Organometallic Photochemistry

Organometallic Photochemistry

Gregory L. Geoffroy

Department of Chemistry
Pennsylvania State University
University Park, Pennsylvania

Mark S. Wrighton

Department of Chemistry
Massachusetts Institute of Technology
Cambridge, Massachusetts

1979

ACADEMIC PRESS

A Subsidiary of Harcourt Brace Jovanovich, Publishers

NEW YORK   LONDON   TORONTO   SYDNEY   SAN FRANCISCO

7262-2295

CHEMISTRY

COPYRIGHT © 1979, BY ACADEMIC PRESS, INC.
ALL RIGHTS RESERVED.
NO PART OF THIS PUBLICATION MAY BE REPRODUCED OR
TRANSMITTED IN ANY FORM OR BY ANY MEANS, ELECTRONIC
OR MECHANICAL, INCLUDING PHOTOCOPY, RECORDING, OR ANY
INFORMATION STORAGE AND RETRIEVAL SYSTEM, WITHOUT
PERMISSION IN WRITING FROM THE PUBLISHER.

ACADEMIC PRESS, INC.
111 Fifth Avenue, New York, New York 10003

United Kingdom Edition published by
ACADEMIC PRESS, INC. (LONDON) LTD.
24/28 Oval Road, London NW1 7DX

Library of Congress Cataloging in Publication Data

Geoffroy, Gregory L
    Organometallic photochemistry.

    Includes bibliographical references and index.
    1. Organometallic compounds. 2. Photochemistry.
3. Transition metal compounds. I. Wrighton, Mark S.,
Date   joint author.   II. Title.
QD411.G46   547'.05   79–6933
ISBN 0–12–280050–8

PRINTED IN THE UNITED STATES OF AMERICA

82   9 8 7 6 5 4 3 2

Contents

QD 708
.2
G82
Chem

Contents

Preface

Investigations into the photochemical properties of coordination compounds have been seriously under way for nearly two decades now, and a fairly high level of understanding of their photophysics and excited-state reactivity has begun to emerge. In direct contrast, the photochemical properties of organometallic complexes, with the exception of the metal carbonyls, are very poorly understood. A number of compounds have been studied, but—as will be obvious from the latter chapters in this book—these studies have not constituted a systematic investigation but rather a relatively random sampling. This situation is likely to change rapidly in the near future, what with the growing interest in organometallics in general and with the realization by photochemists that this is a fertile area to study. It is the primary purpose of this book to provide a firm basis upon which these studies can build. This book only deals with *transition-metal* organometallic complexes.

Chapter 1 presents background material on organometallic excited states and excited-state processes and is designed principally to acquaint students or new workers entering the field with these topics. Chapters 2–8 are detailed reviews of those photochemical studies which have been conducted up to the time of the writing. As such, they should provide a useful guide to what has been done and should point out those areas in need of investigation. We have tried to be as comprehensive as possible in our coverage. However, many photochemical studies of organometallics are buried in papers whose principal focus is not on photochemistry, and it is often difficult to find reference to those investigations. It is thus likely that some previous work has been missed, and a person interested in studying a particular compound should conduct a careful literature search on that species before beginning his research. The review chapters are organized according to type of organometallic. Chapter 2 presents a detailed discussion of metal carbonyls, and Chapters 3–8 discuss, in turn, olefin complexes, arene complexes, cyclopentadienyl complexes, isocyanide complexes, hydride complexes, and alkyl complexes. The organization within each chapter is according to the central metal atom and its group in the periodic table.

G. L. Geoffroy
M. S. Wrighton

1

Electronic Structure of Organometallic Complexes

I. INTRODUCTION

Transition-metal organometallic chemistry has become an extremely important area of chemistry in recent years, principally due to the usefulness of members of this class of compounds for catalyzing or assisting the transformation of organic substrates. The thermal reactions of a large number of transition-metal organometallics have been examined in detail, and guiding principles have begun to emerge concerning thermal reactivity. Although the last decade has seen a maturation of the field of inorganic photochemistry as restricted to classical coordination compounds in aqueous solution [1,2], relatively few photochemical studies have been conducted on true organometallic complexes, excluding the metal carbonyls [3]. It is likely that organometallic photochemistry will prove to be an area of intensive investigation in the coming years, especially since recent results have shown that irradiation of organometallics can lead to catalytically and synthetically useful transformations. It is the purpose of this book to provide a firm basis from which investigators can conduct these studies.

The bulk of the chemistry of transition-element organometallic complexes has concerned the d-block elements. There is increasing interest in the organometallic chemistry of the f-block elements, but little has been done concerning the photochemistry of such systems. Accordingly, we shall restrict our discussion of the electronic structure and photochemistry to the d-block elements, with the exception of a few isolated cases.

In order to discuss the photochemical properties of specific organometallic compounds in detail, it is necessary to develop an appreciation of the bonding and electronic structure of organometallics and a knowledge of the principles of photochemistry. These topics are discussed in this first chapter. The succeeding chapters critically survey the photochemical properties of those organometallic complexes examined to date. These chapters are organized according to the classes of ligands in the order of carbonyl, olefin, arene, cyclopentadienyl, isocyanide, hydride, and alkyl complexes. Each chapter is then subdivided according to the central metal in the complex with the metals presented in order of their left-to-right position in the periodic table.

This first chapter discusses the bonding of representative members of each of these ligand classes with initial emphasis on the descriptive nature of the orbitals involved in the interaction. The 18-valence-electron rule, of special importance for appreciating the stability and reactivity of organometallics, is described and is followed by a detailed discussion of the electronic absorption spectra and excited states of representative organometallic complexes.

An organometallic compound, as defined throughout this book, is any transition-metal complex which contains at least one direct metal–carbon bond, excluding cyanide and carbonate ligands. In addition, we have chosen to discuss the photochemistry of hydride complexes because of their close relationship to organometallics and because of the critical importance of these compounds in assisting organic transformations. An attempt has been made throughout to follow closely the IUPAC nomenclature rules [4] for coordination complexes.

II.   STRUCTURE AND BONDING IN ORGANOMETALLICS

A.   Carbon Monoxide and Organic Isocyanide Ligands

In order to discuss the bonding between a metal and a ligand, it is necessary to consider the metal and ligand orbitals involved and to classify those orbitals according to their symmetry. It is most convenient to illustrate the symmetry properties by referring to the sign of each of the orbital lobes. Orbital interaction then can only occur in those cases which generate a net positive overlap of lobes with the same sign. In bonding to ligands, the metal employs its nd, $(n + 1)$s, and $(n + 1)$p orbitals, and the ligand orbitals can be of various types. In considering the metal–ligand interactions we will focus mainly on the metal d orbitals to illustrate bonding modes. Carbon monoxide and organic isocyanides are quite similar in their bonding properties, and it is appropriate to discuss these ligands together.

Carbon monoxide has a filled σ orbital and two filled π orbitals localized mainly between carbon and oxygen. It also possesses two lone pairs of electrons, localized on the carbon and oxygen atoms but directed away from the molecule. Because of the electronegativity difference between carbon and oxygen, the spatial extent of the carbon lone pair is greater than that of the oxygen lone pair. Carbon monoxide also possesses two mutually perpendicular π-antibonding (π^*) orbitals directed away from the CO internuclear region, and these two orbitals are empty in the ground state. Since oxygen is more electronegative than carbon, the filled orbitals are localized to a greater extent on oxygen than on carbon, and the empty π^* orbitals are more localized on carbon. When CO is bonded to a single transition metal, the M—C—O linkage is invariably linear. If the z direction is chosen along the M—C—O bond, then the d_{z^2}, d_{xz}, and d_{yz} orbitals have the proper symmetry to interact with the carbon lone pair, the π_x^*, and π_y^* orbitals, respectively. This interaction is shown in Fig. 1-1 with positive orbital lobes drawn crosshatched.

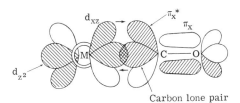

Carbon lone pair

FIG. 1-1.   Bonding interactions of CO with transition metals.

The net effect of the bonding is that carbon monoxide donates some electron density to the metal in a σ fashion from its carbon lone pair and accepts electron density from the metal in a π fashion into its π^* orbitals. Carbon monoxide is hence classified as a σ-donor, π-acceptor ligand. It is the σ bonding which contributes principally to the total bond energy, but the π bonding has important ramifications. Both interactions weaken the carbon–oxygen bond, although the π interaction has the greater effect because it directly populates a C—O antibonding orbital. It is this ability of carbon monoxide to accept electron density from the metal which allows it to stabilize metals in low oxidation states.

Isocyanide ligands of the general formula CNR normally bind to metals in a linear fashion through carbon. The bonding picture is very similar to that of carbon monoxide except that the ≡O portion is replaced by ≡N—R.

Nitrogen is less electronegative than oxygen, and thus isocyanides are in general weaker π acceptors but stronger σ donors than carbon monoxide. The R group of CNR can be varied through a very wide range of organic functional groups, and thus the electronic properties of the ligand can be selectively tuned. Good electron-withdrawing groups such as $p\text{-}NO_2C_6H_4$ make CNR a better π acceptor, whereas good electron-releasing groups such as t-butyl increase its σ-donor ability. Further, the π^*-CN orbitals can conjugate with the π and π^* orbitals of arene substituents; and, as discussed in more detail in Chapter 6, this conjugation has the effect of lowering the energy of one of the π^*-CN orbitals, thus dramatically altering its π-accepting properties.

B.   OLEFIN LIGANDS

The bonding of olefins is most conveniently described using what has come to be called the "Dewar–Chatt" bonding model. The bonding picture is quite analogous to CO and CNR except that different ligand orbitals are employed. Olefins normally bond to metals in an edge-on fashion with the carbon–carbon bond perpendicular to the metal–olefin bond. The simplest olefin, ethylene, possesses filled σ- and π-bonding orbitals which are localized principally between the carbons and also corresponding π^*-antibonding orbitals localized mainly on each carbon atom and directed away from the C—C bond (Fig. 1-2). If the z axis lies along the metal–olefin bond, then the filled π orbital has the proper symmetry to give net overlap with the metal d_{z^2} orbital; alkenes can donate electron density to the metal in a σ fashion. Likewise, the empty π^* orbital can interact with d_{xz} to accept electron density from the metal, and olefins are considered σ-donor, π-acceptor ligands. In general, the π-accepting ability is much less than that of CO or CNR, primarily because of the relatively low electronegativity of the carbons. The electronic properties can be greatly varied, however, by altering the olefin substituents. Electron-withdrawing groups like CF_3 increase the π bonding

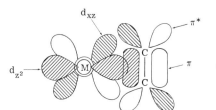

FIG. 1-2.   Bonding interactions of olefins with transition metals.

but decrease the σ bonding, whereas alkyl substituents have the opposite effect. The overall strength of the metal–olefin bond appears to increase as the substituents become more electron-withdrawing.

Alkynes bond similarly to transition metals through an exactly analogous orbital interaction. One significant difference, however, arises from the availability of a second π,π^* orbital set perpendicular to the first which allows alkynes to bond to two metals in a bridging fashion, as illustrated in (I) for $[Co_2(CO)_6(PhC_2Ph)]$.

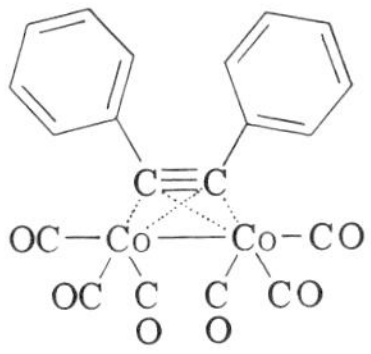

(I)

C. ARENE LIGANDS

Numerous arene complexes have been prepared and characterized [5], and the normal mode of bonding is one in which the planar arene ring lies above the metal and perpendicular to the metal–arene bond. A variety of arenes have been employed in the preparation of complexes ranging from substituted benzenes to naphthalene, anthracene, and benzonorbornene. The bonding in metal arene complexes is best presented by discussing the most representative ligand benzene. The π-molecular orbitals which can be derived for benzene using the Hückel approximation are shown diagramatically in Fig. 1-3 along with an energy level diagram.

If the z direction is from the metal to the center of the arene ligand, the filled benzene a_{2u} orbital has the proper symmetry to interact with d_{z^2}. However, the spatial distribution of d_{z^2} and of a_{2u} are not aligned to give a large overlap since d_{z^2} essentially points at the hole in the center of the a_{2u} orbital. The principal donation of electrons to the metal occurs via π interaction between d_{xz} and d_{yz} and the filled e_{1ga} and e_{1gb} orbitals for which the orbital overlap is large. The e_{2u} orbital set does not have the proper symmetry to give strong overlap with any of the metal orbitals although a weak interaction with d_{xy} and $d_{x^2-y^2}$ is possible. Thus, benzene and other arenes are good electron donors but poor electron acceptors. The electronic properties, however, can be altered to a significant extent by varying the nature of the arene substituents.

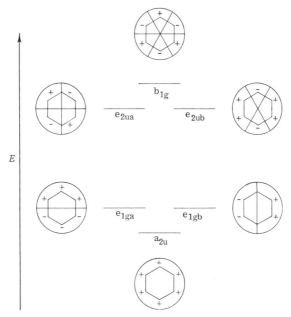

FIG. 1-3.   Benzene π-molecular orbitals. The plus and minus signs designate the signs of the molecular orbitals above the plane of the paper. The portion of the orbitals below the plane of the paper has the opposite sign.

D.   CYCLOPENTADIENYL LIGANDS

The cyclopentadienyl ligand plays a key role in the history of organo-metallic chemistry since it was employed in the synthesis of many of the first characterized organometallics, including ferrocene and $[Re(\eta^5\text{-}C_5H_5)_2H]$. Although the majority of compounds have been prepared using the C_5H_5 ligand, a number have employed substituted derivatives such as $C_5(CH_3)_5$. The C_5H_5 ligand is termed *cyclopentadienyl* because it is the name given the C_5H_5 radical in the accepted organic nomenclature. The ligand is derived for-mally by hydrogen atom abstraction from cyclopentadiene C_5H_6, although most synthetic applications employ the cyclopentadienide ion $[C_5H_5]^-$. The C_5H_5 ligand normally binds to metals in a planar fashion with the plane of the ring perpendicular to the metal–ligand bond and with all five carbon atoms roughly equidistant from the metal. The Greek letter η has been employed to designate ligands bound in this fashion, and the term $\eta\text{-}C_5H_5$ is used [4]. It is more informative, however, to give a superscript to η to indicate the exact number of carbon atoms equidistant from the metal, and $\eta^5\text{-}C_5H_5$ is used throughout this book.

The π-molecular orbitals for the C_5H_5 ligand are shown in Fig. 1-4. If the z axis is the metal–ligand axis, the filled a_2'', e_{1a}'', and e_{1b}'' orbitals can donate electron density to the metal by interacting with d_{z^2}, d_{xz}, and d_{yz}, respectively. The empty e_2'' orbitals do not have the proper symmetry and spatial distribution to interact appreciably with any of the metal d orbitals, except for a weak interaction with d_{xy} and $d_{x^2-y^2}$. The only π-accepting ability comes through interaction of the half-filled e_1'' orbital with d_{xz} or d_{yz}. In most complexes the ligand is considered to bear a formal negative charge in which a complete electron has been transferred to the e_1'' orbital.

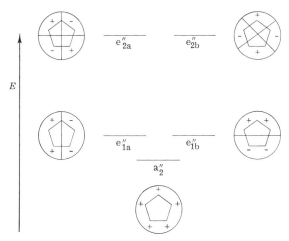

FIG. 1-4. C_5H_5 π-Molecular orbitals.

E. ALKYL, ARYL, AND HYDRIDE LIGANDS

These three ligands are now commonplace in organometallic chemistry, and their bonding properties are well understood. The hydride ligand, formally written as H^-, has a filled s orbital from which electrons are donated in a σ fashion to empty metal orbitals. It, of course, cannot accept electron density from the metal and is simply a σ-donor ligand. It is, however, believed to be the strongest of all σ donors, in part because its small size allows close penetration to the metal (1.55–1.70 Å) and hence gives large orbital overlap. The strong σ donor ability is manifest in the reactivity of hydride complexes since the hydride ligand shows a great tendency to labilize the ligand which is trans to it.

Alkyl ligands are formally considered to be negatively charged, although it is unlikely that a full charge is associated with the ligands in their complexes.

Even so, the bonding is best considered by representation of the ligands as anions such as $CH_3{}^-$. The carbon atom is sp^3-hybridized with a lone pair of electrons in the orbital not bound to a hydrogen. This orbital can interact in a σ fashion with a metal d orbital as illustrated in Fig. 1-5, and alkyl ligands are strong σ-donor ligands but with no π-accepting ability. It must be emphasized, however, that the metal–carbon bond is normally quite covalent in metal complexes, and it is often extremely difficult to deduce the distribution of electron density.

FIG. 1-5.   Bonding of alkyl ligands with transition metals.

Aryl ligands bond through a single carbon of the arene, as illustrated in (II),

$$M-\!\!\left\langle\!\!\bigcirc\right.$$

(II)

and the bonding is similar to that of metal alkyls except that the arene π orbitals can also interact with metal d orbitals of the proper symmetry.

III.   THE 18-VALENCE-ELECTRON RULE

Unlike coordination compounds, organometallic complexes show a high tendency to achieve the inert-gas electron configuration around the central metal, much as organic compounds strive to obtain an inert-gas configuration around carbon. For transition metals, however, the number of electrons needed to reach the configuration is 18, deriving from two s, six p, and ten d electrons. The stability and reactivity of organometallics can often be rationalized in terms of the number of valence electrons, and hence electron counting has been extremely useful for *predicting* structures and reactivity. Organometallics tend to react in such a way so as to obtain the favored 18-electron configuration, including the formation of single or multiple metal–metal bonds when necessary. In general, complexes with 18 valence electrons are usually quite stable whereas 16- and especially 14-valence-electron complexes are quite reactive. A number of 16-electron complexes of the Group VIIIb metals, however, are exceptions and can be isolated, e.g., *trans*-$[RhCl(CO)(PPh_3)_2]$.

There are several methods which can be used to count valence electrons. Of course, it makes no difference how one counts as long as the correct number of electrons is reached. Since it is often difficult to assign a definite oxidation state to the central metal, we prefer to use the counting scheme which always considers the metal neutral *for the purposes of counting*. The rules which can be written for counting by this method are given as follows:

Rule 1: Consider the metal to have an oxidation state of zero and count the number of electrons it contributes.

Rule 2: Each ligand shown in Table 1-1 contributes the number of electrons shown.

TABLE 1-1

Electron-Counting Rules for Selected Ligands

Ligand	Number of valence electrons to be added to the count
H, Cl, Br, I, CN, alkyl, aryl, η^1-allyl	1
CO, PR_3, $P(OR)_3$, AsR_3, olefins, nonbridging alkynes	2
η^3-Allyl, NO	3
Bridging alkynes, η^4-dienes	4
η^5-C_5H_5	5
η^6-C_6H_6, other η^6-arenes	6

Rule 3: Add one electron to the count of each metal for each metal–metal bond that is formed.

Rule 4: For anionic complexes, add to the electron count the number of negative charges the *complex* has; for cationic complexes, subtract from the electron count the number of positive charges the *complex* has.

Rule 5: Bridging hydrides contribute $\frac{1}{2}$ electron to the count of each metal; doubly bridging carbonyls contribute 1 electron to the count of each metal.

It must be emphasized that these are only rules to aid in counting valence electrons; they imply nothing about the actual electron density distribution in the complex. For illustration, several examples are worked as follows:

$[Fe(\eta^5$-$C_5H_5)_2]$	$[IrCl(CO)(PPh_3)_2]$	$[Mn_2(CO)_{10}]$
Fe = 8	Ir = 9	Mn = 7
$2 \times (\eta^5$-$C_5H_5) = 10$	Cl = 1	$5 \times (CO) = 10$
Total = 18	CO = 2	Mn − Mn = 1
	$2 \times (PPh_3) = 4$	Total = 18 for each Mn
	Total = 16	

IV. EXCITED STATES OF ORGANOMETALLIC COMPLEXES

Organometallic complexes have a variety of low-lying excited states. By low-lying states we mean those which can be populated by optical irradiation in the near-infrared, visible, and ultraviolet region. This region corresponds to wavelengths approximately in the range 200–1100 nm. It has been very useful to identify the low-lying excited states according to the one-electron excitation involved. Table 1-2 lists the types of one-electron excitations which have been identified in organometallics. A discussion of key examples in each category follows in the subsequent sections. It is important to recognize that characterization of excited states according to one-electron considerations can be, and often is, a gross oversimplification because of the complete neglect of electron repulsion effects. However, such considerations lead to relatively straightforward chemical expectations.

TABLE 1-2

Types of One-Electron Excited States in Organometallic Complexes

Excited state	Common abbreviation	Example
Ligand field (metal-centered)	"d–d," LF	$[W(CO)_5(piperidine)]$
Intraligand (ligand-centered)	IL	$fac\text{-}[ReCl(CO)_3(3\text{-styrylpyridine})_2]$
Ligand-to-metal charge transfer	LMCT	$[Fe(\eta^5\text{-}C_5H_5)_2]^+$
Metal-to-ligand charge transfer	MLCT	$[W(CO)_5(4\text{-formylpyridine})]$
Metal-to-solvent charge transfer	MSCT	$[Fe_4(CO)_4(\eta^5\text{-}C_5H_5)_4] \cdot CCl_4$
Metal–metal (metal-centered over ≥ 2 metals)	$\sigma_b \rightarrow \sigma^*$ $d\pi \rightarrow \sigma^*$	$[Co_2(CO)_6(PPh_3)_2]$

A. INTRALIGAND EXCITED STATES

Ligands which can be coordinated to metals have their own set of excited states. For organic ligands the study of such excited states comprises the field of organic photochemistry. *A priori*, coordination of an organic ligand to a metal will result in perturbation of its electronic structure. If the free ligand has one-electron transitions which terminate or originate in an orbital centered on the ligating atom, the perturbation can be so great that it is inappropriate to refer to the transition in the complex as an intraligand (IL) transition. On the other hand, coordination of certain ligands involves relatively small perturbation of the electronic structure, and we can expect the complex to have ligand-localized excited states which are not too different from the free ligand itself.

As one extreme, consider the bonding of C_6H_6 in $[Cr(\eta^6\text{-}C_6H_6)(CO)_3]$. The bonding involves the π and π^* levels of C_6H_6. It is readily apparent that the $\pi \to \pi^*$ excitations of the C_6H_6 ligand are going to be drastically changed upon coordination. By way of contrast, bonding of pyridine in $[W(CO)_5(pyridine)]$ is not likely to influence the $\pi \to \pi^*$ excitations, since the principal binding is via the nitrogen atom. However, quite clearly any $n \to \pi^*$ excitation in the pyridine ligand will be altered beyond recognition, since the principal mode of bonding to the metal is via the σ interaction of the nitrogen lone pair. The situation is shown schematically in Fig. 1-6. Note that while the π and π^* levels are unperturbed, the n level is changed considerably and the $n\pi^*$ state in pyridine loses its identity upon coordination.

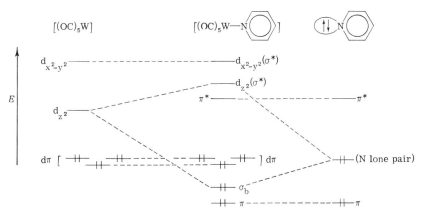

FIG. 1-6. Perturbation of pyridine electronic structure via σ bonding to $[W(CO)_5]$.

Examples of easily identifiable IL absorptions in organometallic complexes of certain organic ligands do exist. One set of such examples can be found among the complexes of styrylpyridine where the binding is through the nitrogen atom of the pyridine ring. For example, the lowest absorption bands of the $fac\text{-}[ReCl(CO)_3L_2]$ (L = $trans$-3- and $trans$-4-styrylpyridine) complexes are at approximately the same position and absorptivity as the $\pi \to \pi^*$ absorptions for free ligands. The vibrational structure revealed in the low-temperature spectrum is also similar to that for the free ligand. On these grounds the lowest absorptions in these complexes have been assigned as IL $\pi\text{-}\pi^*$ absorptions [6]. We expect, then, that the chemistry associated with these excited states will be similar to that for the free ligand. For styrylpyridine, the $\pi \to \pi^*$ excitation results in cis–trans isomerization, and in *fac*-$[ReCl(CO)_3L_2]$ the photoisomerization behavior is attributed to IL $\pi-\pi^*$ excitations [6].

In summary, if the ligand has low-lying transitions involving orbitals which are centered on atoms not directly bonded to the metal, these same transitions survive in the complex as identifiable IL transitions. Naturally, there will always be some perturbation; but when the absorptions are similar to those of the free ligand, this will be an important clue as to the excited-state reactivity of the coordinated ligand.

B.   LIGAND-FIELD EXCITED STATES (METAL-CENTERED EXCITATIONS)

Ligand field (LF) or so-called "d–d" transitions correspond to electronic transitions between the d orbitals on the metal. These are often said to be metal-centered transitions, but organometallics are highly covalent substances and the "metal-centered" d orbitals actually have considerable ligand character. Since metal–ligand bonding involves d orbitals, the electronic transitions originating and/or terminating in d orbitals in organometallics have consequence with respect to metal–ligand binding. In fact, it is now widely believed that LF excited states are the states responsible for efficient photosubstitution, especially when the photosubstitution occurs by a dissociative mechanism. Such excited states have been of particular importance in the photosubstitution of mononuclear metal carbonyls. A number of cases are treated in detail in Chapter 2, but we will consider some ideal cases here.

First, it is important to recognize that organometallics by definition are in the large LF strength regime, since carbon donors are high in the spectrochemical series. Consequently, all organometallics are low-spin systems, and we expect relatively high-energy absorptions. Further, the absorptions associated with the LF transitions in organometallics are relatively intense, owing to the large degree of covalence. Unlike inorganic ions such as Co^{3+} and Cr^{3+} in aqueous solutions for which the LF absorptions have associated absorptivities in the 10–100 range, organometallics often exhibit LF absorptions with absorptivities in the thousands. However, the considerations derived from symmetry, the nature of the ligands, and the central-metal oxidation state effects generally parallel those found in simple inorganic complexes.

Let us now consider in a general way why LF transitions lead to substantial labilization of metal–ligand bonds [7]. We will use simple one-electron arguments to understand the excited-state labilization in much the same way that $\pi \rightarrow \pi^*$ excited states of olefins are rationalized to undergo cis–trans isomerization; depopulation of the π_b level and population of the π^* reduces the C—C bond order to yield a C—C bond having little barrier to rotation. In metal complexes the d orbitals play a role in the π and σ bonding between the metal and ligand. In the d^6, O_h $[M(CO)_6]^-$ (M = V, Nb, Ta), $[M(CO)_6]$ (M = Cr, Mo, W), or $[M(CO)_6]^+$ (M = Mn, Re) complexes, for example, the d orbital diagram in Fig. 1-7 is appropriate. The O_h hexacarbonyls have all been given the LF assignment for the first absorption system. The first LF absorption energy is in the order $[V(CO)_6]^- < [Cr(CO)_6] < [Mn(CO)_6]^+$ as expected for a LF transition. Despite the high symmetry, the spin-allowed LF bands are quite intense ($\varepsilon \approx 10^3$ liter mol^{-1} cm^{-1}), and in going down a group, e.g., Cr to Mo to W, the spin-forbidden transitions have associated absorption intensities which parallel the expected heavy-atom (spin–orbit) effect. The filled "t_{2g}" orbitals are of π symmetry and are π-bonding with respect to the M—CO linkage. The lowest unoccupied set of e_g orbitals is of σ symmetry and is strongly σ-antibonding with respect to the M—CO bond. The LF excited states all involve $t_{2g} \rightarrow e_g$ one-electron excitations; thus, the M—CO binding should be lower upon LF excitation.

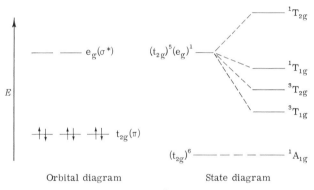

Orbital diagram   State diagram

FIG. 1-7.   Low-spin d^6, O_h LF diagrams.

In fact, the d^6 metal hexacarbonyls that have been studied are extremely photosensitive with respect to dissociative loss of CO. It is interesting that the $t_{2g} \rightarrow e_g$ type one-electron excitation results in extreme labilization. This can be appreciated from the following. Let us assume that the ground- and excited-state substitution of the metal hexacarbonyl proceeds by a dissociative-type mechanism involving the generation of a pentacarbonyl species. The question is: what is the excited-state dissociation rate constant compared to the ground state? Given that (1) the ground-state rate constant can be in the range 10^{-6} sec^{-1}, (2) the excited-state lifetime is known to be $<10^{-10}$ sec, and (3) virtually every metal hexacarbonyl which is excited produces the pentacarbonyl, we estimate the excited-state dissociation constant to be

$\sim 10^{10}$ sec^{-1}. This is some 16 orders of magnitude greater than for the ground state! This reasoning can be applied to numerous organometallic systems, and the general conclusion is that LF excitation can give tremendous increases in ligand dissociation rates.

An important question is whether the extreme increase in ligand lability is due predominantly to depopulation of π-bonding levels or to population of the σ-antibonding levels; or do both contribute substantially. Obviously, both contribute, but it does appear that it is the population of the σ-antibonding levels which is most consequential. This can be appreciated by realizing that the d^5 $[V(CO)_6]$ is relatively substitution-inert, whereas d^6 $[V(CO)_6]^-$ is photosubstitution-labile upon $t_{2g} \to e_g$ type excitation. Similar pairs of d^5/d^6 systems exist among classical coordination complexes, e.g., $[Fe(CN)_6]^{3-/4-}$, and the d^6 system is very photosubstitution-labile. The point is that the d^5 species which has a hole in the t_{2g} set is isolable and does not seem to have the ligand lability of the excited d^6 systems. The widely held belief is that when the orbital of termination is σ-antibonding, then the complex will have greatly enhanced ligand dissociation rates.

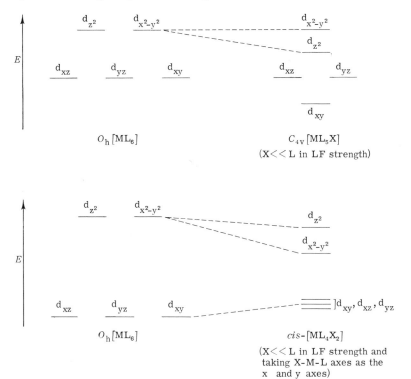

FIG. 1-8.   d-Orbital splittings upon descent in symmetry and LF strength.

Examination of lower symmetry, e.g., C_{4v} and C_{2v}, complexes has proved useful in illustrating the nature and reactivity of the lowest excited states in organometallic complexes. In the d^6 metal hexacarbonyls, consider replacing one or two CO groups by a saturated amine to be typical of C_{4v} $[ML_5X]$ or C_{2v} cis-$[ML_4X_2]$ complexes for X ≪ L in LF strength. As expected, the first absorption system shifts substantially to the red in the C_{4v} compared to O_h compounds; Fig. 1-8 shows the associated d orbital shifts. Note further that the orbital splittings shown lead to fairly clear-cut notions concerning the axis of labilization upon absorption of light. For example, $d\pi \rightarrow d_{z^2}$ excitation in $[ML_5X]$ should tend to most enhance ligand lability along the X—M—L axis (z axis), since the d_{z^2} orbital has directed σ^* character. Likewise, the cis-$[ML_4X_2]$ should be labilized along the X—M—L axes in the lowest one-electron excited state ($d\pi \rightarrow d_{x^2-y^2}$). Such considerations can be used to understand certain aspects of the stereochemistry of ligand photosubstitution, and these are developed in detail in Chapter 2 for certain metal carbonyl derivatives.

C.   METAL-TO-LIGAND CHARGE-TRANSFER EXCITED STATES

Low-valent metal complexes often have low-lying transitions which are termed metal-to-ligand charge transfer (MLCT). The transition originates in some metal-centered orbital and terminates in some ligand-localized orbital, and in the extreme approximation we can view an MLCT transition as one which results in an oxidized metal and a reduced ligand (Eq. 1-1). The energy of an MLCT transition is dependent on the relative ease of oxidation of the

$$M-L \xrightarrow{h\nu} M^+ - L^- \tag{1-1}$$

central metal and the availability of a low-lying acceptor orbital in the ligand. Generally, low-valent organometallics have both an easily oxidized central metal and low-lying acceptor orbitals, and consequently the MLCT transitions are commonplace. The absorption coefficient associated with such transitions is typically in the range of 10^4–10^5 liter mol^{-1} cm^{-1}, and the band positions can be quite solvent-sensitive.

In one-electron terms, the character of MLCT excited states is very different compared to either LF or IL excited states. We expect the central metal to be more susceptible to nucleophilic attack in the MLCT excited state than in the ground state, and at the same time the ligand develops radical-anion character and should be more reactive toward electrophiles than the ground-state complex. Except for associative substitution processes, the MLCT excited states are not typically found to be substitution-labile enough to undergo reaction within the lifetime of the complex. This follows from two considerations: (1) the transition originates in a filled d orbital, which is usually not that consequential with respect to metal–ligand binding; and (2) a ligand-localized orbital is populated, which does not substantially influence metal–ligand bonding. Indeed, the electrostatic attraction generated in an MLCT state may make the M—L bond more inert. Changes in the charge distribution on the central metal may effect significant changes in certain dissociative-type processes. One unimolecular, dissociative process that can be envisioned for MLCT excited states is reductive elimination; i.e., reaction (1-2) is likely to be slower than that from an MLCT excited

$$\textit{Thermal:} \qquad \text{L—M}\begin{smallmatrix}\diagup\text{H}\\[4pt]\diagdown\text{H}\end{smallmatrix} \longrightarrow \text{L—M} + \text{H}_2 \qquad\qquad (1\text{-}2)$$

state (1-3), since in the latter the metal is in a higher oxidation state and can be stabilized by increased electron density via reductive elimination.

$$\textit{MLCT State:} \qquad \text{L}^-\text{—M}^+\begin{smallmatrix}\diagup\text{H}\\[4pt]\diagdown\text{H}\end{smallmatrix} \longrightarrow \text{L—M} + \text{H}_2 \qquad\qquad (1\text{-}3)$$

MLCT excited states are assuming an important role in the development and understanding of inorganic photochemical systems, generally because of their inertness to ligand lability. Simply, this allows other processes to occur. Of particular interest are optical emission and bimolecular chemical processes. For example, the comparison of ground and MLCT excited-state acid–base equilibria in Eqs. (1-4) and (1-5) is a direct measure of the

$$\text{M—L} \underset{-\text{H}^+}{\overset{\text{H}^+}{\rightleftharpoons}} \text{M—LH}^+ \qquad \text{(Thermal, ground state)} \qquad\qquad (1\text{-}4)$$

$$[\text{M}^+\text{—L}^-]^* \underset{-\text{H}^+}{\overset{\text{H}^+}{\rightleftharpoons}} [\text{M}^+\text{—LH}]^* \qquad \text{(MLCT, excited state)} \qquad\qquad (1\text{-}5)$$

consequences of electron redistribution upon the excitation. Another emerging class of bimolecular organometallic photoreactions is electron transfer. Referring to the schematic orbital diagram in Fig. 1-9, the MLCT excited species is simultaneously a more powerful reductant and a more powerful

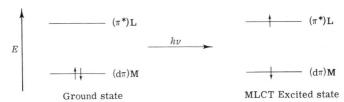

FIG. 1-9.   Schematic molecular orbital diagrams comparing the electron distribution in ground and MLCT states.

oxidant than the ground state. The excited electron in $(\pi^*)L$ is less stable, providing the reducing power; and the "hole" in $(d\pi)M$ is more stable, providing the increased oxidizing power. Ligand-field excited states can be similarly characterized, but the ligand dissociation reaction dissipates the excitation energy prior to most bimolecular electron transfer events. Finally, again owing to their persistence in fluid solutions, MLCT excited organometallic complexes have been used as donors of electronic excitation energy in collisional energy transfer experiments.

Aspects of the simple theory and hypotheses set out for LF and MLCT excited states can be illustrated by a comparison of the spectral and photochemical properties of $[W(CO)_5(piperidine)]$ and $[W(CO)_5(4\text{-formylpyridine})]$. The details are set out in Chapter 2, but the fundamentals can be appreciated by comparison of schematic orbital pictures in Fig. 1-10. The d-orbitals are positioned the same in each case, since the coordination sphere is the same. Indeed, it is found that LF bands present in $[W(CO)_5(piperidine)]$ are also found in the same position and relative intensity in $[W(CO)_5(4\text{-formylpyridine})]$. But unlike the saturated piperidine ligand, 4-formylpyridine has a low-lying $\pi^*(L)$ acceptor orbital, and the $[W(CO)_5(4\text{-formylpyridine})]$ complex exhibits a low-lying, extremely solvent-sensitive, absorption not present in the piperidine complex. The $[W(CO)_5(piperidine)]$ efficiently loses piperidine upon photoexcitation, consistent with the population of the $d_{z^2}(a_1)$ orbital which has directed σ^* character. The lowest excited state in the 4-formylpyridine complex, however, is inert with respect to photosubstitution.

Substituent effects and central metal effects can be used to identify MLCT absorptions. The lower energy MLCT absorptions are found for the more easily reduced L and for the lower valent, more easily oxidized central metal. For example, the MLCT position is $\sim 8000\ cm^{-1}$ higher in energy in

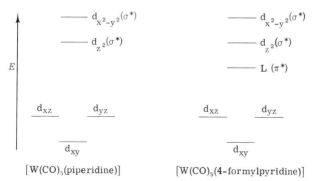

FIG. 1-10.   Comparison of the orbital ordering for two complexes having no low-lying MLCT (left) and having a low-lying MLCT (right).

[W(CO)$_5$(3,4-dimethylpyridine)] compared to [W(CO)$_5$(4-formylpyridine)], and a similar shift is found in the [(η^5-C$_5$H$_5$)M(CO)$_2$L] (M = Mn, Re) complexes. It may be possible to manipulate the nature of the lowest excited state; e.g., (1) fac-[ReCl(CO)$_3$(trans-4-styrylpyridine)$_2$] exhibits a lowest excited state which is trans-4-styrylpyridine π–π* IL in character while the lower valent, but still 5d^6, cis-[W(CO)$_4$(trans-4-styrylpyridine)$_2$] exhibits a lowest excited state which is W $\rightarrow$ π* trans-4-styrylpyridine CT; (2) [W(CO)$_5$L] exhibit a lowest absorption which is LF or W $\rightarrow$ LCT depending on whether L = 3,4-dimethylpyridine or 4-formylpyridine, respectively. Such manipulation of lowest excited state is very consequential with respect to the nature and efficiency of the observed photoinduced reactions.

D.   LIGAND-TO-METAL CHARGE-TRANSFER EXCITED STATES

Excitation of electrons from ligand-centered orbitals to orbitals localized on the central metal are termed ligand-to-metal charge-transfer (LMCT) transitions. The energy of such transitions clearly depends on the availability of low-lying, empty or partially filled metal orbitals and on the relative ease of oxidation of the ligands. Such excited states have been of relatively little importance in organometallic complexes, since the unfilled metal d orbitals are generally not low-lying. This is a fact which follows from the high LF strengths associated with organometallic complexes, placing the unfilled d orbitals at high energy. It also follows from the 18-valence-electron rule that there are relatively few paramagnetic complexes where there are low-lying, partially filled d orbitals. LMCT excited states are likely well within the optical region of the spectrum in many complexes, but they are not often the lowest excited state. Indeed, there are relatively few systems for which unequivocal LMCT assignments have been made in organometallics, and there seem to be few claims that such organometallic excited states are photoreactive.

As an unequivocal lowest LMCT excitation we can point to the 3d^5, [Fe(η^5-C$_5$H$_5$)$_2$]$^+$ ion and its derivatives [8]. Figure 1-11 shows the one-electron level diagram appropriate to this ion, showing that there is a relatively low-lying hole in the e$_{2g}$, a$_{1g}$ set of the Fe^{3+}-localized orbitals. The LMCT transition observed at $\sim$16,200 cm^{-1} in the spectrum of [Fe(η^5-C$_5$H$_5$)$_2$]$^+$ is the (η^5-C$_5$H$_5$)(e$_{1u}$) $\rightarrow$ Fe^{3+} (e$_{2g}$) transition. Consistent with the LMCT character of the transition, the first absorption maximum red-shifts with the introduction of electron-releasing substituents in the (η^5-C$_5$H$_5$) ring: ([Fe(η^5-C$_5$H$_5$)$_2$]$^+$ (λ_{max} = 16,220 cm^{-1}); [Fe(η^5-C$_5$H$_5$) (η^5-C$_5$H$_4$n-Bu)]$^+$ (λ_{max} = 16,000 cm^{-1}); and [Fe(η^5-C$_5$H$_4$n-Bu)$_2$]$^+$ (λ_{max} = 15,390 cm^{-1}). One other compelling fact supporting the LMCT assignment

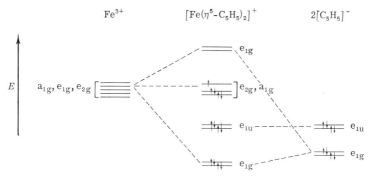

FIG. 1-11.   One-electron levels for $[Fe(\eta^5\text{-}C_5H_5)_2]^+$.

is the absence of such a low-lying absorption in $[Fe(\eta^5\text{-}C_5H_5)_2]$ which does not have a hole in the e_{2g},a_{1g} set. The lowest lying empty d-orbital set in $[Fe(\eta^5\text{-}C_5H_5)_2]$ is the e_{1g} set, and the LMCT $(e_{1g} \rightarrow e_{1g})$ excitation is consequently much higher in energy than in $[Fe(\eta^5\text{-}C_5H_5)_2]^+$. A similar relationship exists between $[Fe(CN)_6]^{3-}$ and $[Fe(CN)_6]^{4-}$ with the former giving a low-lying $CN^- \rightarrow t_{2g}$ LMCT absorption. The pair of organometallics $[V(CO)_6]$ and $[V(CO)_6]^-$ is isoelectronic with the iron hexacyanides, and the lower lying $CO \rightarrow V$ CT is expected in the $[V(CO)_6]$. This expectation is consistent with the fact that $[V(CO)_6]$ exhibits lower energy absorptions than $[V(CO)_6]^-$ [9].

The shift in electron density from ligand to metal in the LMCT excited states leads to the expectation that the ligand will be more susceptible to nucleophilic attack and the metal will be more susceptible to electrophilic attack. Likewise, the LMCT excited species may more readily undergo oxidative addition owing to the increased electron density on the metal. But so far none of the expectations has been realized; rather, LMCT excited states are generally associated with homolytic cleavage of M and L. Consider the one-electron diagram for $[Re(CO)_5CH_3]$ (Fig. 1-12), a complex whose photochemistry has not yet been elucidated. Excitation of an electron from the σ_b level (associated with $-CH_3$) to the $d_{z^2}(\sigma^*)$ level is an example of an LMCT excitation which should lead to labilization of the $Re-CH_3$ bond. However, the orbital of termination $d_{z^2}(\sigma^*)$ also lends lability to the $Re-CO$ σ bond as well, and CO dissociation from Re might compete with formation of CH_3 radicals. In Werner complexes like $[Co(NH_3)_5X]^{2+}$ $(X = Cl, Br, I)$ $X \rightarrow Co^{3+}$ CT excitations are analogous and are well known to be associated with generation of Co^{2+} and the halogen atom. The suggestion implied in Fig. 1-12 is not well substantiated by experiment; $LF(d\pi \rightarrow d_{z^2}(\sigma^*))$ transitions are high in energy and population of $d_{z^2}(\sigma^*)$ alone may be enough to generate alkyl radicals.

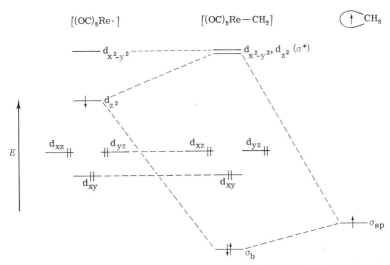

FIG. 1-12.   One-electron level scheme for $[Re(CO)_5CH_3]$.

E.   EXCITED STATES ASSOCIATED WITH METAL–METAL BONDED COMPLEXES

There are large number of structurally well characterized organometallics which have direct metal–metal bonds. Such species exhibit transitions involving metal-centered orbitals which are delocalized over the core of metal atoms present. Such transitions are analogous in some respects to LF transitions in mononuclear complexes, since metal-centered orbitals are involved. In dinuclear and trinuclear metal–metal bonded complexes the metal–metal excitations have been shown to lead to efficient metal–metal bond cleavage producing mononuclear fragments. Ligands coordinated to the metals in polynuclear complexes can still exhibit IL excitations, and MLCT and LMCT excitations will still be present, assuming that the M in MLCT represents not one but n metals over which the metal-centered orbitals are delocalized.

Organometallic complexes containing two metals with a metal–metal bond are the most well-characterized with respect to metal–metal excitations. Of the metal–metal bonded dinuclear complexes $[Mn_2(CO)_{10}]$ represents the cornerstone example. Its electronic structure can be understood with reference to Fig. 1-13. The low-lying one electron transitions are the $\sigma_b \rightarrow \sigma^*$ and the $d\pi \rightarrow \sigma^*$, and these obviously should weaken the metal–metal bond. Referring back to the diagram for $[(OC)_5Re—CH_3]$ (Fig. 1-12), the $\sigma_b \rightarrow \sigma^*$ in $[Mn_2(CO)_{10}]$ is analogous to what one would call the LMCT in the methyl

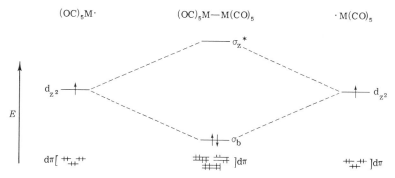

FIG. 1-13.   One-electron level scheme for $[Mn_2(CO)_{10}]$.

compound. However, in $[Mn_2(CO)_{10}]$ the σ_b orbital is clearly a metal–metal orbital. The absorption spectrum of $[Mn_2(CO)_{10}]$ shows an intense ($\varepsilon > 10^4$ liter mol^{-1} cm^{-1}), fairly narrow, solvent-insensitive absorption band in the near-ultraviolet region. Mononuclear Mn carbonyl complexes do not exhibit such a band, and polarized absorption spectral measurements are in accord with the symmetries of the σ_b and σ^* orbitals. A large number of homodinuclear and heterodinuclear metal–metal bonded complexes exhibit a near-ultraviolet absorption band like that for $[Mn_2(CO)_{10}]$. Indeed, a $\sigma_b \to \sigma^*$ absorption is observed for any system where it is appropriate to consider the metal–metal bond as resulting from the coupling of two 17-electron fragments giving a diamagnetic, two-electron metal–metal bonded species. Such is the case for metal–metal bonded dinuclear complexes resulting from any combination of $[Co(CO)_4]$, $[Re(CO)_5]$, $[Mn(CO)_5]$, $[Fe(\eta^5\text{-}C_5H_5)(CO)_2]$, $[Mo(\eta^5\text{-}C_5H_5)(CO)_3]$, $[W(\eta^5\text{-}C_5H_5)(CO)_3]$, etc.

The species $[Mn_2(CO)_8L]$ (M = Mn, Re; L = 1,10-phenanthroline, 2,2'-biquinoline) represent metal–metal bonded species which exhibit IL (L, π–π^*) and $M_2 \to LCT$ transitions. As usual, the IL absorptions are like those for the noncoordinated ligand. The position of the $M_2 \to LCT$ is lower in energy for L = 2,2'-biquinoline than for L = 1,10-phenanthroline in accord with the direction of the charge transfer. Quite interestingly, the $M_2 \to LCT$ results in photoreaction, and the consistent assignment is $\sigma_b \to LCT$, which weakens the metal–metal bond.

For cluster complexes where there are more than two metals bonded to one another there are still metal–metal excited states, but thus far these are relatively poorly characterized. The presence of metal–metal excitations in polynuclear clusters is indicated by low-energy visible absorption. For example, $[Fe_3(CO)_{12}]$ is blue-green. Consistent with the notion that the metal–metal bonds increase in strength down a group, $[Ru_3(CO)_{12}]$ is orange and $[Os_3(CO)_{12}]$ is yellow, reflecting a greater separation between bonding and antibonding metal–metal levels in the heavier species.

Metal–metal excitations in clusters involving three or more metals are not likely to lead to complete fragmentation of the metal–metal framework as is found in the dinuclear species. This follows from the notion that a one-electron excitation will not be capable of breaking more than one bond; and, when it is realized that the orbitals of origin and termination are delocalized over three or more metals, it is possible that such transitions will simply not be sufficient to cause significant labilization. In certain trinuclear clusters, fragmentation results from photoexcitation where Eq. (1-6) probably represents the primary event. As the number of metals increases, the prospects for

$$
\begin{array}{ccc}
\diagup\!\!\!^{M}\!\!\!\diagdown & \xrightarrow{\;h\nu\;} & \diagup\!\!\!^{M}\!\!\!\diagdown \\
M\!\!-\!\!-\!\!-\!\!M & & M\cdot \quad \cdot M
\end{array}
\tag{1-6}
$$

fragmentation decline. The line of reasoning is similar to that in comparing the result of photoexcitation in ethylene and anthracene. In ethylene $\pi_b \to \pi^*$ excitation results in dissociative π-bond cleavage to give a twisted structure, whereas $\pi_b \to \pi^*$ excitation in anthracene is fairly delocalized over several carbons and the excited state is not too distorted. Taking metal clusters to the extreme of a metal surface to which ligands are bonded, we note that "photo-desorption" of species such as CO can result from what seems to be electronic excitations and not heating effects alone. Thus, the unreasonable prospect of metal atom ejection (analogous to cluster fragmentation) does not obtain; cluster complexes may be expected to undergo ligand dissociation reactions, but the excited states for such a process have not yet been characterized.

F.   METAL-TO-SOLVENT CHARGE-TRANSFER EXCITED STATES

Absorptions of complexes in solution have been assigned to transitions which originate in metal-centered orbitals and terminate in solvent orbitals. Such metal-to-solvent charge transfer (MSCT) tends to oxidize the metal complex and reduce the solvent. Obviously, the position of the absorption depends on the solvent and its ease of reduction and on the ease of ionization of the metal complex. The intensity of the absorption can be of the order of 10^3 liter mol^{-1} cm^{-1}.

Good examples of MSCT absorptions among organometallic complexes are not common. However, $[Fe(\eta^5\text{-}C_5H_5)_2]$ and the polynuclear $[Fe(\eta^5\text{-}C_5H_5)(CO)]_4$ both exhibit such a transition in the presence of halocarbon solvent [10,11]. The position of the band maximum depends in the expected manner on the $E_{1/2}$ value of the halocarbon. Further, both $[Fe(\eta^5\text{-}C_5H_5)_2]$ and $[Fe(\eta^5\text{-}C_5H_5)(CO)]_4$ have nearly the same E° for the first oxidation, consistent with the fact that the MSCT is at the same position for these complexes in a common solvent. The MSCT intensity in the two is comparable, but the absorption band is more easily observed in the $[Fe(\eta^5\text{-}C_5H_5)_2]$ case owing to the absence of other intense absorptions in the near-ultraviolet. In both cases irradiation into the MSCT absorption system results in efficient photooxidation of the complex, while irradiation in poor acceptor solvents (e.g., benzene) results in little or no photoreaction.

V. PRINCIPLES OF PHOTOCHEMISTRY

A. INTRODUCTION AND BASIC LAWS

Although transition-element organometallic chemistry is a relatively young field, photochemistry is a fairly well-developed discipline whose concepts and principles apply to all classes of molecules. The purpose of this section is to introduce some definitions, symbols, and terms associated with excited-state chemistry and to outline the basics of light-induced chemical reactions.

The first law of photochemistry states that only optical irradiation that is absorbed by the reacting system can be effective in producing chemical changes. This simple law was formulated in the first half of the nineteenth century—well before quantum theory. In modern terms concerning molecular species, only those molecules which absorb the incident light can undergo primary excited-state events. This crucial fact represents one of the real advantages in photoinduced chemical reactions when compared to heat-driven reactions. The implication of the first law is that it is possible to excite selectively certain absorbing substances in the presence of other molecules which are transparent to the incident irradiation. Thus, molecules which may be a minor constituent of a solution mixture can be selectively excited whereas heating is nonselective.

The second law of photochemistry states that one photon, or light quantum, can excite one molecule. With the advent of lasers, light intensities high enough to detect multiphoton excitations can be generated. However, there have been no demonstrations of the simultaneous excitation of two molecules by one photon. In conventional photochemical experiments we assume that there is a one-to-one correspondence of the number of photons absorbed and the number of excited molecules created.

Complexes of transition elements exhibit absorptions which correspond to electronic transitions from energies of ~ 4000 cm^{-1} to $> 50,000$ cm^{-1}. But, for the most part, we are concerned with electronic transitions resulting from the absorption of light in the ultraviolet and visible region. Generally, excitations of > 40 kcal/mol have been used, which corresponds to red light on the lower end of this scale. Molecules endowed with this amount of excitation energy, hv, are often reactive and short-lived. In studying such species two general types of quantitative measurements have been made: (i) absolute rates of reaction and decay and (ii) efficiency of the decay or reaction process based on the number of photons absorbed. The efficiency of a light-induced process is the quantum yield Φ, defined according to Eq. (1-7). For example, if $\Phi = 1.0$ for a light-induced A to B reaction, this

$$\Phi = \frac{\text{molecules undergoing the process}}{\text{photons absorbed by the molecules}} \tag{1-7}$$

means that every photon absorbed by molecule A yields a B molecule. Generally, the wavelength or energy of the irradiation and the light intensity used are important parameters that should accompany a reported quantum yield.

Knowing that a given excited process occurs with some quantum efficiency says nothing about the rate of the excited-state process. For example, in the A to B conversion the quantum yield gives no details of the rate and mechanism of the decay of the excited A molecule A* to give B. All that can be said concerning the rate of generation of B from A* is that it is fast compared to any other processes, since there is a unit quantum yield for the formation of B; i.e., there are no other (rate) competitive excited-state processes. Determination of rates requires some direct measurement(s) associated with growth of B and consumption of A*. However, Φ can be measured merely by determining the number of molecules of B formed, A molecules consumed, and the number of photons absorbed by the system of A. Thus, conventional chemical techniques can often be used to measure reaction quantum yields, whereas rapid time scale techniques may be required to determine rates of excited-state decay. For organometallics, excited-state lifetimes in the range ~ 1 msec to 10^{-12} sec have been found (*vide infra*). In discussion of photochemical rates one must be careful to distinguish the rate of converting A* to B compared to the amount of B that accumulates per unit time. Obviously, in the present example, the absolute amount of B generated per unit time depends on the rate of A* generation or light intensity. Measurement of the quantum yield serves to standardize comments concerning such matters, and we can regard reactions for which $\Phi \geq 0.1$ as fairly efficient photochemical transformations. Such values are fairly common in organometallic photochemistry.

The lifetime τ of an excited molecule A* is given by Eq. (1-8).

$$\tau = \frac{1}{\sum_{n} k_n} \tag{1-8}$$

when k_n is the rate constant associated with a given excited-state decay process. Further, the quantum efficiency for a given process Φ_i, its associated rate constant k_i, and $\sum_n k_n$ are related as in Eq. (1-9).

$$\Phi_i = \frac{k_i}{\sum_n k_n} \tag{1-9}$$

The value of Φ_i and τ are thus related according to Eq. (1-10).

$$\Phi_i = k_i \tau \tag{1-10}$$

Naturally, the value of τ will be a function of a large number of variables including the nature of the excited-state consumption process(es).

Most photochemists treat an excited species as a unique substance having distinct and characteristic physical, chemical, and structural properties. Some of the properties are measurable in the short lifetime of the excited state; but some properties such as melting point, crystal structure, etc. are not measurements which are likely to be obtained. However, many features of an excited molecule can be measured and correlated with reactivity. In many instances the chemistry and measurements which are possible depend on τ and the rate constant associated with the process of interest. Considerable effort must be expended to completely characterize an excited species.

B. PRIMARY EXCITED-STATE PROCESSES

Excited molecules undergo unimolecular and bimolecular reactions just as do ground-state species. Table 1-3 lists the known classes of photochemical reactions of organometallic substances. It is quite intriguing that we do not yet have excited-state analogs to every ground-state reaction, nor do we have an extensive array of bimolecular reactions. This situation is very likely a consequence of little conscious effort. Unimolecular excited-state processes without net chemical change can also occur as in Eqs. (1-11) and (1-12).

$$A^* \xrightarrow{k_{nr}} A + heat \tag{1-11}$$

$$A^* \xrightarrow{k_r} A + light \tag{1-12}$$

Relaxation of A^* according to Eq. (1-11) is termed nonchemical, nonradiative decay with an associated rate constant k_{nr}. Relaxation according to Eq. (1-12) involves dissipation of excitation energy as light emission and is called radiative decay with an associated constant k_r. Both radiative and nonchemical, nonradiative decay processes contribute to inefficiency in photochemical transformations. Radiative decay, however, is often a welcome

TABLE 1-3

Known Chemical Reactions of Organometallic Excited States

Reaction class	Example
Unimolecular	
Ligand dissociation	$[Cr(CO)_6] \xrightarrow{hv} [Cr(CO)_5] + CO$
Ligand isomerization	$[W(CO)_5(trans\text{-}4\text{-styrylpyridine})] \xrightarrow{hv} [W(CO)_5(cis\text{-}4\text{-styrylpyridine})]$
Reductive elimination	$[IrH_2(diphos)_2]^+ \xrightarrow{hv} [Ir(diphos)_2]^+ + H_2$
Metal–metal bond cleavage	$[Re_2(CO)_{10}] \xrightarrow{hv} 2[Re(CO)_5]$
Rearrangement	$trans\text{-}[W(CO)_4(PPh_3)_2] \xrightarrow{hv} cis\text{-}[W(CO)_4(PPh_3)_2]$
Bimolecular	
Electron transfer	$[ReCl(CO)_3(phen)]^* + PQ^{2+} \rightarrow [ReCl(CO)_3(phen)]^{\dot{+}} + PQ^{\dot{+}}$
Proton transfer	$[Ru(bipy(COO^-)_2)(bipy)_2]^* + 2H^+ \rightarrow [(bipy)_2Ru(bipy(COOH)_2]^{2+*}$
Energy transfer	$[ReCl(CO)_3(phen)]^* + trans\text{-stilbene} \rightarrow stilbene^* + [ReCl(CO)_3(phen)]$
Ligand addition	$[W(CNR)_6]^* + pyridine \rightarrow [W(CNR)_6(pyridine)] \rightarrow$ (intermediate) $[W(CNR)_5(pyridine)] + CNR$

feature in that it provides one easily measurable property of an excited species.

1.   Radiative Decay of Excited Molecules: Luminescence

Excited molecules may undergo spontaneous or stimulated emission. In photochemical experiments we are generally concerned with the spontaneous emission of light from an excited species concomitant with relaxation of the molecule to a more stable electronic state which is generally the ground electronic state. Consequently, a useful piece of information to be gained from measuring the energetic distribution of emitted light (emission spectrum) is the excited-state energy of the molecule. Further, by determining the decay of the emission intensity in time one can directly determine τ, the lifetime of the excited species. In such an experiment the decay is often exponential and the time required for the emission to decay to $1/e$ of its original value is the lifetime τ. Since light is easily detected, emission lifetime and spectral distribution are relatively common measurements. The third general feature associated with emission is the efficiency or luminescence quantum yield, Φ_e (Eq. 1-13). The emission properties of an excited molecule

$$\Phi_e = \frac{\text{photons emitted}}{\text{photons absorbed}} \qquad (1\text{-}13)$$

give important clues as to the nature of the emissive excited state (CT, LF, IL, etc. and its multiplicity). Finally, a determination of the incident wavelengths of light necessary to observe emission (excitation spectrum) reveals important information concerning the relaxation of excited states which are higher in energy than the state from which emission is observed.

Historically, long-lived emission was called *phosphorescence* and prompt (short-lived) emission was termed *fluorescence*. These terms came to be associated with triplet → singlet and singlet → singlet emissions, respectively, in organic systems. In inorganic and organometallic systems the terms have little meaning on two counts. First, organometallics do not typically exhibit long-lived emissions which give an afterglow effect persisting for several seconds after an exciting light is switched off. Second, in many inorganic systems the multiplicity of the excited and ground states is not restricted just to singlet and triplet. Thus, it is generally more appropriate to describe the phenomenon as luminescence or simply emission.

Both absorption and emission are radiative processes and consequently have some factors in common controlling probability. The radiative decay constant k_r in Eq. (1-12) can be approximated by Eq. (1-14)

$$k_r = \frac{\varepsilon_{max}}{10^{-4}} \tag{1-14}$$

where ε_{max} refers to the molar absorptivity associated with the $A \xrightarrow{h\nu} A^*$ transition. This equation has a number of underlying assumptions and applies mainly to near-UV and visible transitions. Radiative decay constants of $\sim 10^9$ sec^{-1} are predicted for fully allowed absorptions ($\varepsilon \sim 10^5$ liter mol^{-1} cm^{-1}) while $\sim 10^3$ sec^{-1} radiative decay constants are predicted for strongly forbidden absorptions ($\varepsilon \sim 10^{-1}$).

The connection between absorption and emission probability is best when the excited state and ground state are structurally very similar. The ideal situation may be as sketched in Fig. 1-14a where the geometries as reflected by the shape of the potential energy curve and position of the energy minimum are very similar in ground and excited state. Very often the excited state relaxes to a thermally equilibrated equilibrium geometry which is quite different than the ground state. Consequently, conclusions concerning k_r drawn from absorptions between the ground state and the excited state are inappropriate. Figure 1-14b shows a situation where distortion so alters the picture that one cannot use the absorptions to predict k_r accurately. As can be seen for the hypothetical cases shown in Fig. 1-14, absorptions are likely to be good indicators of emission properties when the emission spectrum and absorption spectrum overlap, whereas a large distortion may be the cause of an emission spectrum which is significantly red-shifted from the first absorption system.

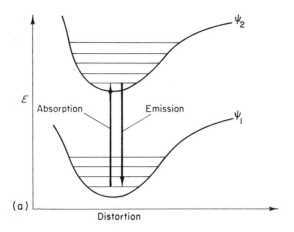

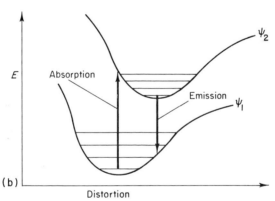

FIG. 1-14.   Potential energy curves for ground and excited states. (a) Ψ_1, Ψ_2 have same geometry; (b) Ψ_1 and Ψ_2 have different geometry.

For emissive molecules (and all molecules are emissive to some degree!) the value of k_r can be measured. From Eq. (1-10) it is clear that two experiments are needed: both τ and Φ_e must be measured. For this specific case then, Eq. (1-15) applies.

$$\Phi_e = k_r \tau \qquad (1\text{-}15)$$

For emissive samples, τ is easily measured and is the most widely used source of such information in organometallics. The typical procedure is to excite a sample with a pulse of light such that the pulse width is much shorter than τ. The emission intensity is then measured as a function of time with the beginning of the pulse representing $t = 0$; τ is taken as the time for the emission intensity to decay to $1/e$. A simple block diagram of a lifetime apparatus is given in Fig. 1-15. The value of Φ_e, the number of photons emitted divided by the number absorbed, is a much more tedious, generally less accurate, measurement. In fact, there are relatively few organometallic systems for which both τ and Φ_e have been measured to allow evaluation of k_r. The difficulty in measuring Φ_e is partly a consequence of the fact that the molecules emit with a spherical distribution and one needs some way of counting all of the photons emitted.

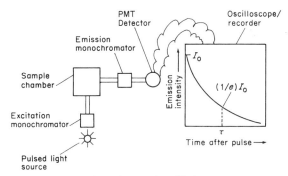

FIG. 1-15.   Block diagram for a lifetime apparatus.

The intensity of emitted light as a function of wavelength or energy is the emission spectrum. Like other spectra, the emission spectrum is a characteristic of the species under investigation. In this case the key species is an excited molecule and the phenomenon is associated with relaxation of the excited species to its ground state or some other lower lying electronic state. Generally, excited molecules are produced by an optical excitation beam and emission from the sample is monitored at right angles. When the emission is detectable, there is a simple way to determine the excited-state lifetime (*vide supra*), but other crucial facts are revealed. First, the energy of the emitted light gives the vertical (Franck–Condon) energy between the emitting state and the termination state. Further, the emission spectrum may exhibit vibrational structure with spacings associated with the termination state, since emission generally originates from the 0th vibrational level of the emitting state. Finally, comparison of the emission spectrum and the absorption spectrum may reveal the nature and degree of structural change upon excitation from the ground state. Emission spectra are easy to measure in the wavelength range of ~ 220 nm to ~ 1100 nm. The unavailability of sensitive detectors at lower energy has precluded significant studies at lower energy. A number of commercial spectrophotometers are available and cover a fairly wide spectral range. A simple block diagram is shown in Fig. 1-16.

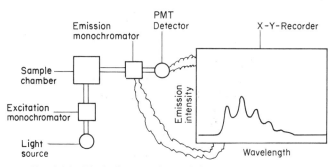

FIG. 1-16.   Block diagram of an emission spectrophotometer.

The excitation spectrum of an emission refers to the spectral distribution of light which is effective in producing emission. Commercial spectrophotometers are often equipped to make this measurement. The procedure is to monitor the emission intensity while varying the excitation wavelength. The typical situation in organometallics is that the only detectable emission is that associated with the lowest excited state decaying to the ground state. Thus, the effective excitation wavelengths will be no lower in energy than the emitted light, and quite naturally, only those exciting wavelengths which are effective in producing excited states will excite the emission. However, excitation wavelengths may directly produce upper excited states which may or may not decay to the lowest excited state from which emission occurs with a detectable yield. Considerable care must be exercised in recording a meaningful excitation spectrum, but the information obtained is quite valuable.

Consider a molecule having excited states Ψ_2, Ψ_3, and Ψ_4 which can be populated by absorption of light by the ground state Ψ_1. It is found that the lowest excited state Ψ_2 is emissive and one wants to determine whether population of Ψ_3 and/or Ψ_4 results in the ultimate population of Ψ_2 by a nonchemical, nonradiative decay process. The excitation spectrum can give an answer to this question because a true excitation spectrum is actually a measure of the *relative* emission quantum yield as a function of wavelength. That is, when Ψ_3 or Ψ_4 is the state of termination in the absorption process from Ψ_1, we can determine the relative number of photons emitted from Ψ_2 per photon absorbed to generate Ψ_4 or Ψ_3. The situation is illustrated schematically in Fig. 1-17. In the state diagram we show a situation where emission is only detectable for the $\Psi_2 \rightarrow \Psi_1$ transition and Ψ_3 and Ψ_4 may decay either by undergoing a reaction or internally converting to Ψ_2 by nonchemical, nonradiative decay paths. The optical absorption spectrum is sketched along with an excitation spectrum for an optically dilute sample of this hypothetical substance. The conclusion is that Ψ_4 must do something

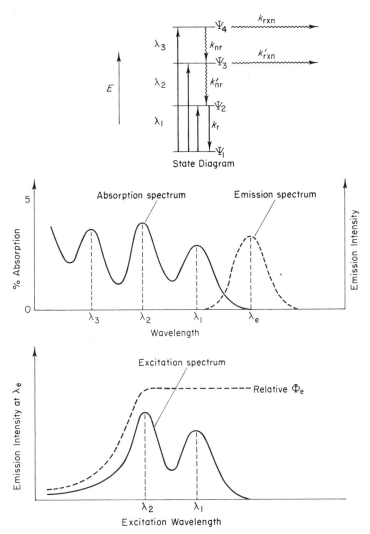

FIG. 1-17. Hypothetical state diagram and absorption, emission, and excitation spectra.

other than decay through Ψ_3 down to Ψ_2 which emits; i.e., from the excitation spectrum we see no maximum at λ_3 as we see at λ_1 and λ_2. Thus, in this example Ψ_4 must be undergoing reaction at such a fast rate that nonchemical, nonradiative decay to Ψ_3 is very inefficient. For the optically dilute sample the excitation spectrum should look just like the absorption spectrum for a situation where Φ_e is independent of excitation wavelength; i.e., a case where all absorbed photons equally efficiently populate the emitting state independent of whether upper excited states are directly populated. By optically dilute we simply refer to a concentration range where the observed emission intensity is proportional to concentration for any exciting wavelength. This arises from geometric/optical considerations. The situation is further complicated by the fact that commercial emission spectrophotometers do not come equipped with excitation systems that can provide monochromatic exciting light at a given resolution with constant intensity independent of wavelength. Thus, in an excitation spectrum which is not corrected for variations in exciting intensity with wavelength the relative emission intensity variations may only reflect variation in lamp output. Electronic, optical, and other ways to "correct" excitation spectra exist and the measurement can become routine.

Table 1-4 lists a number of emissive organometallic complexes. A detailed discussion of these is deferred to an appropriate section of the text, but it is worth noting that a number of organometallics are detectably emissive and that the emission can originate from different types of one-electron excited states.

TABLE 1-4

Representative Emissive Organometallic Complexes

Complex	Emission assignment
[W(CO)$_5$(piperidine)]	LF
[W(CO)$_5$(4-formylpyridine)]	MLCT
[Mo(CNPh)$_6$]	MLCT
[Ru(η^5-C$_5$H$_5$)$_2$]	LF
K[PtCl$_3$(ethylene)]	LF
[Re$_2$(CO)$_8$(1,10-phenanthroline)]	(M—M) LCT
fac-[ReCl(CO)$_3$(3-benzoylpyridine)$_2$]	IL + MLCT
K(diglyme)[Ta(CO)$_6$]	LF
[Rh$_2$(1,3-diisocyanopropane)$_4$]$^{2+}$	M—M

2. Nonradiative Decay

The most important excited-state decay processes are those leading to chemical changes. A list of known unimolecular and bimolecular reactions of organometallic excited states is set out in Table 1-3. A discussion of these by class of complex comprises the bulk of this book. Nonchemical, nonradiative decay which occurs unimolecularly (Eq. 1-11) has received little study at this writing. Likewise, bimolecular nonchemical, nonradiative decay, such as electronic energy transfer (Eq. 1-16), has received little

$$A^* + Q \rightarrow A + Q^* \qquad (1\text{-}16)$$

attention. Such photophysical studies will likely follow the discovery of new fluid solution emissive species. While the observation of emission is no guarantee that the excited-state lifetime will be long enough for efficient bimolecular excited state processes, it is a good indicator that the excited state will persist long enough to at least undergo diffusion controlled processes. So far, the bulk of organometallic photochemistry has resulted from excited states which undergo fast, dissociative-type processes leading to coordinatively unsaturated, ground-state primary products.

REFERENCES

1. V. Balzani and V. Carassiti, "Photochemistry of Coordination Compounds." Academic Press, New York, 1970.
2. A. W. Adamson and P. D. Fleischauer, eds., "Concepts of Inorganic Photochemistry." Wiley, New York, 1975.
3. M. S. Wrighton, *Chem. Rev.* **74**, 401 (1974).
4. *Pure Appl. Chem.* **28**, 39 (1971).
5. W. E. Silverthorn, *Adv. Organomet. Chem.* **13**, 47 (1975).
6. M. S. Wrighton, D. L. Morse, and L. Pdungsap, *J. Am. Chem. Soc.* **97**, 2073 (1975).
7. M. Wrighton, H. B. Gray, and G. S. Hammond, *Mol. Photochem.* **5**, 165 (1973).
8. Y. S. Sohn, D. N. Hendrickson, and H. B. Gray, *J. Am. Chem. Soc.* **93**, 3603 (1970).
9. M. S. Wrighton, D. I. Handeli, and D. L. Morse, *Inorg. Chem.* **15**, 434 (1976); H. Haas and R. K. Sheline, *J. Am. Chem. Soc.* **88**, 3219 (1966).
10. J. C. D. Brand and W. Snedden, *Trans. Faraday Soc.* **53**, 894 (1957).
11. C. R. Bock and M. S. Wrighton, *Inorg. Chem.* **16**, 1309 (1977).

Metal Carbonyls

2

I. INTRODUCTION

Metal carbonyl complexes presently comprise the most important class of organometallic substances which have been the object of photochemical studies. Table 2-1 lists the elements whose carbonyl complexes have established photochemistry. Indeed, metal carbonyls are among the most photoreactive transition-metal complexes, and it may now be said that metal carbonyls can be used to illustrate most of the known types of lowest energy electronic excited states as well as many of the known excited-state reaction pathways. Consequently, this chapter will represent the bulk of the known qualitative and detailed information concerning excited-state chemistry of organometallic species.

There are now a number of reviews concerning the chemical, structural, and physical properties of metal carbonyls [1], and there have been some

TABLE 2-1

Elements Known to Form Photoreactive Carbonyl Complexes

Group V	Group VI	Group VII	Group VIII		
V	Cr	Mn	Fe	Co	Ni
Nb	Mo	—	Ru	Rh	—
Ta	W	Re	Os	Ir	—

reviews of the excited-state chemistry of such complexes [2–8]. Photo-chemistry became an important synthetic tool in the metal carbonyl field in the late 1950s and early 1960s, and special emphasis on photoinduced ligand substitution has been maintained through recent years. However, it is now well established that a large number of other excited-state decay processes obtain and many have been illustrated. Accordingly, this chapter will include many examples of the fundamental aspects of organometallic photochemistry.

Naturally, the key structural unit in the complexes to be discussed in this chapter is the M—CO linkage. Such complexes are extraordinarily diverse: metal carbonyls are known for numerous low-valent metals; four-, five-, six-, and seven-coordination is found in mononuclear complexes; formal d^n configurations include largely $n = 4$–10; an array of structurally novel carbonyl cluster complexes exist; and the known derivatives of binary metal carbonyls can, or have been, used to illustrate virtually every important mode of metal–ligand coordination.

It is widely believed that the bonding between CO and a metal is a combination of σ and π bonding. Delocalization of $d\pi$ electrons from the central metal into the π^*CO orbital gives rise to π back-bonding, and overlap of σ symmetry orbitals of the metal and CO yields a strong σ-donor interaction for the CO as diagrammed in Fig. 1-1. The relative importance of σ and π interactions is difficult to assess, but one generally associates stronger π back-bonding with lower valent metals which have a greater tendency to delocalize electron density into the ligand. Thus, we associate stable carbonyl complexes with low-valent metals. As a consequence of the large degree of delocalization of the electrons from the central metal into the ligand, these compounds are highly covalent. Therefore, electronic transitions involving these electrons should yield substantial changes in bonding, providing a general rationale for the extreme photosensitivity of the compounds. For d^n cases where $n = 1$–9, one expects the possibility of ligand-field (LF) absorptions as well as charge-transfer (CT) transitions involving CO and the other ligands and the central metal. For some ligands one also must contend with the probability that intraligand (IL) excited states could be achieved. In complexes having metal–metal bonds, one finds electronic transitions which can be associated with the metal–metal bond.

An important photoreaction class of mononuclear species $[M(CO)_nL_x]$ is either dissociative loss of CO or L [Eqs. (2-1) and (2-2), respectively].

$$[M(CO)_nL_x] \xrightarrow{h\nu} [M(CO)_{n-1}L_x] + CO \qquad (2\text{-}1)$$

$$[M(CO)_nL_x] \xrightarrow{h\nu} [M(CO)_nL_{x-1}] + L \qquad (2\text{-}2)$$

Chemistry of the photogenerated coordinatively unsaturated species includes scavenging by nucleophiles or oxidative-addition substrates. There are just now examples of photooxidation or photoreduction processes which are primary excited-state processes. A number of metal carbonyl–olefin complexes are photosensitive with respect to olefin chemistry (isomerization, dimerization, oligomerization, hydrogenation, hydrosilation, and hydroformylation). This photochemistry will be presented in Chapter 3. Polynuclear metal carbonyls, especially tri- and dinuclear complexes, undergo photodeclusterification reactions via primary cleavage of direct metal–metal bonds. Finally, optical luminescence has been detected from a number of carbonyl complexes and is beginning to play a key role in elucidating the nature of electronic excited states.

II.   VANADIUM, NIOBIUM, AND TANTALUM COMPLEXES

A.   GEOMETRIC AND ELECTRONIC STRUCTURES

1. d^6 Species

The V, Nb, and Ta carbonyls have proved difficult to obtain, but a new synthesis for the d^6 M(-1) hexacarbonyl species was recently reported [9].

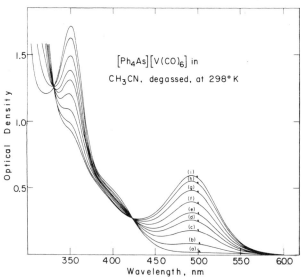

FIG. 2-1.   Electronic spectral changes accompanying 366 nm photolysis of $[Ph_4As][V(CO)_6]$ in CH_3CN to yield $[V(CO)_5(NCCH_3)]^-$. Irradiation times (a) 0 sec, (b) 63 sec, (c) 144 sec, (d) 226 sec, (e) 300 sec, (f) 427 sec, (g) 621 sec, (h) 887 sec, and (i) 1221 sec. Quantum yield for formation of $[V(CO)_5(NCCH_3)]^-$ is 0.51. See Table 2-2 for absorptivities. Reprinted with permission from Wrighton et al. [11], Inorg. Chem. **15**, 434 (1976). Copyright from the American Chemical Society.

The species $[C][M(CO)_6]$ (M = V, Nb, Ta) have been isolated with a number of counterions C^+, and these have received some study in terms of their electronic structure.

The O_h, d^6 $[V(CO)_6]^-$ species was included in an electronic spectral study [10] of d^6 hexacarbonyls. The lowest electronic absorption band has been attributed to an LF $^1A_{1g}(t_{2g}^6) \rightarrow {}^{1,3}T_{1g}(t_{2g}^5e_g)$ type transition. A subsequent study [11] for the complete V, Nb, Ta triad showed the electronic spectrum of the Nb and Ta species to be very similar to that for the V species. Representative electronic spectra for the three complexes are included in Figs. 2-1, 2-2, and 2-3; and spectral assignments are detailed in Table 2-2.

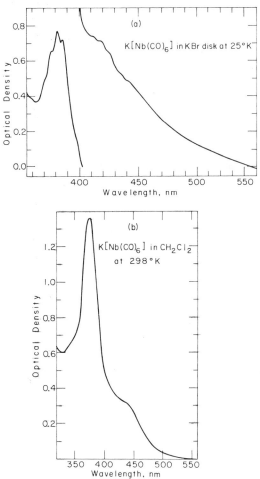

FIG. 2-2. Electronic spectrum of $[K(diglyme)_3][Nb(CO)_6]$. The molar absorptivity of the band near 380 nm is ~ 6000 M^{-1} cm^{-1}; cf. Table 2-2. Reprinted with permission from Wrighton et al. [11], Inorg. Chem. **15**, 434 (1976). Copyright by the American Chemical Society.

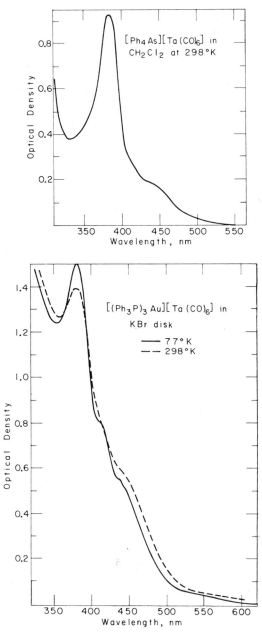

FIG. 2-3.   Electronic spectrum of $[Ph_4As][Ta(CO)_6]$; cf. Table 2-2 for absorptivities. Reprinted with permission from Wrighton *et al.* [11], *Inorg. Chem.* **15**, 434 (1976). Copyright by the American Chemical Society.

TABLE 2-2

Electronic Absorption Spectral Data for Low-Spin d^6 [M(CO)$_6$]$^-$[a,b]

Compound	Solvent	Energy, cm^{-1} × 10^{-3} (ε, M^{-1} cm^{-1})		
		$^1A_{1g}(t_{2g}{}^6) \rightarrow {}^3T_{1g}(t_{2g}{}^5e_g{}^1)$	$^1A_{1g}(t_{2g}{}^6) \rightarrow {}^1T_{1g}(t_{2g}{}^5e_g{}^1)$	M $\rightarrow \pi^*$CO CT
[Ph$_4$As][V(CO)$_6$]	CH$_3$CN	Not observed	25.06 (2090)	28.41 (7150)
	Pyridine	Not observed	25.00 (2340)	28.25 (8160)
	THF	Not observed	24.87 (2170)	28.41 (7560)
	2-MeTHF	Not observed	25.00 (2200)	28.57 (7500)
	KBr	Not observed	24.39 (—)	28.41 (—)
[Et$_4$N][V(CO)$_6$]	CH$_3$CN	Not observed	24.87 (1890)	28.41 (6510)
	Pyridine	Not observed	24.75 (2070)	28.17 (7420)
[K(diglyme)$_3$][Nb(CO)$_6$]	KBr[c]	22.73 (—)	24.39 (—)	26.11 (—)
	CH$_2$Cl$_2$	22.73 (~1500)		26.66 (~6000)
[Ph$_4$As][Ta(CO)$_6$]	CH$_2$Cl$_2$	22.17 (0.31)[d]	24.69 (0.45)	26.18 (1.2)
[(Ph$_3$P)$_3$Au][Ta(CO)$_6$]	KBr	22.22 (0.65)	24.39 (0.8)	25.97 (1.5)
[n-Bu$_4$N][Ta(CO)$_6$]	KBr	22.22	~24.2	26.18
	CH$_2$Cl$_2$	22.22	~24.2	26.18
	EPA	22.22 (920)	24.10 (2400)	26.18 (5800)

[a] Data from Wrighton et al. [11].

[b] At 298°K unless noted otherwise.

[c] 25°K.

[d] Italicized values in parenthesis are the optical densities taken from a spectrum where the concentration of the species is not known.

Note that it is the band in the 400–450 nm region which is attributed to the LF transition and gives the complexes their yellow color. The more intense, sharp band at ~ 380 nm is assigned as the M → π*CO CT band. The spectral features for the hexacarbonyl anions are remarkably similar to those for the isoelectronic, neutral Cr, Mo, and W hexacarbonyls, except that all transitions are red-shifted in the anions. The red-shifted LF and CT transitions in the anions are consistent with the lower oxidation state of the central metal.

In contrast to a number of other organometallic complexes, the low-temperature spectra of $[Nb(CO)_6]^-$ or $[Ta(CO)_6]^-$, but curiously not the $[V(CO)_6]^-$ species, exhibit rich vibrational structure. Spectra in KBr disks or glassy solution show discernible structure logically associated with the M—CO linkage. The systems lend themselves to detailed study, but as yet no detailed analysis has been made.

Complexes of the general formula $[M(CO)_5X]^-$ have not been studied in detail; but for X < CO in LF strength, one finds an absorption at much lower energy than in $[M(CO)_6]^-$ [11]. This is consistent with a C_{4v} complex having lowest $^1A(e^4b_2{}^2) \rightarrow {}^{1,3}E(e^3b_2{}^2a_1)$ LF transitions. The spectrum of $[V(CO)_5(NCCH_3)]^-$ is included in Fig. 2-1, and the ~ 5000 cm^{-1} red shift of the first band position for the C_{4v} compared to the O_h complex is comparable to that found in the isoelectronic neutral complexes like $[Cr(CO)_6]$ compared to $[Cr(CO)_5(NCCH_3)]$ [12].

2.   d^5 Species $[V(CO)_6]$

Since its discovery [13], $[V(CO)_6]$ has been the object of some controversy concerning its electronic and geometric formulation [14]. As a mononuclear d^5 species in solution, it is a unique carbonyl in that it is paramagnetic and does not obey the 18-e^- rule. In contrast to the yellow $[V(CO)_6]^-$ species, $[V(CO)_6]$ is an almost black solid, signaling a possible CO → V dπ CT assignment for the lowest absorptions.

The elusive $[V_2(CO)_{12}]$ has been synthesized by metal atom–ligand co-condensation experiments [15,16]. Spectral studies show that this substance is not that obtained by cooling $[V(CO)_6]$ [15,16], and the unusual magnetic phenomena [17] as a function of temperature cannot therefore be ascribed to dimerization of $[V(CO)_6]$. The low-temperature optical absorption spectrum of $[V(CO)_6]$ [15] has an onset near 400 nm, quite distinct from the room-temperature spectrum. The low-temperature spectrum has been assigned by attributing the lowest energy absorption at 388 nm to an LF transition, $^2B_{2g}(e_g{}^4b_{2g}) \rightarrow {}^2E_g(e_g{}^3b_{2g}a_{1g})$, in the tetragonally distorted $V(CO)_6$. This assignment gains some credibility from the fact that in $[V(CO)_6]^-$ the $^1A_{1g}(t_{2g}{}^6) \rightarrow {}^1T_{1g}(t_{2g}{}^5e_g)$ transition also falls at about 400 nm.

However, the large change in the spectrum from room-temperature solutions to low-temperature remains unexplained.

3.   d⁴ Species $[M(\eta^5\text{-}C_5H_5)(CO)_4]$

The only other set of V, Nb, and Ta carbonyl complexes that has been the object of photochemical studies is the $[M(\eta^5\text{-}C_5H_5)(CO)_4]$ (M = V, Nb, Ta) species where the metal is in the $+1$ oxidation state. These complexes are d⁴ and seven-coordinate, satisfying the $18\text{-}e^-$ rule. Spectral studies have not been reported.

B.   LUMINESCENCE STUDIES

Optical emission has been observed [11] from salts of $[M(CO)_6]^-$ for M = Nb or Ta but not V. Some of the emission spectral features for the pure powders are given in Table 2-3. The emission has been assigned as the emission from the lowest LF excited state. The emission properties do depend somewhat on the cation, but the emission maximum itself is roughly independent of the counterion. The emission spectrum of the $[Ph_4P][Ta(CO)_6]$ exhibits vibrational structure. The lack of emission from the V species suggests that significant spin–orbit coupling is crucial to competitive radiative decay of the lowest excited state.

The absolute emission quantum yield Φ_e and the emission lifetime τ were measured as a function of temperature for the three $[Ta(CO)_6]^-$ salts given in Table 2-3. Emission above $\sim 80°K$ is only barely detectable, and Φ_e increases sharply below this point. Data between 25 and 75°K show that Eq.

TABLE 2-3

Emission Characteristics of $[M(CO)_6]^-$ Salts[a,b]

Compound	Emission maximum, cm^{-1} × 10^{-3} (width at half-height, cm^{-1} × 10^{-3})	Φ_e ±20%	τ × 10^6, sec ±10%
$[n\text{-}Bu_4N][Ta(CO)_6]$	17.20 (3.0)	0.05	23
$[Ph_4As][Ta(CO)_6]$	17.36 (2.2)	0.16	63
$[(Ph_3P)_3Au][Ta(CO)_6]$	17.08 (1.8)	0.25	74
$[K(diglyme)][Nb(CO)_6]$	17.10 (2.6)	—	—

[a] Reprinted with permission from Wrighton et al. [11], Inorg. Chem. **15**, 434 (1976). Copyright by the American Chemical Society.

[b] Data are for the pure powders at 25°K with excitation at 370 nm.

(2-3) obtains; that is, the ratio of the lifetimes at two different temperatures

$$\tau(T_1)/\tau(T_2) = \Phi_e(T_1)/\Phi_e(T_2) \tag{2-3}$$

is equal to the ratio of the quantum yields at those same two temperatures. Using Eqs. (1-9) and (1-10), this means that the radiative decay constant k_r is essentially invariant, and it is the change in the nonradiative decay constant k_{nr} that is responsible for the strong temperature dependence of the emission properties.

C.   PHOTOREACTIONS

1.   Photoreactions of $[M(CO)_6]^-$ and Derivatives

Photosubstitution represents the known photochemistry of $[M(CO)_6]^-$. Most studies have dealt with the V species, and substituted derivatives have been photochemically prepared in synthetic quantities [18]. Table 2-4 shows some of the compounds that have been prepared by the photosubstitution route. The primary photoproduct is probably the coordinatively unsaturated $[M(CO)_5]^-$ which is then scavenged by the nucleophile.

$$[M(CO)_6]^- \xrightarrow{\ h\nu\ } [M(CO)_5]^- + CO \tag{2-4}$$

$$M = V, Nb, Ta$$

TABLE 2-4

Photosubstitution of $[M(CO)_6]^{-a}$

M	Entering ligand	Product
V	$P(OPh)_3$	$[V(CO)_5(P(OPh)_3)]^-$
	PPh_3	$[V(CO)_5(PPh_3)]^-$
	$Ph_2PCH_2CH_2PPh_2$	$[V(CO)_5(Ph_2PCH_2CH_2PPh_2)]$
	$AsPh_3$	$[V(CO)_5(AsPh_3)]^-$
	$SbPh_3$	$[V(CO)_5(SbPh_3)]^-$
	$P(n\text{-}C_4H_9)_3$	$[V(CO)_5(P(n\text{-}C_4H_9)_3)]^-$
	$Ph_2PCH_2CH_2PPh_2$	$[V(CO)_4(Ph_2PCH_2CH_2PPh_2)]^-$
	$(Ph_2AsC_5H_4)_2Fe$	$[V(CO)_4((Ph_2AsC_5H_4)_2Fe)]^-$
	2,2'-Bipyridine	$[V(CO)_4(2,2'\text{-bipyridine})]^-$
Nb	PPh_3	$[Nb(CO)_5(PPh_3)]^-$
	$Ph_2PCH_2CH_2PPh_2$	$[Nb(CO)_4(Ph_2PCH_2CH_2PPh_2)]^-$
Ta	PPh_3	$[Ta(CO)_5(PPh_3)]^-$
	$Ph_2PCH_2CH_2PPh_2$	$[Ta(CO)_4(Ph_2PCH_2CH_2PPh_2)]^-$

^{a} Data from Davison and Ellis [18].

Irradiation of $[M(CO)_6]^-$ in rigid media at low temperature yields optical spectral changes consistent with the loss of CO to generate C_{4v} $[M(CO)_5]^-$ species [11,19]. Owing to the nature of the solvents used, it is likely that the sixth coordination site is filled with a solvent molecule. A degree of reversibility was noted in the Nb case in a KBr matrix at $26°K$ (Eq. 2-5).

$$[Nb(CO)_6]^- \underset{h\nu'}{\overset{h\nu}{\rightleftarrows}} [Nb(CO)_5]^- + CO \qquad (2\text{-}5)$$

$$\lambda_{max} = 380 \text{ nm} \qquad \lambda_{max} = 550 \text{ nm}$$

KBr matrix, $26°K$

The photosubstitution quantum efficiency is very high (Table 2-5), and the yields are apparently independent of the excitation wavelength. The data suggest that the reactive states are the lowest LF states achieved by $t_{2g} \rightarrow e_g$ excitations. High quantum yields (~ 0.5) for loss of CO are consistent with the fact that the t_{2g} levels are π-bonding with respect to M—C interactions, while the e_g levels are strongly σ-antibonding. Since the d^6 species are anions, the t_{2g} levels are probably more π-bonding than in the neutral and cationic electronic analogs. Thus, depopulation of t_{2g} should cause considerable bond weakening.

No studies of the photochemistry of the substituted derivatives of $[M(CO)_6]^-$ have been reported. However, intermediates during the substitution to form $[V(CO)_4L]^-$ (L = bidentate ligand) are consistent with stepwise loss of CO as in Eq. (2-6). Photosubstitution of $[M(CO)_5X]^-$ is thus likely.

$$[V(CO)_6]^- \xrightarrow[Ph_2PCH_2CH_2PPh_2]{h\nu} [V(CO)_5(Ph_2PCH_2CH_2PPh_2)]^- + CO$$

$$\downarrow h\nu \qquad\qquad (2\text{-}6)$$

$$[V(CO)_4(Ph_2PCH_2CH_2PPh_2)]^- + CO$$

2. $[M(\eta^5\text{-}C_5H_5)(CO)_4]$ Complexes

The photochemistry reported for these complexes again involves the substitution of CO; examples are given in Table 2-6 [20]. The primary photosubstitution of two CO groups by acetylenes is unique, and it is curious that two acetylenes in the same molecule can apparently bind differently. One photochemical reaction is reported involving coordinated acetylenes and CO (Eq. 2-7), but no information regarding quantum efficiency is known.

$$(2\text{-}7)$$

TABLE 2-5

Photosubstitution of $[M(CO)_6]^-$ in Solution[a,b]

Compound	Entering group, X	$[M(CO)_5X]^-$ Formation quantum yield			Isosbestic points, nm	$[M(CO)_5X]^-$ Absorption maximum, nm (ε)
		313 nm	366 nm	436 nm		
$[Ph_4As][V(CO)_6]$	CH_3CN	—	0.51	0.56	333, 423	495 (2470)
	Pyridine	—	0.48	0.57	340, 382	520 (5660)
	4-Ethylpyridine	—	—	—	—	520
$[Et_4N][V(CO)_6]$	CH_3CN	0.54	0.60	0.63	327, 430	493 (1740)
	Pyridine	—	0.51	0.58	—	518 (4620)
$[n\text{-}Bu_4N][Ta(CO)_6]$	Pyridine	—	—	—	358, 411	536

[a] Reprinted with permission from Wrighton et al. [11], Inorg. Chem. **15**, 434 (1976). Copyright by the American Chemical Society.
[b] Degassed solutions of the $[M(CO)_6]^-$ salts in a neat solution of X at 298°K.

TABLE 2-6

Photosubstitution of $[M(\eta^5-C_5H_5)(CO)_4]$ and Derivatives[a]

Starting complex	Entering group, L	Product
$[V(\eta^5-C_5H_5)(CO)_4]$	PhC≡CPh	$[V(\eta^5-C_5H_5)(CO)_2L]$
$[V(\eta^5-C_5H_5)(CO)_3PPh_3]$	PPh$_3$	$[V(\eta^5-C_5H_5)(CO)_2PPh_3L]$
$[V(\eta^5-C_5H_5)(CO)_2(PPh_3)_2]$	PhC≡CPh	$[V(\eta^5-C_5H_5)(CO)(PPh_3)_2L]$
$[V(\eta^5-C_5H_5)(CO)_4]$	1,3-Butadiene	$[V(\eta^5-C_5H_5)(CO)_3L]$
$[V(\eta^5-C_5H_5)(CO)_4]$	1,3-Cyclohexadiene	$[V(\eta^5-C_5H_5)(CO)_3L]$
$[V(\eta^5-C_5H_5)(CO)_4]$	2,3-Dimethyl-1,3-butadiene	$[V(\eta^5-C_5H_5)(CO)_3L]$
$[Nb(\eta^5-C_5H_5)(CO)_4]$	PhC≡CPh	$[Nb(\eta^5-C_5H_5)(CO)_2L]$
$[Nb(\eta^5-C_5H_5)(CO)_3PPh_3]$	PPh$_3$	$[Nb(\eta^5-C_5H_5)(CO)_2PPh_3L]$
$[Nb(\eta^5-C_5H_5)(CO)_2(PPh_3)_2]$	PhC≡CPh	$[Nb(\eta^5-C_5H_5)(CO)(PPh_3)_2L]$
$[Nb(\eta^5-C_5H_5)(CO)_2PhC≡CPh]$	PhC≡CPh	$[Nb(\eta^5-C_5H_5)(CO)(PhC≡CPh)L]$
$[Ta(\eta^5-C_5H_5)(CO)_4]$	PhC≡CPh	$[Nb(\eta^5-C_5H_5)(CO)_2L]$

[a] Data from Nesmeyanov, Fischer *et al.* [20].

III. CHROMIUM, MOLYBDENUM, AND TUNGSTEN CARBONYLS

A. GEOMETRIC STRUCTURE

The commonly known carbonyls of Cr(0), Mo(0), and W(0) are six-coordinate octahedral complexes $[M(CO)_6]$ (for structural determinations, see [21]). Other stable complexes containing only the central metal and CO include the dimers $[M_2(CO)_{10}]^{2-}$ having a single M—M bond [22,23]. Numerous compounds of the $[M(CO)_n(L)_{6-n}]$ variety have been prepared, many photochemically [2–7].

Complexes which are formally seven-coordinate are also found.[†] A typical example is the dimeric complex $[Mo_2(CO)_6(\eta^5-C_5H_5)_2]$ (III). Assuming $\eta^5-C_5H_5$ to have a negative charge and to be a six-electron donor occupying three coordination sites, the central metal is in the $+1$ oxidation state. In complex (IV), though, it is appropriate to identify the central metal

$(CO)_3Mo—Mo(CO)_3$

(III)

(IV)

[†] We adopt the formalism that cyclobutadiene dianion, cyclopentadiene anion, and benzene are six-electron donors and hence can occupy three coordination sites.

as being in a $+2$ oxidation state. Other complexes involving the $+2$ oxidation state are clearly seven-coordinate as exemplified by species such as $[W(CO)_2(diars)_2(I)]I$ and $[Mo(CO)_3(L)I_2]$ (L = bidentate ligand) [24]. Finally, seven-coordinate compounds of the type $[W(CO)_3(diars)Br_2]^+$ can be obtained [24]. Thus, for the zerovalent metal complexes, six-coordination is common, while for the $+1$, $+2$, and $+3$ oxidation states seven-coordination is found. Important work in the area of excited-state chemistry involves the six-coordinate compounds and η^5-C_5H_5 complexes like (III) and (IV).

B.   Electronic Structure

Considerable effort has been directed toward understanding the electronic structure of carbonyl complexes of Cr, Mo, and W. Both mononuclear and dinuclear metal–metal bonded complexes have been the object of study, and the results have proved valuable in a number of important photochemical systems. Generally, the complexes exhibit a number of intense ($\varepsilon > 10^2$) transitions in the UV–visible region which are associated with LF and MLCT and LMCT absorptions. Intraligand absorptions are indicated in several cases, and in dinuclear complexes absorptions associated with the M—M bond often dominate the low-energy region of the optical spectrum.

1. $[M(CO)_6]$ Complexes

Spectra for the $[M(CO)_6]$ (M = Cr, Mo, W) compounds were determined early [25], and the lowest energy absorption at $\sim 30{,}000$ cm^{-1} in each was assigned as the $^1A_{1g} \rightarrow {}^1T_{1g}$ LF transition. The band appears only as a shoulder on the more intense M $\rightarrow \pi^*$CO CT absorption at $\sim 35{,}000$ cm^{-1}. The second LF band, $^1A_{1g} \rightarrow {}^1T_{2g}$, predicted for d^6 O_h complexes, can be observed in the vicinity of $\sim 37{,}500$ cm^{-1} for the $[M(CO)_6]$ species. The most intense transition at $\sim 43{,}000$ cm^{-1} is assigned as a second component of the M $\rightarrow \pi^*$CO CT transition. In later studies [10], which included the isoelectronic $[V(CO)_6]^-$, $[Mn(CO)_6]^+$, and $[Re(CO)_6]^+$ complexes, the same assignments were made except for one band at an energy below the $^1A_{1g} \rightarrow {}^1T_{1g}$ absorption having $\varepsilon \approx 1000$ for $[W(CO)_6]$, $\varepsilon \approx 350$ for $[Mo(CO)_6]$, and not present in the $[Cr(CO)_6]$. This low-energy absorption was identified as the lowest LF spin-forbidden singlet $\rightarrow$ triplet transition, $^1A_{1g} \rightarrow {}^3T_{1g}$. Spectra of $[M(CO)_6]$ in the low-energy region are shown in Fig. 2-4, and data are summarized in Table 2-7. The enhanced intensity of the $^1A_{1g} \rightarrow {}^3T_{1g}$ transition with increasing atomic weight of the central metal is expected owing to the larger spin–orbit coupling in the heavier metal. The constancy of the value of $10Dq$ for $[Cr(CO)_6]$, $[Mo(CO)_6]$, and $[W(CO)_6]$

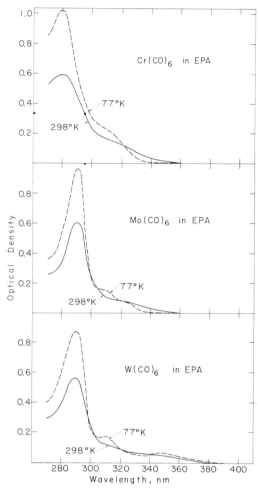

FIG. 2-4. Electronic absorption spectra of [M(CO)₆]. Spectral changes accompanying change in temperature from 298 to 77°K are not corrected for solvent contraction. For molar absorptivities see Table 2-7.

is due to a balancing of diminishing σ bonding and increasing π bonding for the heavier metal systems [10]. While it has been argued [26] that all of the bands in these complexes are CT absorptions, the LF treatment provides the best rationale for the band positions including the lowest singlet → triplet absorption. Further, the LF approach accounts well for the observed spectral changes occurring upon substitution to yield [M(CO)₅X] and [M(CO)₄X₂] (*vide infra*). Finally, the intensities of the LF transitions are uncommonly large because of the high degree of covalence in these molecules; i.e., the molecular orbitals have substantial contribution from both metal and ligand

TABLE 2-7

Electronic Spectral Assignments for $[M(CO)_6]$ (M = Cr, Mo, W)[a,b]

Complex	Bands, cm^{-1} (ε)	Assignment
$[Cr(CO)_6]$	29,500 (700) ⎱ 31,550 (2670) ⎰	$^1A_{1g} \to {}^1T_{1g}$ ($t_{2g}^6 \to t_{2g}^5 e_g^1$)
	35,700 (13,100)	$^1A_{1g} \to c^1T_{1u}$ (M → π*CO)
	38,850 (3500)	$^1A_{1g} \to {}^1T_{2g}$ ($t_{2g}^6 \to t_{2g}^5 e_g^1$)
	43,600 (85,100)	$^1A_{1g} \to d^1T_{1u}$ (M → π*CO)
$[Mo(CO)_6]$	28,850 (350)	$^1A_{1g} \to {}^3T_{1g}$ ⎫
	30,150 (1690) ⎱ 31,950 (2820) ⎰	$^1A_{1g} \to {}^1T_{1g}$ ⎬ ($t_{2g}^6 \to t_{2g}^5 e_g^1$)
	34,600 (16,800)	$^1A_{1g} \to c^1T_{1u}$ (M → π*CO)
	37,200 (7900)	$^1A_{1g} \to {}^1T_{2g}$ ($t_{2g}^6 \to t_{2g}^5 e_g^1$)
	42,800 (138,000)	$^1A_{1g} \to d^1T_{1u}$ (M → π*CO)
$[W(CO)_6]$	28,300 (1000)	$^1A_{1g} \to {}^3T_{1g}$ ⎫
	29,950 (1680) ⎱ 31,850 (3250) ⎰	$^1A_{1g} \to {}^1T_{1g}$ ⎬ ($t_{2g}^6 \to t_{2g}^5 e_g^1$)
	34,650 (17,600)	$^1A_{1g} \to c^1T_{1u}$ (M → π*CO)
	37,100 (7400)	$^1A_{1g} \to {}^1T_{2g}$ ($t_{2g}^6 \to t_{2g}^5 e_g^1$)
	43,750 (208,000)	$^1A_{1g} \to d^1T_{1u}$ (M → π*CO)

[a] Data from Beach and Gray [10].
[b] CH_3CN solution at 300°K.

atomic orbitals, tending to remove restrictions associated with the intensity of "d—d" transitions.

The d-orbital one-electron energy level diagram for $[M(CO)_6]$ is shown in Fig. 2-5. The ground electronic state $^1A_{1g}$ has a t_{2g}^6 electronic configuration and the one-electron excitation to $t_{2g}^5 e_g^1$ yields the $^{1,3}T_{1g}$ and $^{1,3}T_{2g}$ excited states [27]. These one-electron excitations can result in dramatic changes in the substitutional lability of $[M(CO)_6]$ since both σ bonding and π bonding are diminished by depopulation of $t_{2g}(\pi_b)$ and population of $e_g(\sigma^*)$ [28].

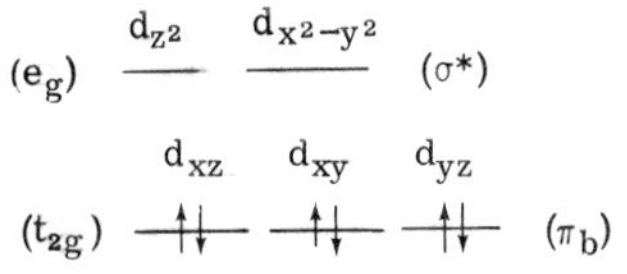

FIG. 2-5. One-electron level diagram for $[M(CO)_6]$ (M = Cr, Mo, W).

2. $[M(CO)_n L_{6-n}]$ Complexes

Complexes of C_{4v} and C_{2v} symmetry have received some attention. Matrix isolation techniques have even allowed the study of the (photogenerated) coordinatively unsaturated $[M(CO)_5]$ species by optical spectroscopy. A

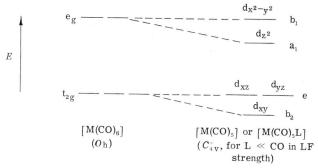

FIG. 2-6.   One-electron level diagram for O_h and C_{4v} complexes.

number of complexes can be understood in terms of LF lowest excited states; others feature lowest MLCT excited states.

Vibrational spectra [29,30] of photogenerated $[M(CO)_5]$ in low-temperature matrices show it to have a square-pyramidal (C_{4v}) structure. One reported metal atom–ligand cocondensation experiment [31] gives $[Cr(CO)_5]$ a D_{3h} structure which has been disputed by others [29,30]. The removal of a CO from $[M(CO)_6]$ to generate the C_{4v} species should affect the d-orbital diagram as shown in Fig. 2-6. The lowest energy spectral feature of $[M(CO)_5]$ has been assigned as the $^1A_1(e^4b_2{}^2) \rightarrow {}^1E(e^3b_2{}^2a_1{}^1)$ LF absorption, [32], and the polarization of the band is consistent with the interpretation [32a]. Data in Table 2-8 show that the spectrum of $[M(CO)_5]$ is very sensitive to the matrix, and this can be ascribed to the geometry-dependent energy of the lowest excited state. But in every case the lowest energy absorption band for $[M(CO)_5]$ is significantly red-shifted when compared to $[M(CO)_6]$, as expected from Fig. 2-6.

TABLE 2-8

Spectral Features of Matrix Isolated $[M(CO)_5]^{a,b}$

Matrix	Band maxima, cm^{-1}		
	$[Cr(CO)_5]$	$[Mo(CO)_5]$	$[W(CO)_5]$
Ne	16,000	—	—
SF$_6$	17,900	22,200	21,700
CF$_4$	18,300	—	—
Ar	18,800	23,300	22,900
Kr	19,300	—	—
Xe	20,300	24,200	24,000
CH$_4$	20,400	24,300	24,200

[a] Data from Perutz and Turner [32].
[b] Only the $^1A_1(e^4b_2{}^2) \rightarrow {}^1E(e^3b_2{}^2a_1{}^1)$ absorption is given.

Absorption spectra for $[M(CO)_5L]$ (L $\ll$ CO in LF strength) have been assigned in the low energy region [33–37]. For ligands L having no low-lying π^* levels (e.g., piperidine) the lowest absorption band system has been associated with the $e^4b_2^2 \rightarrow e^3b_2^2a_1^1$ transition (Fig. 2-6). For M = W the spin-

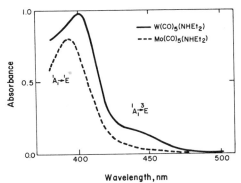

FIG. 2-7. Electronic absorption spectra of $[W(CO)_5(NHEt_2)]$ (——) and $[Mo(CO)_5(NHEt_2)]$ (————) in aliphatic hydrocarbon solution. The intense band ($\varepsilon \cong 5000$) in the vicinity of 400 nm is identified as the $^1A_1(e^4b_2^2) \rightarrow {}^1E(e^3b_2^2a_1^1)$ spin-allowed transition, and the shoulder only observed for the tungsten complex is the corresponding spin-forbidden $^1A_1 \rightarrow {}^3E$ transition.

TABLE 2-9

$[M(CO)_5L]$ Complexes Having LF Lowest Excited States in Absorption

M	L	Absorption band maxima, cm^{-1} (ε)	Assignment	Reference
Cr	Piperidine[a]	23,810 (3500)	LF($^1A_1 \rightarrow {}^1E$)	[34]
		40,000	M $\rightarrow \pi^*$CO CT	
Cr	PPh$_3$[a]	27,855 (1400)	LF($^1A_1 \rightarrow {}^1E$)	[35]
Mo	NH$_3$[b]	25,063 (4200)	LF($^1A_1 \rightarrow {}^1E$)	[36]
	n-PrNH$_2$[b]	25,253 (4200)	LF($^1A_1 \rightarrow {}^1E$)	[36]
	Piperidine[b]	25,381 (5000)	LF($^1A_1 \rightarrow {}^1E$)	[36]
W	NH$_3$[b]	22,727 (600) sh	LF($^1A_1 \rightarrow {}^3E$)	[36]
		24,510 (3800)	LF($^1A_1 \rightarrow {}^1E$)	[36]
	n-PrNH$_2$[b]	22,727 (610) sh	LF($^1A_1 \rightarrow {}^3E$)	[36]
		24,631 (3800)	LF($^1A_1 \rightarrow {}^1E$)	[36]
	Piperidine[b]	22,727 (590) sh	LF($^1A_1 \rightarrow {}^3E$)	[36]
		24,814 (3850)	LF($^1A_1 \rightarrow {}^1E$)	[36]
	Et$_2$O[a]	21,930 —	LF($^1A_1 \rightarrow {}^3E$)	[33]
		23,920	LF($^1A_1 \rightarrow {}^1E$)	[33]
	Br^{-c}	22,030 (500)	LF($^1A_1 \rightarrow {}^3E$)	[37]
		24,150 (2230)	LF($^1A_1 \rightarrow {}^1E$)	[37]

[a] Aliphatic hydrocarbon, ambient temperature.
[b] Benzene solution, 298°K.
[c] EtOH solution, 300°K.

forbidden $^1A_1(e^4b_2{}^2) \rightarrow {}^3E(e^3b_2{}^2a_1{}^1)$ transition has significant intensity; a comparison of the lowest spectral features in $[Mo(CO)_5(NHEt_2)]$ and $[W(CO)_5(NHEt_2)]$ is given in Fig. 2-7. Table 2-9 gives data for C_{4v} complexes having lowest LF excited states in absorption. Generally, the upper excited states have not been definitively assigned.

If L has low-lying π^* orbitals, MLCT may be the lowest energy transition in $[M(CO)_5L]$. A comparison of the absorption spectrum of $[W(CO)_5(piperi-dine)]$ and $[W(CO)_5(4\text{-formylpyridine})]$ (Fig. 2-8) is illustrative [38]. The 4-formylpyridine complex exhibits a low-lying W → L CT absorption, while exhibiting all of the low-lying LF bands present in the $[W(CO)_5(piperdine)]$ system. Variation in the pyridine substituents results in variation in the CT band position (Table 2-10). The more electron-releasing substituents give a higher energy CT absorption, as expected. The absolute value of the shift in W → L CT position is remarkably similar to that found in the analogous series of d^6 $[Ru(NH_3)_5L]^{2+}$ complexes [39]. Finally, concerning this series of complexes, it should be noted that the M → L CT bands are very solvent-sensitive; more polar or polarizable solvents give blue-shifted CT band maxima, while LF bands are little affected. Consequently, when LF and M → L CT bands are overlapping, it is possible to resolve them by varying the solvent [38]. An example is given in Fig. 2-9.

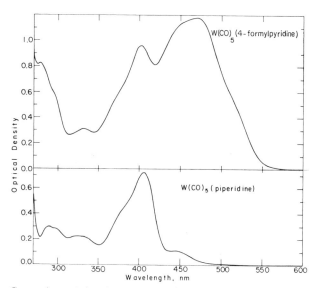

FIG. 2-8. Comparison of the absorption spectra of $[W(CO)_5(piperidine)]$ and $[W(CO)_5$ (4-formylpyridine)] at 298°K in isooctane. Note that all of the bands present in $[W(CO)_5$ (piperidine)] are present in $[W(CO)_5(4\text{-formylpyridine})]$ at very nearly the same position and relative intensity. Reprinted with permission from Wrighton *et al.* [38], *J. Am. Chem. Soc.* **98**, 4105 (1976). Copyright by the American Chemical Society.

TABLE 2-10

Spectral Band Maxima and Assignments for $[W(CO)_5L]$ Complexes[a,b]

L^c	Absorption maximum,[d] nm (ε)	$^1A_1(e^4b_2^2) \to {}^3E(e^3b_2^2a_1^1)$,[e] nm ($\varepsilon$)	$^1A_1(e^4b_2^2) \to {}^1E(e^3b_2^2a_1^1)$, nm ($\varepsilon$)	$W \to L$ CT, nm (ε)
Piperidine	407 (3960)	443 (560)	407 (3960)	None
3,4-DiMe-py	347 (6580)	435 (615)	~390	347 (6580)
4-Me-py	351 (6960), 375 (6730)	437 (655)	~390	351 (6960)
Py	382 (7480)	440 (615)	~390	355 (~6800)
3-Br-py	399 (9670)		Overlapping CT and LF[f]	
3-Acetyl-py	398 (7550)		Overlapping CT and LF[f]	
3-Benzoyl-py	400 (5700)		Overlapping CT and LF[f]	
3,5-DiBr-py	407 (9410)		Overlapping CT and LF[f]	
4-Benzoyl-py	405 (7410), 435 (6600)	g	405 (7410)	435 (6600)
4-Acetyl-py	404 (8400), 440 (8200)	g	404 (8400)	440 (8200)
4-CN-py	404 (5530), 455 (7060)	g	404 (5530)	455 (7060)
4-Formyl-py	402 (5300), 470 (6470)	~445	402 (5300)	470 (6470)

[a] Reprinted with permission from Wrighton et al. [38], J. Am. Chem. Soc. **98**, 4105 (1976). Copyright by the American Chemical Society.

[b] Isooctane solutions at 298°K.

[c] py = pyridine.

[d] Only actual band maxima are given here.

[e] Generally, this band is only a shoulder.

[f] For example, cf. curves D and E in Fig. 2-9.

[g] Obscured by W → L CT absorption.

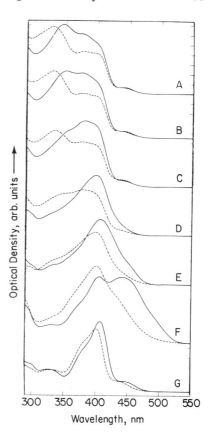

FIG. 2-9. Absorption spectra at 298°K in isooctane (———) and isooctane/EtOH (1:1 volume) (----) for [W(CO)₅X] at the same concentration for both solvents. (A) [W(CO)₅(3,4-diMe-pyridine)]; (B) [W(CO)₅(4-Me-pyridine)]; (C) [W(CO)₅(pyridine)]; (D) [W(CO)₅(3-Br-pyridine)]; (E) [W(CO)₅(3,5-diBr-pyridine)]; (F) [W(CO)₅(4-acetylpyridine)]; (G) [W(CO)₅-piperidine)]. Reprinted with permission from Wrighton *et al.* [38], *J. Am. Chem. Soc.* **98**, 4105 (1976). Copyright by the American Chemical Society.

An MLCT assignment is undoubtedly appropriate in a number of other [M(CO)₅L] complexes, but few detailed studies have been reported. For example, the carbene complexes [M(CO)₅C(OC₂H₅)R] have low-lying absorptions [40] near that for the [M(CO)₅(amine)] complexes. However, the carbene is a carbon-bonded system and should not have LF transitions at such low energies. Thus, the assignment as a CT transition is probable. The situation with [W(CO)₅(CPh₂)] is similar; the lowest absorption in hexane is at 435 nm (ε = 10,400) [41] and an MLCT assignment is likely. A final comparison between two [M(CO)₅L] species is interesting; consider the complexes [W(CO)₅(1-pentene)] and [W(CO)₅(tetracyanoethylene)]. The 1-pentene complex is a pale yellow substance and has no well-defined transitions below ∼ 350 nm [42]. In considerable contrast, the blue tetracyanoethylene complex has an intense transition at ∼ 800 nm [43]. This low-lying band at 800 nm is apparently an MLCT transition. The tetracyanoethylene was reported to be bonded to the metal in the same manner as ethylene

[43]. Thus, the extremely large red-shifted lowest absorption must be a consequence of the exceptional π-acceptor properties of the tetracyanoethylene.

Complexes having ligands with low-lying excited states often exhibit intraligand (IL) transitions. One interesting case is [W(CO)$_5$(styrylpyridine)] [44]; this complex exhibits a low-lying IL($\pi \to \pi^*$) absorption, but there are overlapping LF and MLCT bands. The intense near-UV IL($\pi \to \pi^*$) transition is observed near the position found in the free ligand (Fig. 2-10). The *trans*-4-styrylpyridine complex exhibits an IL transition with an intensity, position, and vibrational structure similar to that found in the free ligand. Complexes of *trans*-2-styrylpyridine have similar spectral features [44].

Complexes of C_{2v} symmetry, [M(CO)$_4$L] (L = bidentate ligand) or *cis*-[M(CO)$_4$L$_2$], have been studied [45–48]. All of the definitive studies have involved nitrogen-donor ligands L, and the situation is basically like that in the C_{4v} systems. For aliphatic amine L the lowest excited states are LF, and Fig. 2-11 can be useful in understanding the rather small changes in the position of the lowest absorption in *cis*-[M(CO)$_4$L$_2$] compared to [M(CO)$_5$L]. As in [M(CO)$_5$L], the W complex, but not the Mo *cis*-[M(CO)$_4$L$_2$] species, exhibits a LF singlet → triplet absorption (Fig. 2-12) [48]. The relatively low

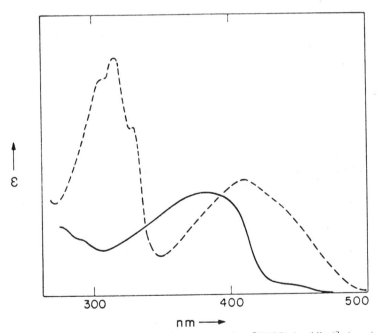

FIG. 2-10. Ultraviolet–visible absorption spectra for [W(CO)$_5$(pyridine)] (———) and [W(CO)$_5$(*trans*-4-styrylpyridine)] (----) in isooctane at 300°K. The maximum molar extinction coefficients for the [W(CO)$_5$(*trans*-4-styrylpyridine)] are 7865 at 408 nm and 16,346 at 316 nm. The band maximum is at 380 nm for [W(CO)$_5$(pyridine)] and molar extinction coefficient is ~6800. Reprinted with permission from Wrighton *et al.* [44].

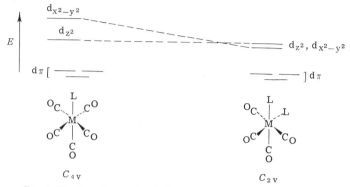

FIG. 2-11.   One-electron level diagrams for C_{4v} and C_{2v} complexes.

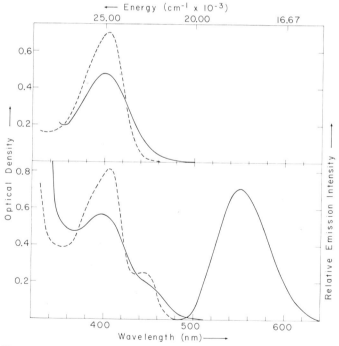

FIG. 2-12.   Absorption (left curves) at 298 (——) and 77°K (----) in EPA for [Mo(CO)$_4$ (ethylenediamine)] (3×10^{-4} M and 1.00 cm pathlength) (upper curves) and [W(CO)$_4$(ethylenediamine)] (at 4×10^{-4} M and 1.00 cm pathlength) (lower curves). Emission of [W(CO)$_4$ (ethylenediamine)] (right, lower curve) in EPA at 77°K. Reprinted with permission from Wrighton and Morse [48].

molar absorptivities, the slight red-shift compared to $[M(CO)_5L]$, and the photochemistry (*vide infra*) are all consistent with the LF assignments for *cis*-$[M(CO)_4L_2]$ or $[M(CO)_4L]$ (L = aliphatic amine). Some data are given in Table 2-11.

In C_{2v} complexes of unsaturated nitrogen-donor ligands MLCT and IL transitions are observable. For L = pyridine, the LF and MLCT transitions in *cis*-$[M(CO)_4L_2]$ are overlapping as in $[M(CO)_5L]$ [48]. But in tetra-carbonyl complexes of 2,2'-bipyridine, 1,10-phenanthroline, and related ligands the MLCT band system is well below the LF band system. [45–48]. Representative spectra are shown in Fig. 2-13, and some data are given in

TABLE 2-11

Spectral Features of $[M(CO)_4L]$ or *cis*-$[M(CO)_4L_2]$ Complexes

Complex	Bands, cm^{-1} (ε)	Assignment	Reference
$[Cr(CO)_4(Ethylenediamine)]^a$	23,600 (1300)	LF($^1A_1 \rightarrow {}^1A_1, {}^1B_2$)	[45,48]
	29,700 (5360)		
	35,599 (6300)		
	40,300 (18,200)	M $\rightarrow \pi^*$CO CT + LF	
	46,200 (21,900)		
$[Mo(CO)_4(Ethylenediamine)]^a$	25,300 (1700)	LF($^1A_1 \rightarrow {}^1A_1, {}^1B_2$)	[45,48]
	32,700 (10,000)		
	38,300 (20,000)	M $\rightarrow \pi^*$CO CT + LF	
	43,400 (20,000)		
cis-$[Mo(CO)_4(Pyridine)_2]^b$	25,320	LF($^1A_1 \rightarrow {}^1A_1, {}^1B_2$) + M $\rightarrow$ py CT	[48]
	33,000	M $\rightarrow \pi^*$CO CT + LF	
$[W(CO)_4(Ethylenediamine)]^a$	22,200 (400)	LF($^1A_1 \rightarrow {}^3A_1, {}^3B_2$)	[45,48]
	25,200 (1400)	LF($^1A_1 \rightarrow {}^1A_1, {}^1B_2$)	
	33,200 (8300)		
	39,200 (27,900)	M $\rightarrow \pi^*$CO CT + LF	
	45,000 (27,700)		
cis-$[W(CO)_4(n\text{-}PrNH_2)_2]^c$	22,200	LF($^1A_1 \rightarrow {}^3A_1, {}^3B_2$)	[48]
	24,270	LF($^1A_1 \rightarrow {}^1A_1, {}^1B_2$)	
	33,670	M $\rightarrow \pi^*$CO CT	
cis-$[W(CO)_4(Piperidine)_2]^c$	22,470	LF($^1A_1 \rightarrow {}^3A_1, {}^3B_2$)	[48]
	24,630	LF($^1A_1 \rightarrow {}^1A_1, {}^1B_2$)	
	33,900	M $\rightarrow \pi^*$CO CT	
cis-$[W(CO)_4(Piperidine)_2]^b$	22,120 (4700)	LF($^1A_1 \rightarrow {}^3A_1, {}^3B_2$) + M $\rightarrow$ py CT	[48]
	24,390 (9400)	LF($^1A_1 \rightarrow {}^1A_1, {}^1B_2$) + M $\rightarrow$ py CT	
	33,000	M $\rightarrow \pi^*$CO CT	

[a] MeOH solution, ambient.
[b] Benzene solution, 298°K.
[c] EPA, 77°K.

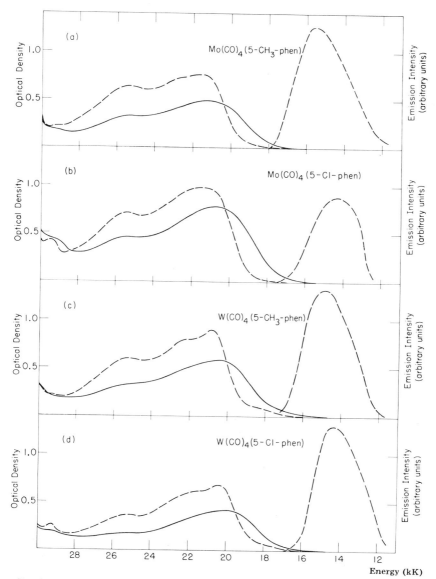

FIG. 2-13. Absorption (left) and emission (right) at 298 (———) and 77°K (----) in EPA solution. The spectral changes accompanying cooling from 298 to 77°K are not corrected for solvent contraction. The emission spectra are corrected, and the excitation source is the 351, 364 nm emission of an argon ion laser. Reprinted with permission from Wrighton and Morse [48].

TABLE 2-12

MLCT Absorption Maxima in $[M(CO)_4L]^a$

| L | Band, cm^{-1} $(\varepsilon)^b$ | | |
	Cr	Mo	W
2,2'-Bipyridine	20,110 (3715)	21,600 (4790)	20,530 (5130)
1,10-Phenanthroline	20,200 (1430)	21,740 (3550)	20,830 (5020)
5-CH$_3$-1,10-Phenanthroline	20,000 (1460)	21,280 (4560)	20,620 (3390)
5-Cl-1,10-Phenanthroline	19,680 (1470)	20,790 (3530)	19,960 (5390)
5-Br-1,10-Phenanthroline	19,530 (3260)	20,620 (8180)	19.960 (4170)
5-NO$_2$-1,10-Phenanthroline	18,690	19,880 (7330)	

^{a} Data from Wrighton and Morse [48].
^{b} CH$_2$Cl$_2$ solution, 298°K.

Table 2-12. The MLCT position depends on the substituents on L in a manner consistent with the direction of the CT [48], and as in the $[M(CO)_5L]$ complexes, the MLCT band position blue-shifts significantly from alkane to more polar or polarizable solvents [45,46]. Note that the LF band in $[M(CO)_4L]$ is still present at ~ 400 nm in those complexes having a lowest MLCT band (Fig. 2-13).

cis-$[W(CO)_4(trans$-4-Styrylpyridine)$_2]$ exhibits an MLCT absorption as the lowest energy absorption and an IL transition in the near-UV (Fig. 2-14)

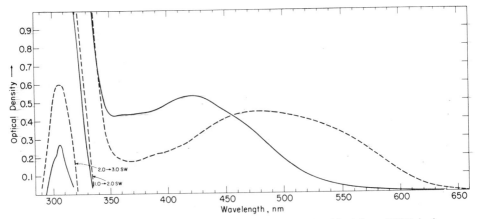

FIG. 2-14. Electronic spectra of cis-$[W(CO)_4(trans$-4-styrylpyridine)$_2]$ at 298°K in iso-octane/benzene (70:30 v/v) (------) and CH$_3$CN/benzene (90:10 v/v) (———). Concentration of the complex is 4.5×10^{-5} M and a 1 cm pathlength cell was used. Note that the OD scale for the IL band is different from the MLCT band. Reprinted with permission from Pdungsap and Wrighton [49].

[49]. The lowest absorption is very solvent-dependent, consistent with the MLCT assignment. The IL band has a molar absorptivity per W of $\sim 45,000$ liter mol^{-1} cm^{-1} or about twice that for the free ligand.

3. $[M(Arene)(CO)_nL_{3-n}]$ Complexes

Absorption spectra of $[M(\eta^6\text{-}C_6H_6)(CO)_3]$ and related complexes that have been studied are dominated by MLCT absorptions. Some spectral data and assignments are given in Table 2-13 [50–54]. For the tricarbonyl species the prominent spectral feature is a sharp, intense band near 31,000 cm^{-1} ($\varepsilon \approx 10^4$ liter mol^{-1} cm^{-1}) which has been assigned as a M $\rightarrow$ arene CT with some M $\rightarrow \pi^*$CO CT character. Though the low-energy region has not been studied in detail, there are probably low-energy LF absorptions. In the Mo and W tricarbonyls, weak, low-energy features may be attributable to such transitions.

The most conclusive assignment is for certain $[M(arene)(CO)_2L]$ complexes. For example, in $[Cr(\eta^6\text{-}C_6(CH_3)_6)(CO)_2L]$ (L = pyridine, *cis*-4-styrylpyridine, *trans*-4-styrylpyridine) the lowest absorption is clearly Cr $\rightarrow$ L CT. This follows from the substantial red-shifted first maximum, compared to the tricarbonyl, and from the variation among the three L (Table 2-13). Since $[Cr(\eta^6\text{-}C_6(CH_3)_6)(CO)_2(piperidine)]$ and related complexes are orange in color, it is likely that there are weak LF transitions in the visible region for such complexes, but no definitive assignments have been made.

4. $[M(\eta^5\text{-}C_5H_5)(CO)_3X]$ and Related Complexes

Spectral data have been reported for certain M(II) complexes (Table 2-14) [55–57]. The formally d^4 complexes exhibit a rather weak visible absorption which appears to be essentially unaffected by solvent. An LF assignment is appropriate and consistent with the photochemistry. The complex $[Mo(\eta^5\text{-}C_5H_5)(CO)_2Br_2]^-$ exhibits a red-shifted absorption compared to $[Mo(\eta^5\text{-}C_5H_5)(CO)_3Br]$, consistent with the LF interpretation [57].

A series of cationic complexes $[M(\eta^5\text{-}C_5H_5)(CO)_2L]^+$ (M = Mo, W; L = unsymmetrical α-diimine) have been prepared and their UV–vis spectra suggest an MLCT lowest absorption band [58,59]. Bands are in the ~ 550 nm range with absorptivities in the range of 5000 liter mol^{-1} cm^{-1}. A related set of neutral complexes $[M(\eta^3\text{-}C_3H_4R)(CO)XL]$ (M = Mo, W; X = Cl, Br, I, NCS; L = α-diimine) also shows a very low-energy visible absorption which blue-shifts with more polar solvents [60]. An MLCT assignment is suggested.

TABLE 2-13

Spectral Data for [M(η⁶-Arene)(CO)₂L]

Arene	M	L	Bands, cm⁻¹ (ε)	Assignment	Reference
C_6H_6	Cr	CO[a]	~26,620	LF?	[50,51]
			31,220	M → C_6H_6 CT	
			38,500	M → π*CO CT	
		Pyridine[b]	20,200	M → π*CO CT	[52]
		trans-Styrylpyridine[b]	16,390	M → L CT	[52]
				M → L CT	
$1,3,5\text{-}(CH_3)_3C_6H_3$	Cr	CO[a]	~32,000 (12,600)	M → arene, π*CO CT	[51–53]
			38,500	M → π*CO CT	
			45,450	M → π*CO CT	
		Pyridine[b]	19,230 (4900)	M → L CT	[53]
		trans-4-Styrylpyridine[b]	15,625	M → L CT	[52]
$(CH_3)_6C_6$	Cr	CO[b]	30,960 (8660)	M → arene, π*CO CT	[52]
			38,900 (7200)	M → π*CO CT	
			45,450 (25,700)		
		Pyridine[c]	19,420 (4000)	M → L CT	[52]
		cis-4-Styrylpyridine[b]	15,870 (6000)	M → L CT	[52]
		trans-4-Styrylpyridine[b]	14,810 (10,000)	M → L CT	[52]
$1,3,5\text{-}(CH_3)_3C_6H_3$	Mo	CO[c]	31,750 (8300)	M → arene, π*CO CT	[51]
			34,700	M → π*CO CT	
			39,500		
			43,500		
$(CH_3)_6C_6$	W	CO[c]	24,400 (560)	LF(?)	[54]
			30,800 (15,000)	M → arene, π*CO CT	
			34,700 (6500)	M → π*CO CT	
			42,000 (~1500)		

[a] Methanol solution, ambient. [b] Isooctane, 298°K. [c] Isooctane, 298°K. [b] Isooctane, 298°K. [c] Isooctane, 1 M pyridine, 298°K.

TABLE 2-14

Spectral Data for $[M(\eta^5\text{-}C_5H_5)(CO)_3X]$ Complexes[a]

M	X	Solvent	Bands, cm^{-1} (ε)	Assignment	Reference
Mo	—SCN	CCl$_4$	21,050 (389)	LF	[55]
			31,250 sh	—	
		CH$_2$Cl$_2$	21,050	LF	
			31,250	—	
		Acetone	21,140	LF	
			30,300	—	
	Br	CH$_2$Cl$_2$	20,920 (503)	LF	[56]
		THF	21,010 (490)	LF	[57]
	—SCN	CH$_2$Cl$_2$	22,470 (—)	LF	[56]
W	Cl	CCl$_4$	21,740 (510)	LF	[55]
			31,750 (2200)	—	

[a] Room-temperature data.

5. Metal–Metal Bonded Complexes

The dinuclear $[M_2(CO)_{10}]^{2-}$ complexes feature a direct M—M bond. The bonding is viewed according to Fig. 2-15, and the interesting spectral feature is that there are low-lying absorptions associated with the M—M bond [61,62]. A sharp, intense near-UV absorption is associated with the $\sigma_b \to \sigma^*$ transition in these complexes, and a lower-energy, weaker band is associated with a $d\pi \to \sigma^*$ excitation. The heterodinuclear M—M' bonded species $[MM'(CO)_{10}]^-$ (M = Cr, Mo, W; M = Mn, Re) are also "d^7–d^7" systems, and the near-UV–vis spectrum has been assigned similarly [61,63]. Some spectral data for the dianions and monoanions are included in Table 2-15 [55, 61–67].

The spectra of the M—M bonded $[M_2(\eta^5\text{-}C_5H_5)_2(CO)_6]$ complexes have been recorded. Figure 2-16 shows representative UV–vis absorptions for the Mo and W species. Again the intense near-UV band is assigned as a $\sigma_b \to \sigma^*$

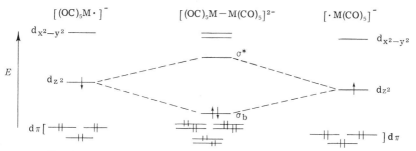

FIG. 2-15. One-electron diagram for $[M_2(CO)_{10}]^{2-}$ (M = Cr, Mo, W).

TABLE 2-15

Spectral Features of Dinuclear Cr, Mo, W Complexes

Complex	Solvent	Bands, cm^{-1} (ε)		Reference
		$\sigma_b \rightarrow \sigma^*$	$d\pi \rightarrow \sigma^*$	
$[Cr_2(CO)_{10}]^{2-}$	CH$_2$Cl$_2$	27,200 (13,500)	24,000 sh (5300)	[61]
$[W_2(CO)_{10}]^{2-}$	CH$_2$Cl$_2$	28,500 (7210)	25,500 sh (4840)	[62]
$[MnCr(CO)_{10}]^-$	EPA	29,240 (13,800)	24,100 (2100)	[61]
$[MnMo(CO)_{10}]^-$	THF	30,400	26,900	[63]
$[MnW(CO)_{10}]^-$	THF	30,700	26,600	[63]
$[Mo_2(\eta^5\text{-}C_5H_5)_2(CO)_6]$	Alkane	25,770 (20,400)	19,530 (1720)	[55]
$[Mo_2(\eta^5\text{-}C_5(CH_3)_5)_2(CO)_6]$	Alkane	24,450 (9500)	20,200 (2500)	[64]
$[W_2(\eta^5\text{-}C_5H_5)_2(CO)_6]$	CCl$_4$	27,620 (20,200)	20,280 (2450)	[55]
$[Cr_2(\eta^5\text{-}C_5H_5)_2(CO)_6]$	Alkane	20,490	16,950 (—)	[64]
$[Cr_2(\eta^5\text{-}C_5H_5)_2(CO)_4]$	Alkane	25,640 (19,200)^a?	16,260 (300)^a?	[64]
$[Mo_2(\eta^5\text{-}C_5H_5)_2(CO)_4]$	Alkane	20,770 (13,800)^a?	26,300 sh^a?	[64]
$[Mo_2(\eta^5\text{-}C_5(CH_3)_5)_2(CO)_4]$	Alkane	30,670 (13,200)^a?	25,320 sh^a?	[64]
$[Mo_2(\eta^5\text{-}C_5H_5)_2(CO)_4(C_2H_2)]$	Alkane	27,780 (7740)	18,870 (845)	[64]
$[W_2(\eta^5\text{-}C_5H_5)_2(CO)_4(C_2H_2)]$	Alkane	28,740 (6180)	19,230 (950)	[64]
$[Mo(\eta^5\text{-}C_5H_5)(CO)_3Mn(CO)_5]$	Alkane	26,810 (14,000)	22,220 (1400)	[65]
$[Mo(\eta^5\text{-}C_5H_5)(CO)_3Re(CO)_5]$	Alkane	27,400 (6500)	22,030 (640)	[65]
$[W(\eta^5\text{-}C_5H_5)(CO)_3Mn(CO)_5]$	Alkane	27,620 (14,900)	22,220 (1460)	[65]
$[W(\eta^5\text{-}C_5H_5)(CO)_3Re(CO)_5]$	Alkane	30,210 (9000)	24,330 (1400)	[65]
$[Mo(\eta^5\text{-}C_5H_5)(CO)_3Co(CO)_4]$	CCl$_4$	28,010 (11,600)	19,010 (460)	[66,67]
$[W(\eta^5\text{-}C_5H_5)(CO)_3Co(CO)_4]$	CCl$_4$	20,240 (10,700)	19,490 (500)	[66,67]
$[Mo(\eta^5\text{-}C_5H_5)(CO)_3Fe(CO)_3(\eta^5\text{-}C_5H_5)]$	CCl$_4$	25,130 (11,800)	18,350 (1230)	[66]

^{a} Assignments questionable.

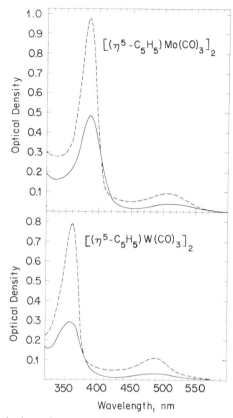

FIG. 2-16. Electronic absorption spectra in EPA at 298 (———) and 77°K (————). The intense band ($\varepsilon \approx 1.7 \times 10^4$) is the $\sigma_b \to \sigma^*$. Reprinted with permission from Wrighton and Ginley [55], *J. Am. Chem. Soc.* **97**, 4246 (1975). Copyright by the American Chemical Society.

transition and the visible band is attributed to a $d\pi \to \sigma^*$ excitation (Table 2-15). Though the symmetry is low, the spectra of these "d^5–d^5" dimers are assignable. A comparison of the spectra of the M—M bonded complexes with mononuclear, but analogous, complexes lends credence to the $\sigma_b \to \sigma^*$ and $d\pi \to \sigma^*$ assignments. In complexes like $[W(CO)_5Br]^-$ or $[W(\eta^5-C_5H_5)(CO)_3Br]$, one does not observe an intense, near-UV absorption band (cf. Tables 2-9 and 2-14). However, each exhibits a low-energy LF band, just as found in the M—M bonded species. The intense $\sigma_b \to \sigma^*$ transition in the M—M bonded complexes is a consequence of the M—M bond; a related transition must obtain in species like $[W(CO)_5Br]^-$, but it is likely to be found at much higher energies and would be termed $Br^- \to W$ CT (Fig. 2-17) owing to the disparate orbital electronegativities in Br compared to a metal radical like $[Mo(\eta^5-C_5H_5)(CO)_3]$ or $[W(CO)_5]^-$. Following this vein, other

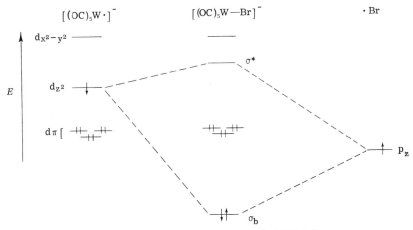

FIG. 2-17.   One-electron diagram for $[W(CO)_5Br]^-$.

metal radicals like $[Co(CO)_4]$, $[Mn(CO)_5]$, $[Fe(\eta^5\text{-}C_5H_5)(CO)_2]$, etc. can be bonded to Cr-, Mo-, and W-centered radicals to give M—M′ species which exhibit $\sigma_b \rightarrow \sigma^*$ and $d\pi \rightarrow \sigma^*$ absorptions. Various M—M′, d^n–$d^{n'}$, systems involving Cr, Mo, and W are included in Table 2-15. It has been noted [67] that the $\sigma_b \rightarrow \sigma^*$ transition of an M—M′ species is generally situated between the $\sigma_b \rightarrow \sigma^*$ positions associated with the M—M and M′—M′ bonded species. This results from the similar orbital electronegativities involved and signals a lack of significant ionic character in the M—M′ bond. This follows from the finding that the $\sigma_b \rightarrow \sigma^*$ position can be correlated with the M—M dissociation energy [68]; higher energy $\sigma_b \rightarrow \sigma^*$ was associated with stronger M—M bonds. The ability to "predict" the M—M′ $\sigma_b \rightarrow \sigma^*$ position from M—M and M′—M′ suggests that the M—M′ bond energy can be predicted from that of M—M and M′—M′. Such can be the case only when there is little ionic contribution. As an example, the "$\sigma_b \rightarrow \sigma^*$" absorption in $[W(CO)_5Br]^-$ is not situated between that for $[W_2(CO)_{10}]^{2-}$ and Br_2; the W—Br bond has significant ionic character.

A comparison of the spectra of $[M_2(\eta^5\text{-}C_5H_5)_2(CO)_6]$ [55], $[M_2(\eta^5\text{-}C_5H_5)_2(CO)_4]$ [64], and $[M_2(\eta^5\text{-}C_5H_5)_2(CO)_4(C_2H_2)]$ [64] reveals an interesting trend. First, consider the comparison between the single-bonded complexes $[M_2(\eta^5\text{-}C_5H_5)_2(CO)_6]$ and the triple-bonded species $[M_2(\eta^5\text{-}C_5H_5)_2(CO)_4]$. The $\sigma_b \rightarrow \sigma^*$ transition in the single-bonded species is at significantly lower energy than the first intense absorption in the triple-bonded species. The Mo—Mo bond in $[Mo_2(\eta^5\text{-}C_5H_5)_2(CO)_4]$ [69] is ~ 0.8 Å shorter than the 3.235 Å Mo—Mo bond [70] in the single-bonded hexacarbonyl. Thus, the higher energy absorption in the triple-bonded species *may* be a $\sigma_b \rightarrow \sigma^*$ excitation associated with the shorter (stronger) M—M bond. The acetylene complexes $[M_2(\eta^5\text{-}C_5H_5)_2(CO)_4C_2H_4]$ also have a direct M—M bond and exhibit spectral features [64] which are similar to those for $[M_2(\eta^5\text{-}C_5H_5)_2(CO)_6]$. A structural characterization of the Mo [71] and the W [72] acetylene complexes reveals that the M—M bond is 0.3 Å shorter than in the hexacarbonyl. A comparison of the $\sigma_b \rightarrow \sigma^*$ bands shows that the shorter M—M bonded species has the higher energy absorption [72].

C.   LUMINESCENCE STUDIES

Only a few luminescent Cr, Mo, or W carbonyl complexes are known. The first reports [33] of luminescence from any metal carbonyl involved C_{4v} complexes of the general formula [W(CO)$_5$(n-electron donor)]. Emission was observed at 77°K in solution or as the pure solid. The corresponding Cr and Mo complexes were not observed to luminesce under the same conditions used for the W species. The lack of detectable emission from the Cr and Mo species was associated with lack of an observable $^1A_1(e^4b_2{}^2) \rightarrow {}^3E(e^3b_2{}^2a_1{}^1)$ absorption band, and the emission was attributed to the LF triplet → singlet transition. Ligand-field emission spectral features of some C_{4v} complexes are set out in Table 2-16 [33,73], and the emission of several [W(CO)$_5$L] complexes are shown in Fig. 2-18. The emission lifetimes are in the 10^{-6}–10^{-5} sec

TABLE 2-16

Emission Properties of [W(CO)$_5$L] Complexes[a,b]

Complex	Emission maximum, nm	Rel Φ_e	$\tau_e \times 10^6$, sec
[W(CO)$_5$(EtNH$_2$)]	533		0.92
[W(CO)$_5$(t-BuNH$_2$)]	533		1.2
[W(CO)$_5$(i-PrNH$_2$)]	533		0.65
[W(CO)$_5$(c-HxNH$_2$)]	533	1.0	1.1
[W(CO)$_5$(Et$_2$NH)]	533	6.0	5.1
[W(CO)$_5$(Me$_2$NH)]	533		2.6
[W(CO)$_5$(Et$_2$NMe)]	533		25.5
[W(CO)$_5$(Et$_3$N)]	533		9.7
[W(CO)$_5$(n-Bu$_3$N)]	533		6.9
[W(CO)$_5$(Me$_2$NCH(CH$_3$)Ph)]	533	12.0	15.5
[W(CO)$_5$(EtOH)]	~535		11.7
[W(CO)$_5$(i-PrOH)]	~535		6.6
[W(CO)$_5$(t-BuOH)]	~535		6.3
[W(CO)$_5$(Et$_2$O)]	533		7.1
[W(CO)$_5$(Acetone)]	538		5.3
[W(CO)$_5$(Acetone-d$_6$)]	~535		5.7
[W(CO)$_5$(Cyclohexanone)]	~535		3.6

[a] Data from Wrighton et al. [33,73].
[b] Methycyclohexane solutions at 77°K.

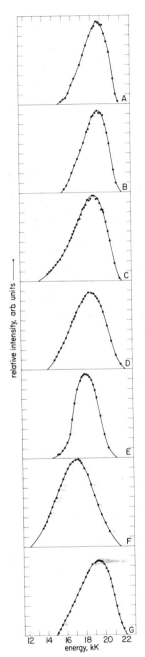

Fig. 2-18.   Corrected 351, 364 nm excited emission spectra of [W(CO)$_5$L] complexes at 77°K in EPA. For (A–G) L is 3,4-diMe-pyridine, 4-Me-pyridine, pyridine, 3-Br-pyridine, 3,5-diBr-pyridine, 4-acetylpyridine, and piperidine, respectively. Reprinted with permission from Wrighton *et al.* [38], *J. Am. Chem. Soc.* **98**, 4105 (1976). Copyright by the American Chemical Society.

range, typical of heavy-metal complexes exhibiting emission which has spin-forbidden character [74].

The emission of the series of $[W(CO)_5(pyridine-X)]$ complexes, given in Table 2-10 and Fig. 2-18, reveals that the MLCT states are emissive. Other data are given in Table 2-17 [38], showing that for those complexes which have lowest MLCT excited states the lifetimes are longer and the emission maxima are red-shifted compared to $[W(CO)_5(piperdine)]$ which exhibits LF emission. For L = 4-cyanopyridine, 4-acetylpyridine, 4-benzoylpyridine, and 4-formylpyridine, the MLCT excited state is the lowest excited state. In the other complexes the lowest excited state was assigned as the LF $^3E(e^3b_2{}^2a_1{}^1)$ state.

TABLE 2-17

Emission Properties of $[W(CO)_5L]$ Complexes[a,b]

L	Emission maximum ($cm^{-1} \times 10^{-3}$)	Emission width,[c] ($cm^{-1} \times 10^{-3}$)	$\tau_e \times 10^6$
3,4-Dimethylpyridine	19.7	3.1	0.96
4-Methylpyridine	19.6	3.1	1.07
Pyridine	19.1	4.0	0.86
3-Bromopyridine	18.9	4.1	1.00
3-Acetylpyridine	18.7	4.4	1.17
3-Benzoylpyridine	18.5	4.4	1.25
3,5-Dibromopyridine	18.1	2.9	1.25
4-Cyanopyridine	16.6	4.1	33.0
4-Acetylpyridine	17.0	4.5	38.3
4-Benzoylpyridine	16.8	4.2	29.5
4-Formylpyridine	15.2	4.0	15.0
Piperidine	18.3	4.1	0.82

[a] Reprinted with permission from Wrighton *et al.* [38], *J. Am. Chem. Soc.* **98**, 4105 (1976). Copyright by the American Chemical Society.

[b] Corrected for variation in detector response as a function of wavelength. All data for EPA solutions at 77°K.

[c] Width of emission band at half-height.

Ligand-field and MLCT states are also emissive at low temperature for the C_{2v} complexes [47,48]. Some emission spectra are included in Fig. 2-12 and 2-13, and some data are given in Table 2-18. For the aliphatic amine complexes and for the pyridine complexes, it appears that, as in the C_{4v} complexes, the emission is from an LF state. For these, emission is only observable from the W complexes. But for the 2,2′-bipyridine and related complexes, emission is from the MLCT excited state, and emission is found for both Mo and W complexes. One final interesting fact is that emission is observable at room

TABLE 2-18

Emission Data for $[M(CO)_4L]$ and cis-$[M(CO)_4L_2]$ Complexes

M	L or L_2	Emission maximum[a] (kK $\pm$ 0.05)	Emission half-width[a]	Emission τ (μsec $\pm$ 10%)	Emission Φ ($\pm$20%)
Mo	2,2'-bipy	15.25	3440		
	1,10-phen	15.66	3150	11.6	0.09
	5-CH$_3$-1,10-phen	15.55	3260	13.2	0.08
	5-Cl-1,10-phen	14.40	2940	13.3	0.04
	5-Br-1,10-phen	14.40	3325	9.5	
W	2,2'-bipy	15.10	3500		
	1,10-phen	15.30	3210	11.6	0.05
	5-CH$_3$-1,10-phen	14.90	3200	12.5	0.04
	5-Cl-1,10-phen	14.40	2950	7.9	0.02
	5-Br-1,10-phen	14.40	3340	7.9	
	(Pyridine)$_2$	18.31		48.0	
	Ethylenediamine	18.32		24.0	
	(n-PrNH$_2$)$_2$	18.52		0.9	
	(Piperidine)$_2$	18.35		29.0	

[a] From corrected emission spectra; emission half-width is the width of the emission band at half-height; excitation wavelength is 351.1, 363.4 nm from an argon ion laser. EPA solution, 77 K. Data are from Wrighton and Morse [48].

temperature from the pure solid $[M(CO)_4L]$ (L = 2,2'-bipyridine, 1,10-phenanthroline, etc.) although fluid solution emission is not detectable at room temperature.

D.   PHOTOREACTIONS

Carbonyl complexes of Cr, Mo, and W have extensive photochemistry. Though substitution is the dominant reaction, examples of isomerization, linkage isomerization, and intraligand isomerization are known. A reaction of growing importance is M—M bond cleavage. The photochemical reactions are organized according to reaction types, beginning with photosubstitution.

1.   Photosubstitution

a. $[M(CO)_nL_{6-n}]$ Complexes.   Chemistry involving ligand exchange and substitution dominates the excited-state processes of $[M(CO)_6]$ complexes. It appears certain that the photochemical formation of $[M(CO)_5L]$ is obtained by the sequence outlined in Eqs. (2-8)–(2-10). Several lines of evidence support very efficient generation of the coordinatively unsaturated intermediate $[M(CO)_5]$ which has a substantial lifetime.

$$[M(CO)_6] \xrightarrow{\ hv\ } [M(CO)_5] + CO \tag{2-8}$$

$$[M(CO)_5] + CO \xrightarrow{\ k_9\ } [M(CO)_6] \tag{2-9}$$

$$[M(CO)_5] + L \xrightarrow{\ k_{10}\ } [M(CO)_5L] \tag{2-10}$$

It was found that a reversible photoreaction occurs upon photolysis of $[M(CO)_6]$ in a methyl methacrylate polymer [75]. The slow thermal bleaching of the yellow intermediate formed during photolysis is thought to be due to reaction (2-9). Infrared characterization of the $[M(CO)_5]$ intermediate was first gained by Sheline and co-workers [76] who obtained IR spectra after photolysis of $[M(CO)_6]$ at 77°K in methylcyclohexane glasses. The IR spectra supported assignment of the primary photoproduct as C_{4v}, $[M(CO)_5]$. However, evidence obtained upon thawing the $[Mo(CO)_5]$ sample implicated isomerization from a species of C_{4v} symmetry to one of D_{3h} symmetry. Strohmeier and his colleagues advanced chemical evidence [77] supporting the mechanism in Eqs. (2-8)–(2-10). The high initial quantum yield of ~ 1.0 for $[M(CO)_5L]$ formation was found to be independent of M (M = Cr, Mo, W) and L. If a substantial contribution to the substitution process is an associative mechanism, one expects a dependence on the entering group L. It should be emphasized, however, that the *lack* of an effect by the entering group is not itself conclusive proof of the dissociative mechanism. The absolute value of the CO dissociation quantum yield now seems to be somewhat less than 1.0 [78], but this does not alter the fact that the reactions are efficient and probably mainly dissociative.

An interesting series of papers have appeared concerning the photochemistry of matrix isolated $[M(CO)_n]$ species [29,30,32,32a,79]. Beginning with $[M(CO)_6]$, the species $[M(CO)_5]$, $[M(CO)_4]$, and $[M(CO)_3]$ have been observed upon irradiation. The structure of photogenerated $[M(CO)_5]$ is apparently C_{4v} in the matrices. Quite interestingly, $[Cr(CO)_5]$ can be photo-oriented with plane polarized light [322]. The characterization of $[M(CO)_5]$ by flash photolysis in fluid solutions has been attempted [80–83]. Transients have been observed, and some quantitative information concerning the rate of reaction of $[Cr(CO)_5]$ and L has been reported. Though not all of the data are totally reproducible from laboratory to laboratory, it is clear that irradiation of $[M(CO)_6]$ leads to some coordinatively unsaturated species which is scavengeable with virtually any nucleophile. Indeed, it is the apparent reactivity of $[M(CO)_5]$ with virtually everything (N_2, impurities, excess $[M(CO)_6]$, etc.) which has led to false conclusions in the flash studies. But with the low-temperature matrix studies, flash studies, quantum yields vs. L, and high absolute quantum yields, it is certain that reaction (2-8) is efficient in solution. Further, though the question of the interconvertibility of C_{4v} and D_{3h} [M(CO)$_5$] in solution remains, the species generated in matrices has a C_{4v} geometry.

Identification and characterization of the reactive excited state in [M(CO)$_6$] complexes have not been pursued, probably because it suffices to say that the main decay path is dissociative loss of CO. The [M(CO)$_6$] species have not been found to luminesce, and it is likely that spectroscopic excited states are extremely short-lived. The triplet-sensitized reaction of [Cr(CO)$_6$] has been carried out [84], and the quantum yield for loss of CO was found to be unity. The fact that the direct-irradiation and triplet-sensitized yields are the same is consistent with decay of the excited states proceeding through a low-lying triplet state, but this point is clearly not proved. The $t_{2g}^6 \rightarrow t_{2g}^5 e_g$ one-electron excitation in these systems gives rise to the $^1T_{1g}$, $^3T_{1g}$, $^1T_{2g}$, and $^3T_{2g}$ excited states, all of which should be substantially more reactive than the ground state. Depopulation of the t_{2g} level diminishes π back-bonding, and concomitant population of e_g diminishes σ bonding regardless of the spin multiplicity of the excited state achieved. Our ability to resolve the question of relative reactivity of different spin states involved in these reactions may ultimately depend on our ability to observe the reactive state prior to its decay.

The synthetic utility of the sequence (2-8)–(2-10) has had considerable impact on systematic studies of chemical properties of the [M(CO)$_n$(L)$_{6-n}$] complexes. From numerous early successes we conclude that derivatives of [M(CO)$_6$] can be prepared by irradiation in the presence of almost any coordinating agent L. Our objective here is to account generally for the degree of substitution ultimately obtained and to explain how to control it. It was recognized from the outset that photolysis of [M(CO)$_5$L] could result in the loss of another CO molecule (Eq. 2-11) or loss of L (Eq. 2-12).

$$[M(CO)_5L] \xrightarrow{h\nu} [M(CO)_4L] + CO \qquad (2\text{-}11)$$

$$[M(CO)_5L] \xrightarrow{h\nu} [M(CO)_5] + L \qquad (2\text{-}12)$$

Reaction (2-11) leads to potentially two geometric isomers of [M(CO)$_4$L$_2$], and reaction (2-12) leads simply to ligand exchange in the presence of added L. The relative efficiencies of processes (2-11) and (2-12) were found to be very sensitive to the nature of L. In fact, for certain L, such as tetrahydrofuran (THF), process (2-11) is fairly insignificant, and nearly complete conversion of [M(CO)$_6$] to [M(CO)$_5$(THF)] can be achieved. The THF is weakly bound, and a pure [M(CO)$_5$L] species is obtained by addition of L to the solution of [M(CO)$_5$(THF)], as in reaction (2-13).

$$[M(CO)_5(THF)] \xrightarrow[L]{\Delta} [M(CO)_5L] + THF \qquad (2\text{-}13)$$

The relative importance of reaction (2-11) was found to increase with increasing strength of the M—L bond [85]. It is not obvious that such a correlation should exist since the excitation energies are high enough to yield loss of either the CO or L. If a common excited state is responsible for both reaction (2-11) and (2-12), the correlation could be rationalized by merely assuming that photoexcitation causes the same relative increase in substitution rate for L and CO. In such a case, comparison of ground-state binding strength may yield the correlation observed: when L and CO are more comparable in binding strength, release of CO is competitive with release of L, though in the ground state both undergo substitution slowly. Thus, we note crudely, though quite generally, that when L resembles CO in its bonding properties, sequential substitution of CO by L is possible until every CO has been replaced; e.g., $[Mo(CO)_6]$ can be converted to $[Mo(P(OCH_3)_3)_6]$.

A second parameter also affects the relative efficiencies of reaction (2-11) and (2-12). The reaction quantum yield for (2-11) was found to be sensitive to the wavelength of the exciting light, as evidenced by data like those shown in Table 2-19 [86,87]. Higher energy irradiation yields more efficient loss of CO. Such an effect can be attributed to at least two reactive excited states or to differences in the reactivity of one excited state depending on the vibrational level directly achieved. The latter alternative is not likely since the reactions are carried out in condensed media. Additional data [44] (Table 2-20) reveal that both reaction (2-11) and (2-12) are wavelength-dependent, with reaction (2-12) having attenuated importance upon higher excitation energy. The opposite wavelength dependence for the two processes can be rationalized by invoking two reactive LF excited states. The situation is

TABLE 2-19

Wavelength Dependence for Quantum
Efficiency of $[M(CO)_5(Pyridine)]$ to
$[M(CO)_4(Pyridine)_2]$ Conversion[a]

Central metal	Solvent	$\Phi_{366\,nm}$	$\Phi_{436\,nm}$
Cr	C_6H_6	0.21	0.13
Cr	THF	0.28	0.17
Mo	C_6H_6	0.16	0.11
Mo	THF	0.35	0.22
W	C_6H_6	0.11	0.08
W	THF	0.06	0.02

[a] Data from Strohmeier and von Hobe [86,87].

TABLE 2-20

Wavelength Dependence for Processes (2-11) and (2-12) for
M = W and L = Pyridine[a]

Irradiation, λ (nm)	$\Phi_{\text{Eq. (2-11)}}$	$\Phi_{\text{Eq. (2-12)}}$	Irradiation, λ (nm)	$\Phi_{\text{Eq. (2-11)}}$	$\Phi_{\text{Eq. (2-12)}}$
436	0.00_2	0.63	313	0.03_9	0.38
366	0.01_3	0.50	254	~ 0.04	0.34

[a] Quantum yields may depend on conditions used; cf. Wrighton *et al.* [44].

detailed in Fig. 2-19. Low-energy excitation yields population of the $d_{z^2}(\sigma_z{}^*)$ orbital with σ-antibonding character directed principally along the z axis, strongly labilizing the σ donor, pyridine. Higher energy excitation populates the $d_{x^2-y^2}(\sigma_{xy}{}^*)$ orbital with strong labilizing effects for the equatorial CO groups. Internal conversion of the upper state to the lower state with rate constant k_{nd} adequately accounts for the fact that reaction (2-12) occurs upon high-energy excitation. Impressive support of the rationale of the reactivity of [W(CO)$_5$(pyridine)] is found in the observation of selective incorporation of ^{13}CO into equatorial positions of [M(CO)$_5$(piperidine)] (M = Mo, W) [88,89]. The results with the pyridine complexes, however, are somewhat ambiguous, owing to the subsequently established fact that such complexes exhibit M $\rightarrow$ py CT [38] in the wavelength region where the CO lability is observed [44,86,87].

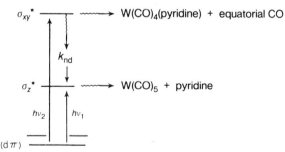

FIG. 2-19. Correlation of excited state and observed reactivity for [W(CO)$_5$pyridine].

Detailed study of the C_{4v} [M(CO)$_5$(aliphatic amine)] and [M(CO)$_5$(NH$_3$)] complexes [36], which only have LF lowest excited states, confirms the essential picture given in Fig. 2-19. Table 2-21 [36] shows some quantum yield data for the substitution of CO or L. Upon excitation to the LF $^{1,3}E(e^3b_2{}^2a_1{}^1)$ states, the principal reaction is loss of the aliphatic amine, consistent with the directed σ^* character of the state. Higher energy excita-

TABLE 2-21

Quantum Yields for Photosubstitution of $[M(CO)_5L]$ Complexes[a,b]

M	L	Entering group, X	Product	$\Phi_{366\,nm}$	$\Phi_{405\,nm}$	$\Phi_{436\,nm}$
Mo	NH_3	CO	$[Mo(CO)_6]$		0.64	0.58
W	Piperidine	CO	$[W(CO)_6]$		0.65	
Mo	Piperidine	Piperidine	$[Mo(CO)_4L_2]$	0.13	0.11	0.04_8
Mo	n-PrNH$_2$	n-PrNH$_2$	$[Mo(CO)_4L_2]$	0.24	0.20	0.05_7
W	Piperidine	Piperidine	$[W(CO)_4L_2]$	0.03_4	0.008_1	0.006_1
W	n-PrNH$_2$	n-PrNH$_2$	$[W(CO)_4L_2]$	0.05_7	0.02_2	0.01_6
Mo	NH_3	1-Pentene	$[Mo(CO)_5X]^c$	0.74	0.84	0.71
W	NH_3	1-Pentene	$[W(CO)_5X]^d$	0.49	0.66	0.56
W	n-PrNH$_2$	1-Pentene	$[W(CO)_5X]^d$	0.60	0.65	0.73
W	Piperidine	1-Pentene	$[W(CO)_5X]^d$	0.45	0.51	0.49

[a] Reprinted with permission from Wrighton [36], *Inorg. Chem.* **13**, 905 (1974). Copyright by the American Chemical Society.

[b] All quantum yields are $\pm 10\%$ in benzene solution at 298°K.

[c] Constitutes $> 70\%$ of primary photoproducts.

[d] Constitutes $> 90\%$ of primary photoproducts.

tion, presumably producing population of the $d_{x^2-y^2}$ orbital, results in a higher efficiency for CO substitution. One curious fact is that the CO substitution yield for the Mo complexes is generally higher than for the W species as found for L = pyridine (Table 2-19).

Study of the photosubstitution of L in a series of $[W(CO)_5L]$ (L = pyridine and substituted pyridine) showed rather conclusively that lowest MLCT excited states do not lead to either L or CO substitution [38]. Substitution quantum yields as a function of L are given in Table 2-22 [38] and show that those complexes exhibiting a lowest MLCT, rather than LF, excited state do not undergo efficient substitution of L. Spectral data for the complexes are given in Tables 2-10 and 2-17 and Figs. 2-8, 2-9, and 2-18. The situation for the C_{4v} pyridine and substituted pyridine complexes is given in Fig. 2-20. In another series of C_{4v}, d^6 complexes $[Ru(NH_3)L]^{2+}$, involving a similar range of L, the same trend was observed and the same conclusion was drawn [90].

One final interesting experiment concerning $[W(CO)_5(pyridine)]$ should be mentioned. Irradiation of $[W(CO)_5(pyridine)]$ at 12°K in an Ar matrix results in the generation of coordinatively unsaturated species [91]. At 366 nm, $[W(CO)_5]$ is generated, suggesting a dissociative mechanism for loss of L. At 254 nm, free CO is detectable, consistent with the wavelength-dependent photochemistry in solution at 298°K.

Photosubstitution of cis-$[M(CO)_4L_2]$ or $[M(CO)_4L]$ (L = nitrogen-donor ligand) has been carried out in fluid solution at 298°K [48]. For L = aliphatic

TABLE 2-22

Photosubstitution of L in $[W(CO)_5L]^{a,b}$

	$\Phi \pm 10\%$		Lowest excited state
L	436 nm	514 nm^c	
3,4-Dimethylpyridine	0.53		LF
4-Methylpyridine	0.55		LF
Pyridine	0.62		LF
3-Bromopyridine	0.66		LF
3-Acetylpyridine	0.75		LF
3-Benzoylpyridine	0.73		LF
3,5-Dibromopyridine	0.82		LF
4-Benzoylpyridine	0.12	0.02	W → L CT
4-Cyanopyridine	0.12	0.02	W → L CT
4-Acetylpyridine	0.15	0.02	W → L CT
4-Formylpyridine	0.05	0.002	W → L CT
Piperidine	0.58		LF

^{a} Reprinted with permission from Wrighton *et al.* [38], *J. Am. Chem. Soc.* **98**, 4105 (1976). Copyright by the American Chemical Society.

^{b} Irradiation carried out at 298°K in isooctane/1-pentene (2:1 by volume). The only photoproduct is $[W(CO)_5(1\text{-pentene})]$.

^{c} Data at 514 nm are given only for those complexes that absorb sufficiently to give accurate quantum yields.

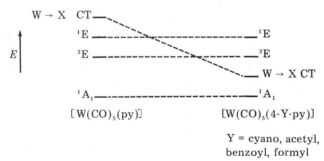

Y = cyano, acetyl, benzoyl, formyl

Fig. 2-20. Correlation of lowest excited state and substitution lability; cf. Table 2-22.

amine or pyridine, LF lowest excited states obtain; and, as in the $[M(CO)_5L]$ complexes, the photochemistry is dominated by loss of L (Table 2-23). For such complexes the lowest excited state involves destabilization of the σ bonding along the OC—M—N— donor axes, and the observed chemistry is thus reasonable. Loss of CO from $[M(CO)_4L]$ (L = 2,2′-bipyridine and

TABLE 2-23

Photosubstitution of Pyridine in
cis-$[W(CO)_4(Pyridine)_2]^{a,b}$

Irradiation, λ, nm	$\Phi \pm 10\%$
436	0.23
405	0.23
366	0.23

[a] Reprinted with permission from Wrighton and Morse [48].
[b] 298°K in benzene.

related ligands) does occur upon photoexcitation (Table 2-24) [48]. The quantum efficiencies are small and wavelength-dependent, and the MLCT state is not reactive. The upper excited states which are reactive are probably LF states. Consistent with these findings, cis-$[W(CO)_4(trans$-4-styrylpyridine)]$ is not found to undergo photosubstitution from the lowest excited state [49].

Consider now the photosubstitution behavior of complexes $[M(CO)_nL_{6-n}]$ where L is more similar to CO in its bonding properties. For $[Mo(CO)_5 \cdot P(C_6H_{11})_3]$, it was found that irradiation in a hydrocarbon or 2-methyltetrahydrofuran glass leads to loss of CO, not $P(C_6H_{11})_3$ [92]. This is in contrast to the results for $[W(CO)_5(pyridine)]$ [91]. Elegant studies have also been carried out on $[M(CO)_5CS]$ complexes at low temperature [93], revealing that CO is photodissociated rather than CS. In these systems a square-pyramidal, coordinatively unsaturated $[M(CO)_4L]$ species is formed and is found to exist in two isomeric forms as in Eq. (2-14). Thus, these experiments suggest little selectivity in the axis of labilization. However, the

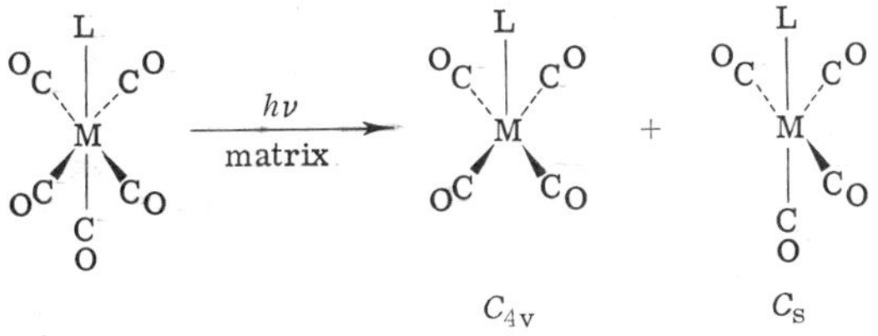

$$\xrightarrow[\text{matrix}]{h\nu} \qquad + \qquad \qquad (2\text{-}14)$$

C_{4v} $\qquad\qquad$ C_s

lack of a statistical ratio (4:1) of the C_s and C_{4v} products seemingly demands some preferential loss of axial CO, if the observed products actually are due to primary photoproducts. Study of $trans$-$[(^{13}CO)W(CO)_4CS]$ revealed that the loss of CO is essentially statistical, but the primary photoproduct relaxes to a mixture of C_s and C_{4v} isomers [93]. These data suggest that interpreting the stereospecificity of CO loss from products alone may be an error.

TABLE 2-24

Photosubstitution of [W(CO)$_4$L] Complexes to Yield [W(CO)$_3$LX] Complexes[a,b]

L	X (M)	Solvent	$\Phi_{436\,nm}$	$\Phi_{405\,nm}$	$\Phi_{366\,nm}$	$\Phi_{313\,nm}$
1,10-phen	CH$_3$CN (neat)	CH$_3$CN	1.6×10^{-4}	1.2×10^{-3}	9.2×10^{-3}	2.2×10^{-2}
2,2'-bipy	CH$_3$CN (neat)	CH$_3$CN	5.3×10^{-5}			2.1×10^{-2}
5-CH$_3$-1,10-phen	CH$_3$CN (neat)	CH$_3$CN	1.5×10^{-5}			2.0×10^{-2}
5-Br-1,10-phen	CH$_3$CN (neat)	CH$_3$CN	1.2×10^{-4}			2.4×10^{-2}
1,10-phen	Pyridine (0.02)	Benzene	0.9×10^{-4}	1.5×10^{-3}		1.0×10^{-2}
1,10-phen	trans-4-Styrylpyridine (0.02)	Benzene		1.4×10^{-3}		

[a] Reprinted with permission from Wrighton and Morse [48].
[b] All quantum yields are $\pm 10\%$; 298°K.

For CO-like ligands, multiple L photosubstitution products, beginning with $[M(CO)_6]$, have been made (Table 2-25) [94–97]. The isolation of such photoproducts depends on several factors including final product stability and photolabilization of CO at intermediate stages of substitution. All of the cases for which $[ML_6]$, $[M(CO)L_5]$, and $[M(CO)_2L_4]$ are found have L as a good π-acceptor ligand. That is, the loss of CO to yield stable low-valent complexes requires entering ligands capable of stabilizing the low-valent

TABLE 2-25

Photosubstitution of Chromium, Molybdenum, and Tungsten Carbonyl by
Ligands Capable of Stabilizing Low-Valent Metals

Starting complex	Entering ligand, L	Product	Reference
$[Cr(CO)_6]$	$(n\text{-}C_3H_7)OPF_2$	$[CrL_6]$	[94]
	$P(OCH_3)_2F$	$[CrL_6]$	[94]
	$P(OCH_3)_3$	$[Cr(CO)L_5]$	[94]
		$cis\text{-}[Cr(CO)_2L_4]$	[95]
	$(CH_3)P(OCH_3)_2$	$[Cr(CO)L_5]$	[94]
	$(CH_3O)P(CH_3)_2$	$[Cr(CO)L_5]$	[94]
	$P(CH_3)_3$	$cis\text{-}[Cr(CO)_2L_4]$	[95]
$[Mo(CO)_6]$	PF_3	$[Mo(CO)_5L]$	[96]
		$cis\text{-}[Mo(CO)_5L]$	[96]
		$trans\text{-}[Mo(CO)_4L_2]$	[96]
		Isomers of $[Mo(CO)_3L_3]$	[96]
		$cis\text{-}[Mo(CO)_2L_4]$	[96]
		$trans\text{-}[Mo(CO)_2L_4]$	[96]
		$[Mo(CO)L_5]$	[96]
		$[MoL_6]$	[96]
	$P(OCH_3)_3$	$[MoL_6]$	[94]
	$(n\text{-}C_3H_7O)PF_2$	$[MoL_6]$	[94]
	$P(OCH_3)_2F$	$[MoL_6]$	[94]
	$P(CH_3)_3$	$[Mo(CO)L_5]$	[94]
		$cis\text{-}[Mo(CO)_2L_4]$	[95]
	$(CH_3)P(OCH_3)_2$	$[Mo(CO)L_5]$	[94]
$[W(CO)_6]$	$P(CH_3)_3$	$fac\text{-}[W(CO)_3L_3]$	[95]
		$mer\text{-}[W(CO)_3L_3]$	[95]
		$cis\text{-}[W(CO)_4L_2]$	[95]
		$[W(CO)_4L_2]$	[95]
		$cis\text{-}[W(CO)_2L_4]$	[95]
		$[W(CO)L_5]$	[94]
	$P(OCH_3)_3$	$[W(CO)L_5]$	[94]
		$cis\text{-}[W(CO)_2L_4]$	[95]
	$(n\text{-}C_3H_7O)PF_2$	$[WL_6]$	[94]
	$P(OCH_3)_2F$	$[WL_6]$	[94]
	$(CH_3)P(OCH_3)_2$	$[W(CO)L_5]$	[94]
	1,3-Butadiene	$[Mo(CO)_2L_2]$	[95]

metal. Ligands having this quality should also tend to make possible the photolabilization of the CO. The substitution of CO by ligands which are like CO will not lead to substantial changes in the electronic structure; and, thus, even though the symmetry may be quite low, the excited states are likely to be O_h-like. Consequently, CO photosubstitution can occur since the $t_{2g}^6 \rightarrow t_{2g}^5 e_g$ type excitation is indiscriminate when all six ligands are good π acceptors. This situation is to be contrasted to that of $[W(CO)_5(piperidine)]$ where there are excited states which yield labilization of either the z axis or x–y axes.

Olefin ligands are CO-like, and metal carbonyl olefin complexes have been implicated as important in a number of photoassisted and photocatalyzed olefin reactions (cf. Chapter 3). Photosubstitution in such systems is crucial. Since the olefins are CO-like, it is not surprising that multiple substitution is found. For example, it was reported that $[W(CO)_5(1\text{-pentene})]$ undergoes loss of CO in the presence of 2.3 M 1-pentene with a quantum yield of 0.31 $\pm$ 0.05 and 0.44 $\pm$ 0.05 at 313 and 366 nm, respectively [42]. Recently, there have been some very nice studies [98–100] on $[M(CO)_4(norbornadiene)]$. These complexes apparently undergo loss of a CO which is trans to an olefin [98,100]. The quantum efficiency for reaction (2-15) was found to be ~ 0.1 [99]. Finally, $[Cr(CO)_3(cycloheptatriene)]$ undergoes CO substitution

$$[Cr(CO)_4(norbornadiene)] \xrightarrow[PPh_3]{hv} mer\text{-}[Cr(CO)_3(PPh_3)(norbornadiene)] \quad (2\text{-}15)$$

upon 366 nm excitation [101], in contrast to the thermal reaction which gives cycloheptatriene substitution [102].

   b. $[M(\eta^6\text{-Arene})(CO)_n L_{3-n}]$ *Complexes.* Early reports on the $[M(\eta^6\text{-}$ arene$)(CO)_3]$ complexes claimed reactions (2-16) [103] and (2-17) [104] as general photoreactions [103,104]. Recent studies [53,105] have shown that

$$[M(\eta^6\text{-arene})(CO)_3] \xrightarrow[L]{hv} [M(\eta^6\text{-arene})(CO)_2 L] + CO \quad (2\text{-}16)$$

$$[M(\eta^6\text{-arene})(CO)_3] \xrightarrow[arene^*]{hv} [M(\eta^6\text{-arene}^*)(CO)_3] + arene \quad (2\text{-}17)$$

the quantum yield for (2-16) for M = Cr is very high. If reaction (2-17) obtains at all, it is very inefficient. The quantum yield for (2-16) was measured for L = pyridine [53], dodecylmaleimide [105]; arene = benzene [53], mesitylene [53,105], and the results are summarized in Table 2-26. The data show that the quantum efficiency is independent of wavelength, arene, entering group, and entering-group concentration. Numerous qualitative studies show that virtually any $[Cr(\eta^6\text{-arene})(CO)_2 L]$ complex can be made according to reaction (2-16), and Table 2-27 [106–112] gives representative photosubstitution products. Some data have been included for the Mo species, but no detailed studies have been reported. Curiously, the W species does

TABLE 2-26

Photosubstitution of $[Cr(\eta^6\text{-Arene})(CO)_3]$

Arene	Entering group, L (M)	Irradiation, λ, nm	Φ	Solvent
Benzene	Pyridine (0.167)	313	0.79[a]	Isooctane
	(0.084)		0.81[a]	
	(0.008)		0.78[a]	
Benzene	Pyridine (0.167)	313	0.62[a]	Benzene
	(0.084)		0.65[a]	
	(0.008)		0.63[a]	
Mesitylene	Pyridine (0.167)	313	0.62[a]	Isooctane
		366	0.72[a]	
		436	0.80[a]	
	Dodecylmaleimide (0.007)	313	0.68[b]	Cyclohexane
	Dodecylmaleimide (0.003)	313	0.91[b]	Benzene

[a] Wrighton and Haverty [53].
[b] Nasielski and Denisoff [105].

not undergo photoinduced CO exchange [104]. This is surprising in view of the similarity of the electronic spectra of $[M(\eta^6\text{-arene})(CO)_3]$ and the similarity in the photochemistry of $[M(CO)_6]$. For W the primary photoreaction may be isomerization of the η^6-arene bonding to a η^4-arene bonding.

At this point the photochemistry of $[Cr(\eta^6\text{-arene})(CO)_3]$ represents one of the more dramatic illustrations of a difference in thermal and photochemical reactivity. While CO is apparently exclusively lost from the excited state [53], thermal reactions are dominated by arene exchange [113]. The differences in ground vs. excited-state reactivity in this case may be ascribed to the fact that one-electron excitation is likely to change M—L binding more dramatically for a two-electron donor compared to the six-electron donor.

Further substitution of CO in $[Cr(\eta^6\text{-arene})(CO)_2L]$ is not common. There have been no detailed studies on the photochemistry of such species, though one should expect further CO loss for situations in which L is CO-like. Substitution of L itself is to be expected for L = amine, THF, etc. where there should be low-lying excited states. Perhaps surprisingly, $[Cr(\eta^6\text{-}$ arene)(CO)$_2$(trans-4-styrylpyridine)], which has an MLCT absorption at very low energies, undergoes efficient, clean photosubstitution of the styrylpyridine ligand. Apparently, the relaxed reactive LF states must be below the MLCT state [52].

As with other metal carbonyls, the coordinatively unsaturated intermediates generated by photolysis of $[M(\eta^6\text{-arene})(CO)_3]$ can be involved as intermediates in redox reactions. These can result in an oxidized central metal and complete loss of all ligands. Examples are represented by Eqs.

TABLE 2-27

Photosubstitution of $[M(\eta^6\text{-Arene})(CO)_3]$

Initial complex	Entering ligand, L	Product	Reference
$[Cr(\eta^6\text{-}C_6H_6)(CO)_3]$	$[Fe(\eta^5\text{-}C_5H_5)C_5H_4]_3P$	$[Cr(\eta^6\text{-}C_6H_6)(CO)_2L]$	[106]
$[Cr(\eta^6\text{-}C_6H_6)(CO)_3]$	2,3-Diazabicyclo[2.2.1]hept-2-ene	$[Cr(\eta^6\text{-}C_6H_6)(CO)_2L]$	[107]
$[Cr(\eta^6\text{-}CH_3OC_6H_5)(CO)_3]$	PPh_3	$[Cr(\eta^6\text{-}CH_3OC_6H_5)(CO)_2L]$	[108]
$[Cr(\eta^6\text{-}(CH_3)_6C_6)(CO)_3]$	Cyclopentene, maleic acid, PhCCPh	$[Cr(\eta^6\text{-}(CH_3)_6C_6)(CO)_2L]$	[109]
$[Mo(\eta^6\text{-Mesitylene})(CO)_3]$	C_2H_4	$[Mo(\eta^6\text{-Mesitylene})(CO)_2L]$	[110]
$[Mo(\eta^6\text{-}CH_3C_6H_5)(CO)_3]$	^{13}CO	$[Mo(\eta^6\text{-}CH_3C_6H_5)(CO)_2(^{13}CO)]$	[111]
$[Cr(\eta^6\text{-}C_6H_6)(CO)_3]$	Tetrahydrofuran	$[Cr(\eta^6\text{-}C_6H_6)(CO)_2L]$	[112]

(2-18) [114] and (2-19) [115]. Nondestructive, oxidative addition to 16-e^-

$$[Cr(\eta^6\text{-arene})(CO)_3] \xrightarrow[\text{MeOH}]{h\nu} Cr(OMe)_3 + \text{arene}$$
$$+ \qquad \text{(2-18)}$$
$$3CO$$

$$[Cr(\eta^6\text{-arene})(CO)_3] \xrightarrow[R_2S_2]{h\nu} Cr(SR)_3 + \text{arene}$$
$$+ \qquad \text{(2-19)}$$
$$3CO$$

intermediates like $[M(\eta^6\text{-arene})(CO)_2]$ is expected; an example is represented by Eq. (2-20) [116]. It is interesting to note that there seem to be no such

$$[Cr(\eta^6\text{-}C_6H_6)(CO)_3] \xrightarrow[Cl_3SiH]{h\nu}$$

$$\text{(2-20)}$$

simple oxidative addition products from the parent $[M(CO)_6]$ complex!

   c. $[M(\eta^5\text{-}C_5H_5)(CO)_3X]$ *and Related Compounds.*   Photosubstitution of the $[M(\eta^5\text{-}C_5H_5)(CO)_3X]$ complexes is dominated by reactions involving primary dissociative loss of CO. Some examples of photosubstitution of mononuclear complexes are given in Table 2-28 [56,57,117–124]. All three CO groups can be substituted using a tridentate phosphorus donor. The restriction that the entering groups replacing CO be extremely good π-acceptor ligands may be relieved somewhat since the central metal in these cases is in either a +1 or +2 oxidation state. The loss of CO induced by photolysis can apparently be achieved regardless of X or the central metal, but the photolability of the ($\eta^5\text{-}C_5H_5$) or X group has not been evaluated.
   The complexes $[Mo(\eta^3\text{-}C_3H_5)(\eta^5\text{-}C_5H_5)(CO)_2]$ [122] and $[M(\eta^5\text{-}C_5H_5)_2(CO)]$ [123] also undergo loss of CO subsequent to excitation. The latter complex is particularly interesting, since reaction (2-21) obtains even

$$[W(\eta^5\text{-}C_5H_5)_2(CO)] \xrightarrow[C_6H_6]{h\nu} [W(\eta^5\text{-}C_5H_5)_2(H)(C_6H_5)] \qquad \text{(2-21)}$$

in the presence of an alkyne. Interestingly, similar reactivity, implicating common intermediates, is found by irradiating $[W(\eta^5\text{-}C_5H_5)_2H_2]$ (cf. Chapter 7).
   Work with these systems can be used to demonstrate other interesting consequences of CO lability. For example, σ to π rearrangements like that in reaction (2-22) [125] have been observed, and it is likely that such reactions

$$[W(\eta^1\text{-}C_3H_5)(\eta^5\text{-}C_5H_5)(CO)_3] \xrightarrow{h\nu} [W(\eta^3\text{-}C_3H_5)(\eta^5\text{-}C_5H_5)(CO)_2] \qquad \text{(2-22)}$$

TABLE 2-28

Photosubstitution of $[M(\eta^5\text{-}C_5H_5)(CO)_3X]$ and Related Complexes

Starting complex	Entering group, L	Product	Reference
$[Mo(\eta^5\text{-}C_5H_5)(CO)_3Cl]$	$(Ph_2PCH_2CH_2)_2PPh$	$[Mo(\eta^5\text{-}C_5H_5)LCl]$	[117]
		$[Mo(\eta^5\text{-}C_5H_5)(CO)LCl]$	[118]
	PEt_3	$[Mo(\eta^5\text{-}C_5H_5)(CO)L_2Cl]$	[119]
	$Ph_2PCH_2CH_2PPh_2$	$[Mo(\eta^5\text{-}C_5H_5)(CO)LCl]$	[118]
	None	$[Mo(\eta^5\text{-}C_5H_5)(CO)_2Cl]_2$	[120]
$[Mo(\eta^5\text{-}C_5H_5)(CO)_3X]$	X^-	$[Mo(\eta^5\text{-}C_5H_5)(CO)_2X_2]^-$	[56,57]
	$X = Cl, Br, I, NCS$		
$[Mo(\eta^5\text{-}C_5H_5)(CO)_2\text{-}$ $(Ph_2PCH_2CH_2PPh_2)]Cl$	None	$[Mo(\eta^5\text{-}C_5H_5)(CO)\text{-}$ $(Ph_2PCH_2CH_2PPh_2)Cl]$	[118]
$[M(\eta^5\text{-}C_5H_5)(CO)_2(CH_3)]$	PPh_3	$[M(\eta^5\text{-}C_5H_5)(CO)_2(L)(CH_3)]$	[121]
	$M = Mo, W$		
$[Mo(\eta^3\text{-}C_3H_5)(\eta^5\text{-}C_5H_5)$ $(CO)_2]$	Et_2NPF_2	$[Mo(\eta^3\text{-}C_3H_5)(\eta^5\text{-}C_5H_5)$ $(CO)L]$	[122]
$[M(\eta^5\text{-}C_5H_5)_2(CO)]$	C_2H_2	$[M(\eta^5\text{-}C_5H_5)_2(C_2H_2)]$	[123]
	$M = Mo, W$		
$[Mo(\eta^5\text{-}C_5H_5)(CO)_2NO]$	PPh_3	$[Mo(\eta^5\text{-}C_5H_5)(CO)(PPh_3)NO]$	[124]

begin with the dissociative loss of CO. Another interesting fact is that ligands containing multiple ligating sites can be metallated with more than one group, e.g., reaction (2-23) [121].

$$[Mo(\eta^5\text{-}C_5H_5)(CO)_3CH_3] \xrightarrow[Ph_2PCH_2CH_2PPh_2]{h\nu}$$

$$\{[Mo(\eta^5\text{-}C_5H_5)(CO)_2(CH_3)]_2(Ph_2PCH_2CH_2PPh_2)\} \quad (2\text{-}23)$$

Irradiation of $[Mo(\eta^5\text{-}C_5H_5)(CO)_2(COCH_3)E(C_6H_5)_3]$ proceeds to give net decarbonylation producing $[Mo(\eta^5\text{-}C_5H_5)(CO)_2(CH_3)E(C_6H_5)_3]$ for $E = As, Sb$ [126]. The reaction undoubtedly involves dissociative rupture of a Mo—CO bond followed by alkyl migration.

A final interesting observation [127] is that $[Cr(\eta^5\text{-}C_5H_5)(CO)_3(CH_3)]$ undergoes photolysis in the absence of entering groups to give the dimers $[Cr_2(\eta^5\text{-}C_5H_5)_2(CO)_6]$ and $[Cr_2(\eta^5\text{-}C_5H_5)_2(CO)_4]$. But in the presence of $P(OCH_3)_3$ or PPh_3, simple substitution products result. Whether the Cr—CH_3 is homolytically cleaved as a primary photoprocess remains unanswered.

2. Metal–Metal Bond Cleavage and Related Reactions

As discussed previously complexes having a direct metal–metal bond exhibit absorptions in the UV–vis region which are attributable to transitions between metal–metal bonding and antibonding orbitals. The complexes $[M_2(\eta^5\text{-}C_5H_5)_2(CO)_6]$ (M = Mo, W) have received considerable study, and it is now well appreciated that the M—M bond can be efficiently cleaved by photochemical means.

The first indication that the bond could be ruptured photochemically came from reaction (2-24) [128]. However, irradiation of $[M_2(\eta^5\text{-}C_5H_5)_2(CO)_6]$

$$[Mo_2(\eta^5\text{-}C_5H_5)_2(CO)_6] \xrightarrow[PPh_3]{h\nu} [Mo(\eta^5\text{-}C_5H_5)(CO)_2(PPh_3)_2{}^+][Mo(\eta^5\text{-}C_5H_5)(CO)_3{}^-]$$

$$(2\text{-}24)$$

in CCl_4 proceeds according to Eq. (2-25) [55]. Further, simultaneous

$$[M_2(\eta^5\text{-}C_5H_5)_2(CO)_6] \xrightarrow[CCl_4]{h\nu} 2[M(\eta^5\text{-}C_5H_5)(CO)_3Cl] \qquad (2\text{-}25)$$

$$M = Mo, W$$

irradiation of $[M_2'(CO)_{10}]$ (M' = Mn, Re) and $[M_2(\eta^5\text{-}C_5H_5)_2(CO)_6]$ in hydrocarbon solutions yields the M—M' bonded complexes (Eq. 2-26).

$$[M_2(\eta^5\text{-}C_5H_5)_2(CO)_6] + [M_2'(CO)_{10}] \xrightarrow[\substack{\text{near-UV, } N_2\text{-purged} \\ \text{benzene solution}}]{h\nu} [M(\eta^5\text{-}C_5H_5)(CO)_3M'(CO)_5]$$

$$M = W, Mo; M' = Mn, Re \qquad (2\text{-}26)$$

Quantum yields for reaction (2-25) given in Table 2-29 show that the M—M bond cleavage is a fairly efficient process and that the stoichiometry is clean.

TABLE 2-29

Photoreaction of $[M_2(\eta^5\text{-}C_5H_5)_2(CO)_6]$ in $CCl_4{}^{a,b}$

M	Photolysis λ, nm	$\Phi_{dis}{}^c$	$\Phi_{appear}{}^d$	N^e
Mo	550	0.35	0.77	2.20
	405	0.42	0.94	2.25
	366	0.45	1.04	2.30
W	550	0.12	0.25	2.05
	366	0.21	0.41	1.93

[a] Reprinted with permission from Wrighton and Ginley [55], *J. Am. Chem. Soc.* **97**, 4246 (1975). Copyright by the American Chemical Society.

[b] At 298°K; degassed solutions.

[c] Quantum yield for disappearance of starting complex.

[d] Quantum yield for appearance of $[M(\eta^5\text{-}C_5H_5)(CO)_3Cl]$.

[e] Ratio of $[M(\eta^5\text{-}C_5H_5)(CO)_3Cl]$ appearance to $[M_2(\eta^5\text{-}C_5H_5)_2(CO)_6]$ disappearance.

That Eq. (2-25) is clean and efficient and Eq. (2-26) requires irradiation of both complexes seemingly demands homolytic M—M bond cleavage as the primary photoprocess (Eq. 2-27) [55]. The 17-electron, d^5 radical undergoes

$$[M_2(\eta^5\text{-}C_5H_5)_2(CO)_6] \xrightarrow{h\nu} 2[M(\eta^5\text{-}C_5H_5)(CO)_3] \qquad (2\text{-}27)$$

subsequent reaction with CCl_4 to give $[M(\eta^5\text{-}C_5H_5)(CO)_3Cl]$ or with another radical such as $M'(CO)_5$ to generate new M—M' species. Consistently, in the absence of radical traps, flash photolysis of $[Mo_2(\eta^5\text{-}C_5H_5)_2(CO)_6]$ yields intermediates which regenerate the starting material [129]. The data were interpreted in terms of photogenerated $[M(\eta^5\text{-}C_5H_5)(CO)_3]$ radicals which back-react bimolecularly at nearly a diffusion-controlled rate.

Irradiation of $[M_2(\eta^5\text{-}C_5H_5)_2(CO)_6]$ in the presence of other halocarbons yields additional information concerning stoichiometry and reactivity. First, irradiation in the presence of Ph_3CCl yielded ESR-detectable $Ph_3C\cdot$ radicals, and irradiation in the presence of $PhCH_2Cl$ produced bibenzyl in good yield [55]. Irradiation of $[W_2(\eta^5\text{-}C_5H_5)_2(CO)_6]$ in the presence of several chlorocarbons established the rate constants of Cl abstraction by $[W(\eta^5\text{-}C_5H_5)(CO)_3]$; the ordering of reactivity for the chlorocarbons was found to be: $CCl_4 > CHCl_3 > PhCH_2Cl > CH_2Cl_2$ [130].

Irradiation of $[M(\eta^5\text{-}C_5H_5)(CO)_3M'(CO)_5]$ leads to homolytic M—M' cleavage [Eqs. (2-28) and (2-29)] [65]. Quantum yields for reaction (2-29) are

$$2[M(\eta^5\text{-}C_5H_5)(CO)_3M'(CO)_5] \xrightarrow[\text{hydrocarbon}]{h\nu} [M_2(\eta^5\text{-}C_5H_5)_2(CO)_6] + [M'_2(CO)_{10}]$$

$$M = Mo, W; \qquad M' = Mn, Re \qquad (2\text{-}28)$$

$$[M(\eta^5\text{-}C_5H_5)(CO)_3M'(CO)_5] \xrightarrow[CCl_4]{h\nu} [M'(CO)_5Cl] + [M(\eta^5\text{-}C_5H_5)(CO)_3Cl]$$

$$M = Mo, W; \qquad M' = Mn, Re \qquad (2\text{-}29)$$

given in Table 2-30. Interestingly, there is a rather substantial decline in the quantum efficiency at 436 compared to 366 nm. This may be a consequence of $d\pi \rightarrow \sigma^*$ (436 nm) vs. a $\sigma_b \rightarrow \sigma^*$ (366 nm) absorption. A similar, but smaller, wavelength dependence obtains for the $[M_2(\eta^5\text{-}C_5H_5)_2(CO)_6]$ species (Table 2-29).

Under certain conditions, irradiation of M—M' bonded species yields M—M and M'—M' bonded complexes in addition to the halogen abstraction products [131]. In such a case it is possible to determine the relative reactivity of the M- and M'-centered radicals toward a given halogen donor. Further, if M—M' bonded species are irradiated in the presence of two different halogen donors, the relative reactivity of the two donors can be obtained. Such studies have been used to establish that the reactivity of the photogenerated $[Mo(\eta^5\text{-}C_5H_5)(CO)_3]$ radical is independent of its source; that is, the

TABLE 2-30

Disappearance Quantum Yields for
$[M(\eta^5\text{-}C_5H_5)(CO)_3M'(CO)_5]$ in $CCl_4{}^{a,b}$

M	M'	$\Phi_{366\,nm} \pm 10\%$	$\Phi_{436\,nm} \pm 10\%$
Mo	Mn	0.51	
Mo	Re	0.56	
W	Mn	0.35	0.05_6
W	Re	0.23	0.05_1

^{a} Reprinted with permission from Ginley and Wrighton [65], *J. Am. Chem. Soc.* **97**, 4908 (1975). Copyright by the American Chemical Society.
^{b} Photolysis in degassed CCl_4 solution at 298°K. Analysis by UV and IR spectral changes. The products of the photolysis are $[M(CO)_5Cl]$ and $[M(\eta^5\text{-}C_5H_5)(CO)_3Cl]$.

photogenerated radicals from a series of metal–metal bonded complexes all involving the $[Mo(\eta^5\text{-}C_5H_5)(CO)_3]$ unit give a species with the same properties toward two halocarbons [131]. Second, the reactivity of a series of metal-centered radicals toward CCl_4 and 1-iodopentane show the following ordering: $[Re(CO)_5] > [Mn(CO)_5] > [W(\eta^5\text{-}C_5H_5)(CO)_3] > [Mo(\eta^5\text{-}C_5H_5)(CO)_3] > [Fe(\eta^5\text{-}C_5H_5)(CO)_2] > [Co(CO)_4]$. The species $[MoW(\eta^5\text{-}C_5H_5)_2(CO)_6]$, $[M(\eta^5\text{-}C_5H_5)(CO)_3Co(CO)_4]$ (M = Mo, W), and $[M(\eta^5\text{-}C_5H_5)(CO)_3Fe(\eta^5\text{-}C_5H_5)(CO)_2]$ (M = Mo, W) have all been shown to undergo efficient, photoinduced, homolytic metal–metal bond cleavage [66,131,136].

Irradiation of $[M_2(\eta^5\text{-}C_5H_5)_2(CO)_6]$ in the presence of other radical traps has led to some interesting chemistry. For example, irradiation of the Mo complex in the presence of NO gives the product indicated in Eq. (2-30) [132]. It has not been determined whether $[Mo(\eta^5\text{-}C_5H_5)(CO)_3NO]$ is an

$$[Mo_2(\eta^5\text{-}C_5H_5)_2(CO)_6] \xrightarrow[\text{NO}]{hv} [Mo(\eta^5\text{-}C_5H_5)(CO)_2NO] \qquad (2\text{-}30)$$

intermediate in the formation of the dicarbonyl product. Irradiation of either $[Mo_2(\eta^5\text{-}C_5H_5)_2(CO)_6]$ or the derivative $[Mo_2(\eta^5\text{-}C_5H_5)_2(CO)_4(PPh_3)_2]$ in the presence of nitrosodurene leads to a stable radical resulting from the reaction of the Mo-centered radical and the radical trap [133].

Homolytic cleavage of direct M—M bonds dominates the photochemistry of such species, but photoreactions of $[M_2(\eta^5\text{-}C_5H_5)_2(CO)_6]$ carried out in media such as pyridine or THF [56,134] give rise to net heterolytic cleavage products. However, it is likely that such chemistry occurs as a consequence of disproportionation reactions of the $[M(\eta^5\text{-}C_5H_5)(CO)_3]$ radicals which are the primary products.

It is worth noting that the photochemistry of $[Mo(\eta^5\text{-}C_5H_5)(CO)_3$ $Sn(CH_3)_3]$ has been investigated and has been shown to involve loss of CO (Eq. 2-31) [135]. Though the complex is an M—M' system, it is likely that

$$[Mo(\eta^5\text{-}C_5H_5)(CO)_3Sn(CH_3)_3] \xrightarrow[P(OC_6H_5)_3]{hv}$$

$$[Mo(\eta^5\text{-}C_5H_5)(CO)_2(P(OC_6H_5)_3)Sn(CH_3)_3] + CO \quad (2\text{-}31)$$

the bond is more ionic than for transition metal–transition-metal bonded systems. Thus, an energy level scheme as in Fig. 2-17 may obtain, and LF lowest excited state reactivity is found.

While M—M homolytic bond cleavage in $[M_2(\eta^5\text{-}C_5H_5)_2(CO)_6]$ is the only detectable photoreaction in CCl_4 [55,135], there are several reports of simple photosubstitution (Eq. 2-32) [121,135]. However, prompt CO

$$[M_2(\eta^5\text{-}C_5H_5)_2(CO)_6] \xrightarrow[L\,=\,P(OC_6H_5)_3]{\Delta} [M_2(\eta^5\text{-}C_5H_5)_2(CO)_4L_2] \quad (2\text{-}32)$$

loss is not efficient [129]; the mechanism for substitution may be as in

$$[M_2(\eta^5\text{-}C_5H_5)_2(CO)_6] \underset{\Delta}{\overset{hv}{\rightleftharpoons}} 2[M(\eta^5\text{-}C_5H_5)(CO)_3] \quad (2\text{-}33)$$

$$[M(\eta^5\text{-}C_5H_5)(CO)_3] + L \xrightarrow{\Delta} [M(\eta^5\text{-}C_5H_5)(CO)_2L] + CO \quad (2\text{-}34)$$

$$2[M(\eta^5\text{-}C_5H_5)(CO)_2L] \xrightarrow{\Delta} [M_2(\eta^5\text{-}C_5H_5)_2(CO)_4L_2] \quad (2\text{-}35)$$

$$[M(\eta^5\text{-}C_5H_5)(CO)_2L] + [M(\eta^5\text{-}C_5H_5)(CO)_3] \xrightarrow{\Delta} [M_2(\eta^5\text{-}C_5H_5)_2(CO)_5L] \quad (2\text{-}36)$$

Eqs. (2-33)–(2-36). The substitution occurs at the radical stage, and isolated products result from radical coupling. This mechanism demands that the disubstituted product be a primary product, perhaps accompanied by the monosubstituted species. In this connection it is appropriate to note that reaction (2-37) is known [64,137] and that the triple-bonded species reacts

$$[M_2(\eta^5\text{-}C_5H_5)_2(CO)_6] \xrightarrow{hv} [M_2(\eta^5\text{-}C_5H_5)_2(CO)_4] + 2CO \quad (2\text{-}37)$$

rapidly with ligands (Eq. 2-38) [69]. This provides another possible alternative for the photochemical formation of substituted products. However,

$$[Mo_2(\eta^5\text{-}C_5H_5)_2(CO)_4] \xrightarrow[L\,=\,PPh_3,\,P(OCH_3)_3]{\Delta} [Mo_2(\eta^5\text{-}C_5H_5)_2(CO)_4L_2] \quad (2\text{-}38)$$

the photogeneration of the triple-bonded complex (Eq. 2-37) very likely implicates the sequence Eq. (2-33), (2-39), (2-40). In the presence of L reaction

$$[M(\eta^5\text{-}C_5H_5)(CO)_3] \underset{}{\overset{\Delta}{\rightleftharpoons}} [M(\eta^5\text{-}C_5H_5)(CO)_2] + CO \quad (2\text{-}39)$$

$$2[M(\eta^5\text{-}C_5H_5)(CO)_2] \xrightarrow{\Delta} [M_2(\eta^5\text{-}C_5H_5)_2(CO)_4] \quad (2\text{-}40)$$

$$[M(\eta^5\text{-}C_5H_5)(CO)_2] + L \xrightarrow{\Delta} [M(\eta^5\text{-}C_5H_5)(CO)_2L] \quad (2\text{-}41)$$

(2-41) should compete with reaction (2-40). Apparently, all observations concerning $[M_2(\eta^5\text{-}C_5H_5)_2(CO)_6]$ are consistent with homolytic M—M cleavage as the principal result of excited-state decay.

3. Rearrangement and Intraligand Reactions

Several interesting rearrangement reactions of Cr, Mo, and W carbonyls have been reported. First, linkage isomerization has been photoinduced (Eq. 2-42) [138]. Also, the related complexes $[M(\eta^5\text{-}C_5H_5)(CO)_2(PPh_3)X]$

$$[Mo(\eta^5\text{-}C_5H_5)(CO)_3NCS] \underset{hv \text{ or } \Delta}{\overset{hv}{\rightleftarrows}} [Mo(\eta^5\text{-}C_5H_5)(CO)_3SCN] \qquad (2\text{-}42)$$

(M = Mo, W; X = Br, I) undergo photoinduced cis–trans interconversion along with substitution reactions apparently involving CO and PPh_3.

Isomerization of $cis\text{-}[M(CO)_4L_2]$ (M = Cr, Mo, W; L = 1,3-dimethyl-4-imidazoline-2-ylidene and related ligands) to the thermodynamically un-stable trans isomer can be photoinduced [139]. A competing photoprocess is the loss of CO. Photoinduced isomerizations of this sort have also been noted in low-temperature matrices [32,32a,92].

As noted in Table 2-28, photosubstitution of CO occurs in $[M(\eta^5\text{-}C_5H_5)(CO)_2NO]$ [124]. Additionally, products which appear to arise from a photogenerated nitrene are found Eq. (2-43). The existence of the nitrene

$$[Mo(\eta^5\text{-}C_5H_5)(CO)_2NO] \xrightarrow[PPh_3]{hv} [Mo(\eta^5\text{-}C_5H_5)(CO)(PPh_3)_2(NCO)] +$$
$$[Mo(\eta^5\text{-}C_5H_5)(CO)_2(PPh_3)(NCO)] \quad (2\text{-}43)$$

gains support from the observation that irradiation of an azido species gives similar product yields [124], (Eq. 2-44). Photogeneration of coordinated

$$[Mo(\eta^5\text{-}C_5H_5)(CO)(PPh_3)_2(N_3)] \xrightarrow{hv} [Mo(\eta^5\text{-}C_5H_5)(CO)(PPh_3)_2(NCO)] +$$
$$[Mo(\eta^5\text{-}C_5H_5)(CO)_2(PPh_3)(NCO)] \quad (2\text{-}44)$$

nitrenes from metal azide complexes has been unambiguously established for species such as $[Ir(NH_3)_5(N_3)]^{2+}$ [140].

Study [44,49] of the complexes $[W(CO)_5L]$ and $cis\text{-}[W(CO)_4L_2]$ (L = 4-styrylpyridine, 2-styrylpyridine) has provided some interesting results concerning the reactivity and interconvertibility of IL and metal-centered excited states. The key point is that the styrylpyridines have intense, spin-allowed $\pi \rightarrow \pi^*$ absorptions as well as low-lying triplet states, and population of either state results in cis–trans isomerization [141]. For the C_{4v} and C_{2v} complexes, substitution and IL isomerization were studied. For the C_{4v} complexes, fairly efficient substitution and IL isomerization are found (Table 2-31) [44]. Quite interestingly, irradiation in the lowest absorption

TABLE 2-31

Quantum Yields for Photoreactions of $[W(CO)_5L]$ Complexes[a]

Reaction[b]	$\Phi_{436\,nm}$	$\Phi_{366\,nm}$	$\Phi_{313\,nm}$	$\Phi_{254\,nm}$
A[c]	0.63	0.50	0.38	0.34
B[d]	0.00_2	0.01_3	0.03_9	~0.04
C[c]	0.16	0.08	0.05	
D	0.49	0.34	0.26	0.21
E	0.08 $(t \to c)$			
	0.31 $(c \to t)$			
F[c]	0.16	0.07	0.13	

[a] Data from Wrighton *et al.* [44].

[b] (A) $[W(CO)_5(py)] \xrightarrow{h\nu} [W(CO)_5(1\text{-pent})]$

(B) $[W(CO)_5(py)] \xrightarrow{h\nu} cis\text{-}[W(CO)_4(py)_2]$

(C) $[W(CO)_5(trans\text{-}4\text{-stpy})] \xrightarrow{h\nu} [W(CO)_5(1\text{-pent})]$

(D) $[W(CO)_5(trans\text{-}4\text{-stpy})] \xrightarrow{h\nu} [W(CO)_5(cis\text{-}4\text{-stpy})]$

(E) $[W(CO)_5(trans\text{-}2\text{-stpy})] \xrightarrow{h\nu} [W(CO)_5(cis\text{-}2\text{-stpy})]$

(F) $[W(CO)_5(trans\text{-}2\text{-stpy})] \xrightarrow{h\nu} [W(CO)_5(1\text{-pent})]$

[c] Φ measured at room temperature in the presence of 3.66 M 1-pentene–isooctane solvent.

[d] Φ for formation of $cis\text{-}[W(CO)_4(py)_2]$ at room temperature in presence of 0.25 M pyridine in isooctane.

system (LF + W → L CT) yields very efficient IL isomerization, and photosubstitution of L occurs but with a lower efficiency than for L = pyridine. These results could be interpreted in terms of reaction from an IL triplet state. This conclusion is supported by the study of $cis\text{-}[W(CO)_4L_2]$ [49]. In this case, the lowest state in absorption is purely W → L CT, and this relaxed CT state may even be lower than the IL triplet. Excitation into the lowest absorption gives little IL isomerization, but direct irradiation into the IL absorption ~313 nm gives isomerization with a quantum yield of 0.1 for *trans* → *cis*-4-styrylpyridine conversion. This shows that the IL reactions are competitive with internal conversion through the CT and LF manifolds. Biacetyl-sensitized isomerization of the coordinated ligands in $cis\text{-}[W(CO)_4L_2]$ was found, and this can be interpreted as due to population of the reactive IL triplet. The excited-state processes in $cis\text{-}[W(CO)_4(trans\text{-}4\text{-styrylpyridine})_2]$ are summarized in Fig. 2-21 [49]. Even for L = cis-4-styrylpyridine the W → L CT was not found to give efficient IL isomerization. From earlier studies [142] of Ru(II) complexes it was expected that the IL cis → trans, but not trans → cis, isomerization would occur, owing to the increase of electron density in the $L\pi^*$ orbital in the W → L CT state.

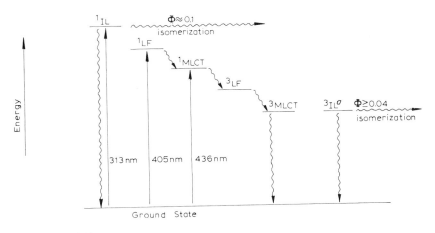

FIG. 2-21. Excited-state processes in *cis*-[W(CO)$_4$(*trans*-4-styrylpyridine)$_2$]. Reprinted with permission from Pdungsap and Wrighton [49].

IV. MANGANESE AND RHENIUM CARBONYLS

A. GEOMETRIC STRUCTURE

The commonly available carbonyls of Mn and Re have the formula [M$_2$(CO)$_{10}$] with the structure shown in (V). The [M$_2$(CO)$_{10}$] species are isostructural with [M$'_2$(CO)$_{10}$]$^{2-}$ [M$'$ = W(−1), Mo(−1), Cr(−1)] [143,144]. A number of six-coordinate M(I) compounds are known including [M(CO)$_6$]$^+$ and [M(CO)$_5$X]. Numerous substituted derivatives have been prepared including metal carbonyl–alkyl compounds where it appears still appropriate to assign the +1 oxidation state to the central metal. The [M(η^5-C$_5$H$_5$)(CO)$_3$] complexes are well-studied and can be viewed as having the substitutionally inert d^6 electronic configuration with a +1 central metal oxidation state.

$$
\begin{array}{ccc}
& \text{OC} \quad\quad \text{CO} & \\
& \quad | \;\; \text{CO} \quad | \;\; \text{CO} & \\
\text{OC} - & \text{M} \longrightarrow \text{M} & - \text{CO} \\
\text{OC}\diagup & |\quad\; \text{OC}\diagup \;\; | & \\
& \text{OC} \quad\quad \text{CO} &
\end{array}
$$

(V)

The M(I) species $[M_2(CO)_8X_2]$ (X = halogen) having structure (VI) can be understood without invoking a direct M—M bond. A large number of polynuclear complexes are known which do contain direct M—M bonds,

$$(CO)_4M \underset{X}{\overset{X}{\diagdown\diagup}} M(CO)_4$$

(VI)

e.g., $[Re_3H_3(CO)_{12}]$ and $[Re_4H_4(CO)_{12}]$; and, in contrast to the situation for Cr, Mo, and W, clusters containing more than two metals are important. The -1 oxidation state has some importance; e.g., $[M(CO)_5]^-$ are known and are isostructural with $[Fe(CO)_5]$.

B. Electronic Structure

1. $[M(CO)_nL_{6-n}]$ Species

Assignment of the spectra of the d^6 $[M(CO)_6]^+$ complexes follows that of the isoelectronic Cr, Mo, and W species. The lowest absorptions are LF and the $M \rightarrow \pi^*CO$ CT absorptions fall at higher energies (Table 2-32) [10]. For the set of isoelectronic, isostructural complexes $[V(CO)_6]^-$, $[Cr(CO)_6]$, $[Mn(CO)_6]^+$ the value of $10Dq$ varies as expected, with the higher oxidation state central metal giving the highest energy LF transitions (cf. Tables 2-2,

TABLE 2-32

Electronic Spectral Features of $[M(CO)_6]^+$ Species[a]

Complex	Bands, cm^{-1} (ε)	Assignment	
$[Mn(CO)_6{}^+][BF_4{}^-]$	49,900 (27,000)	$^1A_{1g} \rightarrow d^1T_{1u}$	$\left.\right\}$ $M \rightarrow \pi^*CO$
	44,500 (16,000)	$^1A_{1g} \rightarrow c^1T_{1u}$	
	39,600 (2200)	$^1A_{1g} \rightarrow {}^1T_{1g}$	$\left.\right\}$ $t_{2g}{}^6 \rightarrow t_{2g}{}^5e_g{}^1$
	37,300 (1100)		
	33,250 (600)	$^1A_{1g} \rightarrow {}^3T_{1g}$	
$[Re(CO)_6{}^+][AlCl_4{}^-]$	51,200 (77,900)	$^1A_{1g} \rightarrow d^1T_{1u}$	$M \rightarrow \pi^*CO$
	47,100 (4600)	$^1A_{1g} \rightarrow {}^1T_{2g}$	$t_{2g}{}^6 \rightarrow t_{2g}{}^5e_g{}^1$
	44,500 (20,000)	$^1A_{1g} \rightarrow c^1T_{1u}$	$M \rightarrow \pi^*CO$
	40,700 (2900)	$^1A_{1g} \rightarrow {}^1T_{1g}$	$\left.\right\}$ $t_{2g}{}^6 \rightarrow t_{2g}{}^5e_g{}^1$
	38,500 (1500)		
	36,850 (708)	$^1A_{1g} \rightarrow {}^3T_{1g}$	

[a] Data from Beach and Gray [10].
[b] CH$_3$CN solution at 298°K.

2-7, and 2-32). At the same time the higher positive charge also leads to higher-energy MLCT transitions. This central charge effect has consequence with respect to the relative position of the IL states and the CT and LF states in that it is possible to find IL absorptions below the metal-centered ones, e.g., fac-[ReX(CO)$_3$(styrylpyridine)$_2$], $vide$ $infra$.

The C_{4v} complexes of Mn and Re have received considerable attention, and we can again make analogy to the isoelectronic, isostructural Cr, Mo, and W complexes. The lowest transitions have been ascribed [37,145] to LF transitions (Table 2-33) [37,146,147]. The spectral comparison given in Fig. 2-22 for C_{4v} W and Re at least is convincing with respect to the fact that a common spectral assignment is appropriate. The Re, but not the Mn, complexes exhibit a prominent spin-forbidden absorption feature attributable to the $^1A_1(e^4b_2{}^2) \rightarrow {}^3E(e^3b_2{}^2a_1{}^1)$ transition. As with the Cr, Mo, and W species, the shift in the position of the first band from the O_h to the C_{4v} complexes is substantial for replacement of CO by ligands which are low in the spectrochemical series. From variation in the position of the first band in the C_{4v} complexes of Cr, Mo, W, Mn, and Re, the spectrochemical series is that expected:

$$CO \sim H \sim alkyl > \text{P-donor} > \text{N-donor} > \text{O-donor} > Cl > Br > I$$
$$\longleftarrow 10Dq$$

The LF spectroscopy of further derivatives (beyond C_{4v}) of Mn and Re is not well developed. The spectra of a few lower symmetry species have been reported (Table 2-34) [146,148,149], but much work remains to be done in this area. The main point here is that for the complexes examined so far, the weak LF bands are not too far red-shifted from those for the C_{4v} species.

More substantive studies of lower symmetry complexes have dealt with species of the general formula fac-[ReX(CO)$_3$L] and fac-[ReX(CO)$_3$L$_2$] where L = 1,10-phenanthroline and related ligands or L = pyridine or substituted pyridine. Such complexes exhibit lowest optical absorptions which are attributable to Re $\rightarrow$ L CT or IL transitions. Some spectral data are given in Table 2-35 [150–152]. For the 1,10-phenanthroline and related complexes the Re $\rightarrow$ L CT absorption is the lowest energy absorption system. The assignment follows from a comparison with the bispyridine complex which has much higher Lπ^* acceptor orbitals and does not exhibit low-lying absorptions. Further, among the particular set of complexes the substituent effect is that expected for an MLCT transition. The lower energy absorption in the 2,2′-biquinoline compound compared to the 1,10-phenanthroline complex is especially convincing. The 1,10-phenanthroline and related complexes also exhibit a relatively nonperturbed IL absorption.

The bispyridine and bis-substituted pyridine complexes of Re also exhibit low-lying Re $\rightarrow$ L CT absorptions. Again, the energetic ordering of the bands with substitution in the pyridine confirms the CT assignment and its

TABLE 2-33

Spectral Features of C_{4v} Mn and Re Carbonyls

Complex	Solvent	Bands, cm^{-1} × 10^{-3} $(\varepsilon)^a$					Reference
		$^1A_1 \rightarrow {}^3E$	$^1A_1 \rightarrow {}^1E$	LF	LF	M → π*CO	
[Re(CO)$_5$NH$_2$CH$_2$CH$_3$]Cl	EtOH	31.45 (1250)	34.13 sh 35.34 (3270)	—	—	42.55 sh 37.17 (55,600)	[37]
[Re(CO)$_5$Cl]	EPA	29.07 (750)	31.06 (1940)	~33.0 (~1300)	36.36 (1600) 37.45 (1600)	Not recorded —	[37]
	CH$_3$CN	29.33 (0.22)	31.55 (0.693)	~33.7 (0.38)	~37.04 (0.45)	Not recorded	
	EtOH	29.24 (0.33)	31.45 (0.10)	~33.9 (0.06)	~37.0 (0.08)	~44.4 (0.70) ~48.08 (1.80)	
	Isooctane	28.74 (0.13)	30.77 (0.36)	~32.8 (0.21)	~36.1 (0.29) ~37.2 (0.29)	42.92 (1.33) 47.17 (5.10)	
[Re(CO)$_5$Br]	EPA	28.33 (600)	30.67 (1780)	~32.6 (~1300)	~36.4 (2400)	Not recorded	[37]
	Isooctane	27.93 (0.17)	30.21 (0.48)	~32.3 (0.28)	Not recorded	Not recorded	
	EtOH	28.73 (0.03)	30.95 (0.09)	~33.1 (0.05)	~37.0 (0.08) ~32.3 (0.28)	~42.19 (0.26) 46.51 (1.60)	
[Re(CO)$_5$I]	EPA	26.18 (220)	29.59 (2150)	~31.7 (~1200)	~34.5 (~2400) ~37.0 (~3200)	Not recorded	[37]
	CH$_2$Cl$_2$	26.24 (0.11)	29.67 (0.99)	~31.6 (0.5)	~35.7 (1.12)	Not recorded	
	EtOH	26.81 (215)	29.76 (2000)	~32.1 (940)	~35.1 (2000) ~36.6 (2800)	~46.9 (~35,000) ~49.5 (60,000)	
	Isooctane	26.04 (0.02)	28.99 (0.30)	30.77 (0.18)	~34.1 (0.27) ~36.0 (0.33)	40.32 sh (0.70) 44.84 sh (4.20) 44.85 (7.15)	

Compound	Solvent						Ref
[Re(CO)₅H]	C₆H₁₂		36.70 (—)			46.51 (—)	[146]
[Mn(CO)₅NH₂CH₃][PF₆]	EtOH	Not observed	29.41 (—)		~37.0 (—)	>50.00	[37]
[Mn(CO)₅NH₂CH₃]Cl	CH₂Cl₂	Not observed	29.59 (1420)		37.0 sh (3730)	—	[37]
	EtOH	Not observed	29.67 (—)		~37.0	—	[37]
[Mn(CO)₅Cl]	EPA	Not observed	29.41 (1200)		37.0 sh (3100)	—	
	Cyclohexane	Not observed	26.67 (0.01)		—	44.05 (0.5), 50.50 (0.6), >52.60 (0.45)	[146]
	CH₃OH	Not observed	26.52 (600)		~37.0 (~1500)	45.05 (13,000)	
[Mn(CO)₅Br]	Cyclohexane	Not observed	25.94 (0.02)		~37.0 (0.07)	43.01 (0.2), ~50.00 (0.75), 53.00	
	CH₃OH	Not observed	26.07 (420)		37.0 (1700)	43.95 (15,000), 50.50 (7000)	
[Mn(CO)₅I]	Cyclohexane	Not observed	23.53 (380)	32.80 (2310)	—	42.00 (15,300), 50.80 (79,000)	
	CH₃OH	Not observed	25.00 (360)	33.60 (3000)	36.40 (1500)	42.70 (20,000), 52.63 (88,000)	
[Mn(CO)₅NO₃]	CHCl₃	Not observed	26.67 (1200)		—	—	[147]
[Mn(CO)₅CF₃]	C₆H₁₂		~37.00 (~3000)			46.380 (30,000), >52.6 (>12,000)	
[Mn(CO)₅CH₃]	C₆H₁₂		~35.50 (~3000)			~45.05 (28,000), 51.02 (36,000)	[146]
[Mn(CO)₅H]	C₆H₁₂		~30.00, ~33.00, 37.00			47.28, 50.63	

ª Values in parentheses which are italicized are the relative absorptivities. Data are for 300°K solutions.

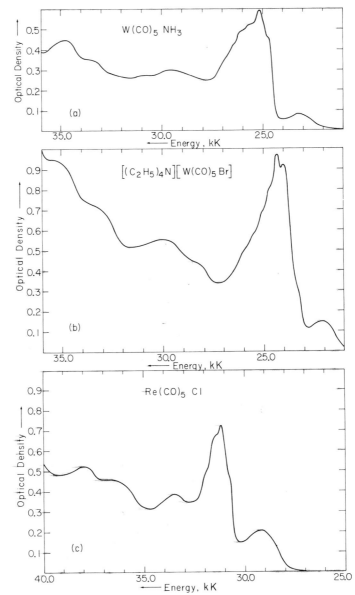

FIG. 2-22. Absorption spectra at 25°K in EtOH. See Tables 2-9 and 2-33 for molar absorptivities and band positions at 298°K. Reprinted with permission from Wrighton *et al.* [37], *J. Am. Chem. Soc.* **98**, 1111 (1976). Copyright by the American Chemical Society.

TABLE 2-34

Spectral Features of Low-Symmetry Mn and Re Complexes
Exhibiting LF Lowest Absorptions[a]

Complex	Solvent	Bands, cm^{-1} (ε)	Assignment	Reference
$[Mn(CO)_4Cl]_2$	C_6H_{12}	23,100 ($\sim$1000)	LF	
		$\sim$29,400 (100–1000)	?	
$[Mn(CO)_4Br]_2$	C_6H_{12}	23,100 (2300)	LF	[146]
		$\sim$29,400 (700)	?	
$[Mn(CO)_4I]_2$	C_6H_{12}	22,800 (100–1000)	LF	
		$\sim$28,600 (100–1000)		
$[Re(CO)_4Cl]_2$	C_6H_{12}	$\sim$27,000 sh	LF	[148]
$[MnCl(CO)_3(NCCH_3)_2]$	C_6H_{12}	26,600 ($\sim$1200)	LF	
$[MnBr(CO)_3(NCCH_3)_2]$	C_6H_{12}	26,500 (1500)	LF	[149]
$[MnCl(CO)_3(Et_2O)_2]$	Et_2O	25,000	LF	

[a] Room-temperature data.

TABLE 2-35

Spectral Features of fac-$[ReX(CO)_3L]$ and fac-$[ReX(CO)_3L_2]$[a]

L	X	Bands, cm^{-1} (ε)	Assignment
1,10-Phenanthroline[b]	Cl	24,400 sh	Re → L CT
		26,530 (4000)	Re → L CT
		37,310 (31,000)	IL
5-CH_3-1,10-Phenanthroline[b]	Cl	23,800 sh	Re → L CT
		26,320 (4100)	Re → L CT
		37,310 (27,000)	IL
4,7-Diphenyl-1,10-phenanthroline[b]	Cl	26,530 (—)	Re → L CT
		34,970 (—)	IL
5-Cl-1,10-Phenanthroline[b]	Cl	23,800 sh	Re → L CT
		25,910 (4100)	Re → L CT
		36,760 (30,300)	IL
5-Br-1,10-phenanthroline[b]	Cl	24,000 sh	Re → L CT
		25,840 (3900)	Re → L CT
		36,500 (27,500)	IL
5-NO_2-1,10-phenanthroline[b]	Cl	22,000 sh	Re → L CT
		25,190 (3800)	Re → L CT
		36,630 (27,700)	IL
2,2′-Biquinoline[b]	Cl	20,800 sh	Re → L CT
		22,830 (1920)	Re → L CT
		26,670 (20,900)	IL
Pyridine[c]	Cl	33,900 (8000)	Re → L CT
		38,460 (10,000)	IL

(*continued*)

TABLE 2-35 (*continued*)

L	X	Bands, cm^{-1} (ε)	Assignment
4-Phenylpyridine[c]	Cl	33,000 (22,600)	Re → L CT
		37,310 (36,500)	IL
	Br	32,790 (22,500)	Re → L CT
		37,310 (37,250)	IL
	I	32,790 (21,050)	Re → L CT
		37,040 (38,500)	IL
3-CN-Pyridine[c]	Cl	32,470 (8900)	Re → L CT
		39,530 (15,600)	IL
3,5-DiCl-pyridine[c]	Cl	32,260 (7000)	Re → L CT
		35,460 (9000)	IL
4-CN-Pyridine[c]	Cl	29,590 (11,100)	Re → L CT
		39,680 (13,000)	IL
4,4'-Bipyridine[c]	Cl	31,550 (13,200)	Re → L CT
		40,810 (35,950)	IL
	Br	31,060 (13,450)	Re → L CT
		40,810 (36,250)	IL
	I	31,250 (12,500)	Re → L CT
		42,020 (35,000)	IL
2,2'-Bipyridine-4,4'-dicarboxylic-acid[d]	Cl	26,670 (—)	Re → L CT
		33,330 (—)	IL
trans-3-Styrylpyridine[e]	Cl	33,670 (45,300)	IL
trans-4-Styrylpyridine[e]	Cl	30,490 (50,300)	IL
	Br	30,300 (53,200)	IL

[a] CH$_2$Cl$_2$ solutions, 298°K unless noted otherwise.
[b] Wrighton and Morse [150].
[c] Giordano and Wrighton [151].
[d] MeOH solution, 298°K
[e] Wrighton *et al.* [152].

direction. The absorption spectrum of *fac*-[ReCl(CO)$_3$(4-phenylpyridine)$_2$] (Fig. 2-23) is representative of this series of complexes.

Medium effects on the Re → L CT absorptions are such that more polar or polarizable solvents give the higher energy band maxima. An example of the magnitude of the effect is given in Table 2-36. The data illustrate another interesting effect: there are large, reversible spectral changes associated with the equilibrium given in Eq. (2-45). Spectra of the limiting forms are given in

$$fac\text{-}[ReCl(CO)_3(4,4'\text{-bipyridine})_2] \xrightleftharpoons[-2H^+]{+2H^+} fac\text{-}[ReCl(CO)_3(4,4'\text{-bipyridineH})_2]^{2+} \quad (2\text{-}45)$$

Fig. 2-24. A similar equilibrium obtains for *fac*-[ReCl(CO)$_3$(2,2'-bipyridine-4,4'-dicarboxylic acid)], and the limiting spectra are shown in Fig. 2-25. In both of these cases the main point is that protonation of the charge-acceptor

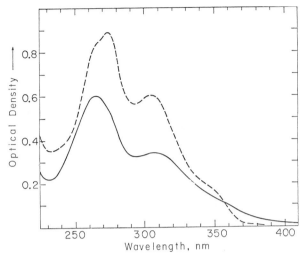

Fig. 2-23.   Absorption spectrum of *fac*-[ReCl(CO)₃(4-phenylpyridine)₂] at 298 (——) and 77°K (---) in EPA. The band at ~ 310 nm is the Re → 4-phenylpyridine CT and the ~ 260 nm band is the IL (4-phenylpyridine) absorption; cf. also Table 2-35 and Giordano and Wrighton [151].

TABLE 2-36

Solvent Dependence of Re → L CT Absorption Band
Maxima in *fac*-[ReCl(CO)₃L₂][a]

L	Solvent	Re → L CT Maximum, cm⁻¹
4,4′-Bipyridine	CH_3CN	32,790
	CH_3OH	31,750
	CH_2Cl_2	31,550
	Dioxane	31,250
	C_6H_6	30,770
4,4′-Bipyridine H⁺	CH_3CN	29,850
	CH_3OH	29,410

[a] Data from Giordano and Wrighton [151]; 298°K.

ligand lowers the energy of the Re → L CT absorption. This reveals that the excited state of the base form is more basic than the ground state, consistent with the increased electron density on L in the excited state. Similar effects have been noted in d^6 Ru(II) complexes of diazines and certain bipyridines [153–156].

It can be seen from Fig. 2-23, 2-24, and 2-25 that IL absorptions can be a prominent feature of the optical spectrum of the *fac*-[ReX(CO)₃L₂] complexes. For L = styrylpyridine the situation is even more extreme: the lowest absorption system is dominated by IL absorption. Figs. 2-26, 2-27,

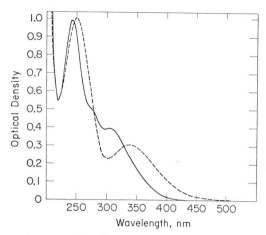

FIG. 2-24. Electronic spectra of *fac*-[ReCl(CO)$_3$(4,4'-bipyridine)$_2$] (—) and *fac*-[ReCl(CO)$_3$(4,4'-bipyridineH)$_2$]$^{2+}$ (----) in CH$_3$CN at 298°K; cf. Tables 2-35 and 2-36 and Giordano and Wrighton [151].

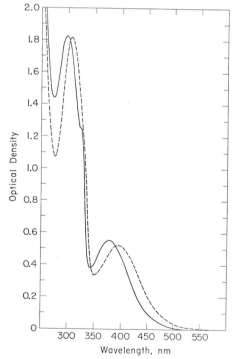

FIG. 2-25. Electronic spectra of *fac*-[ReCl(CO)$_3$(2,2'-bipyridine-4,4'-dicarboxylate)]$^{2-}$ (—) and *fac*-[ReCl(CO)$_3$(2,2'-bipyridine-4,4'-dicarboxylic acid)] in MeOH at 298°K; cf. Table 2-35 and Giordano, Fredericks, and Wrighton [170].   (----)

FIG. 2-26.   Spectra of (a) *trans*-4-styrylpyri-
dine, (b) [ReCl(CO)₃(*trans*-4-styrylpyridine)₂],
and (c) [ReBr(CO)₃(*trans*-4-styrylpyridine)₂] in
EPA at 298 (——) and 77°K (– – – –). The spectra
at 77°K are obtained after cooling the 298°K
solutions and have not been corrected for solvent contraction. The molar absorptivities at
the band maxima are (a) 28,000 M^{-1} cm^{-1},
(b) 50,300 M^{-1} cm^{-1}, (c) 53,300 M^{-1} cm^{-1}
at 298°K. Reprinted with permission from
Wrighton *et al.* [152], *J. Am. Chem. Soc.* **97**,
2073 (1975). Copyright by the American Chemical Society.

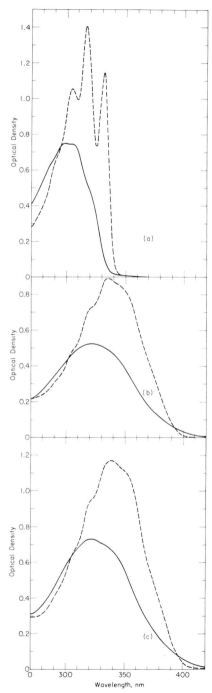

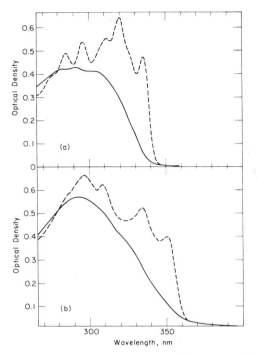

Fig. 2-27. (a) Absorption of *trans*-3-styrylpyridine at 298 (——) and 77°K (----) in EPA; absorptivity at 298°K is 19,000 M^{-1} cm^{-1}. (b) Absorption of *fac*-[ReCl(CO)₃(*trans*-3-stpy)₂] at 298 (——) and 77°K (----) in EPA; absorptivity at 298°K is 45,300 M^{-1} cm^{-1}. Reprinted with permission from Wrighton *et al.* [152], *J. Am. Chem. Soc.* **97**, 2073 (1975). Copyright by the American Chemical Society.

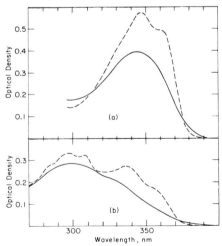

Fig. 2-28. Spectra at 298 (——) and 77°K (----) in EtOH at pH 2 for (a) *trans*-4-styryl-pyridine and (b) *trans*-3-styrylpyridine. At this pH each species is fully protonated. Absorptivities at 298°K are 28.000 and 19,000 M^{-1} cm^{-1} for (a) and (b), respectively. Reprinted with permission from Wrighton *et al.* [152], *J. Am. Chem. Soc.* **97**, 2073 (1975). Copyright by the American Chemical Society.

and 2-28 show spectra of *trans*-3- and *trans*-4-styrylpyridine, their protonated form, and of their *fac*-[ReX(CO)$_3$L$_2$] complexes. The position of the lowest band, its intensity, its structure at low temperature, and the red-shift of the lowest maximum demand an IL assignment for the complexes. Further, it is interesting to note that the spectrum of the complexed ligand most closely resembles the properties of the protonated, rather than the deprotonated, free ligand.

2.   [M(η^5-C$_5$H$_5$)(CO)$_n$L$_{3-n}$] and Related Complexes

The mononuclear complexes [M(η^5-C$_5$H$_5$)(CO)$_n$L$_{3-n}$] (M = Mn, Re) have been studied to some extent with respect to electronic structure. Some electronic spectral data are set out in Table 2-37, and key spectra are shown in Figs. 2-29–2-32 [157]. The spectra can be interpreted with reference to the one-electron level schemes in Fig. 2-33. The splittings in the highest occupied

TABLE 2-37

Electronic Absorption Spectral Band Maxima in Low-Energy Region for Mn and Re Complexes[a,b]

L	[Re(η^5-C$_5$H$_5$)(CO)$_2$L] Band max, nm (ε)	[Mn(η^5-C$_5$H$_5$)(CO)$_2$L] Band max, nm (ε)	[Mn(η^5-C$_5$H$_4$CH$_3$)(CO)$_2$L] Band max, nm
CO	255 (2700)	330 (1100)	330
3,4-DiMe-py	425 (2740); 375 (4740)	470 (1160); 385 (5430)	480; 395
4-Me-py	430 (2780); 380 (4800)	485 (1250); 390 (5800)	490; 400
py	445 (2880); 393 (5000)	495 (1280); 412 (5100)	505; 420
3-Br-py	475 (3040); 415 (5260)		
4-Ph-py	475 (3600); 417 (5570)		
3,5-DiCl-py	505 (2950); 435 (5180)	555 (1880); 470 (5610)	560; 477
3-Acetyl-py	510 (1860); 410 (4650)		
3-Benzoyl-py	510 (1950); 410 (4780)		
4-Acetyl-py	555 (3000); 480 (5310)	640 (1990); 525 (6790)	650; 535
4-Benzoyl-py	555 (3070); 480 (4870)	640 (1760); 525 (6380)	
4-Formyl-py	585 (3100); 495 (5180)	670 (2240); 555 (6530)	680; 565
n-PrNH$_2$	320	455 (825); 380 (1300)	
Piperidine	320	455; 380	355; 380
THF	335	510 (990); 400 (1400)[c]	
NH$_3$	320 (1000); 260 (4000)[d]	455	

[a] Reprinted with permission from Giordano and Wrighton [157], *Inorg. Chem.* **16**, 160 (1977). Copyright by the American Chemical Society.

[b] All spectra recorded in isooctane solution (unless noted otherwise) at 298°K using a Cary 17 instrument; py = pyridine; Me = methyl; Pr = propyl.

[c] Recorded in THF solution.

[d] Recorded in EtOH solution.

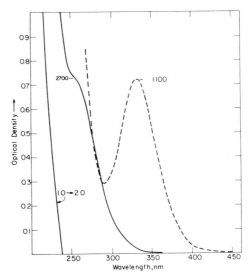

FIG. 2-29. Electronic absorption spectra of $[Re(\eta^5\text{-}C_5H_5)(CO)_3]$ (—), 2.63×10^{-4} M, and $[Mn(\eta^5\text{-}C_5H_5)(CO)_3]$ (–––), 6.67×10^{-4} M, in isooctane solution at $298°K$. The pathlength is 1.00 cm. Reprinted with permission from Giordano and Wrighton [157], *Inorg Chem.* **16**, 160 (1977). Copyright by the American Chemical Society.

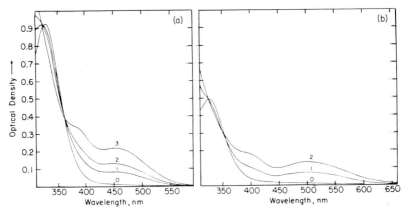

FIG. 2-30. (a) Initial electronic spectral changes accompanying the photoconversion of $[Mn(\eta^5\text{-}C_5H_5)(CO)_3]$ (8.5×10^{-4} M) to $[Mn(\eta^5\text{-}C_5H_5)(CO)_2(n\text{-}PrNH_2)]$ in a N_2-deoxygenated 3.0-ml isooctane solution of 0.05 M n-PrNH$_2$. Curves 0, 1, 2, and 3 correspond, respectively, to 0, 25, 50, and 200 sec irradiation times using a GE black light. Curve 3 corresponds to ~32% conversion. Note the two low-energy bands at 455 and 380 nm for the n-PrNH$_2$ complex. (b) Initial electronic spectral changes accompanying the photoconversion of $[Mn(\eta^5\text{-}C_5H_5)$ $(CO)_3]$ (4.45×10^{-4} M) to $[Mn(\eta^5\text{-}C_5H_5)(CO)_2THF]$ in 3.0 ml of N_2-deoxygenated THF. Curves 0, 1, and 2 correspond, respectively, to 0, 25, and 100 sec irradiation times using the GE black light. Curve 2 corresponds to ~27% conversion. Note the two low-energy bands for the THF complex. Reprinted with permission from Giordano and Wrighton [157], *Inorg. Chem.* **16**, 160 (1977). Copyright by the American Chemical Society.

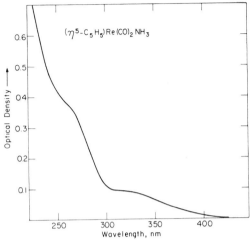

FIG. 2-31. Electronic spectrum of 8.0×10^{-5} M $[Re(\eta^5-C_5H_5)(CO)_2NH_3]$ in deoxygenated EtOH solution using a 1.00 cm pathlength cell. Reprinted with permission from Giordano and Wrighton [157], *Inorg. Chem.* **16**, 160 (1977). Copyright by the American Chemical Society.

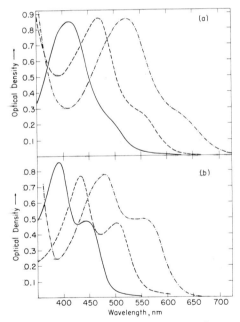

FIG. 2-32. Electronic spectra of $[M(\eta^5-C_5H_5)(CO)_2L]$ ($\sim 10^{-4}$ M) in isooctane solution at 298°K using 1.00 cm pathlength cells for M = Mn (a) and M = Re (b): L = pyridine (—), 3,5-dichloropyridine (– – –), and 4-acetylpyridine (–.–). Band maxima and molar absorptivity are given in Table 2-37. Reprinted with permission from Giordano and Wrighton [157], *Inorg. Chem.* **16**, 160 (1977). Copyright by the American Chemical Society.

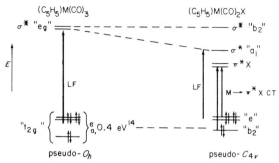

FIG. 2-33.   Correlation diagrams. Reprinted with permission from Giordano and Wrighton [157], *Inorg. Chem.* **16**, 160 (1977). Copyright by the American Chemical Society.

TABLE 2-38

Spectral Properties of $[Mn(\eta^5\text{-}Y)(CO)_3]^a$

Y	Bands, nm (ε)	Assignment	Reference
C_5H_5	330 (1100)	Mn → Y CT	[157]
	216 (12,000)	?	[159]
$C_5H_4CH_3$	330 (—)	Mn → Y CT	[157]
$C_5H_4COCH_3{}^b$	338 (1640)	Mn → Y CT	[51]
$C_5H_4COC_6H_5$	344 (1750)	Mn → Y CT	[159]
	282 (2340)	?	
	247 (14,600)	$\pi \to \pi^*$ IL	
	209 (25,900)	?	
$C_5H_4(o\text{-}CH_3C_6H_4CO)$	339 (1620)	Mn → Y CT	[159]
	285 (4680)	?	
	243 (9800)	$\pi \to \pi^*$ IL	
	211 (23,500)	?	
$C_5H_4(m\text{-}CH_3C_6H_4CO)$	339 (1580)	Mn → Y CT	[159]
	289 (4500)	?	
	251 (11,700)	$\pi \to \pi^*$ IL	
	215 (17,300)	?	
$C_5H_4(p\text{-}CH_3C_6H_4CO)$	339 (1970)	Mn → Y CT	[159]
	257 (14,900)	$\pi \to \pi^*$ IL	
	215 (22,100)	?	
$C_5H_4(o\text{-}FC_6H_4CO)$	342 (2000)	Mn → Y CT	[159]
	278 (2870)	?	
	239 (15,800)	$\pi \to \pi^*$ IL	
	216 (21,900)	?	
$C_5H_4(m\text{-}FC_6H_4CO)$	347 (2140)	Mn → Y CT	[159]
	285 (3310)	?	
	243 (17,000)	$\pi \to \pi^*$ IL	
	218 (20,000)	?	

(*continued*)

TABLE 2-38 (continued)

Y	Bands, nm (ε)	Assignment	Reference
$C_5H_4(p\text{-}FC_6H_4CO)$	343 (1930)	Mn → Y CT	[159]
	294 (2120)	?	
	249 (19,100)	$\pi \to \pi^*$ IL	
	213 (31,800)	?	
$C_5H_4(o\text{-}ClC_6H_4CO)$	341 (1950)	Mn → Y CT	[159]
	285 (2510)	?	
	238 (14,500)	$\pi \to \pi^*$ IL	
	216 (30,400)	?	
$C_5H_4(p\text{-}ClC_6H_4CO)$	342 (2160)	Mn → Y CT	[159]
	298 (2230)	?	
	257 (18,100)	$\pi \to \pi^*$ IL	
	217 (24,500)	?	
$C_5H_4(p\text{-}OCH_3C_6H_4CO)$	340 (2130)	Mn → Y CT	[159]
	280 (15,700)	$\pi \to \pi^*$ IL	
	209 (34,200)	?	
$C_5H_4CH_2C_6H_5$	331 (970)	Mn → Y CT	[159]
	246 (12,600)	?	
$C_5H_4CHCHC_6H_5$	284 (9510)	$\pi \to \pi^*$ IL	[159]
	212 (16,100)	?	
$C_5H_4COCH_2C_6H_5$	336 (1410)	Mn → Y CT	[159]
	280 (2110)	?	
	236 (7050)	$\pi \to \pi^*$ IL	
	218 (12,800)	?	
Indenyl[b]	375 (1415)	Mn → Y CT	[51]

[a] All data for alkane solutions at 298°K unless noted otherwise.
[b] C_2H_5OH solution, 298°K.

levels have been determined from photoelectron spectra [158]. For simplicity, the $[M(\eta^5\text{-}C_5H_5)(CO)_3]$ species are treated as pseudo-O_h, taking the $(\eta^5\text{-}C_5H_5)$ group as equivalent to three two-electron carbon donors. The $[M(\eta^5\text{-}C_5H_5)(CO)_2L]$ complexes, then, are viewed as pseudo-C_{4v} species.

The lowest observable transitions in the tricarbonyls (Fig. 2-29) are probably CT transitions with substantial M → $(\eta^5\text{-}C_5H_5)$ CT character, but the LF transitions are very likely in the same energy region. The M → $(\eta^5\text{-}C_5H_5)$ CT assignment is substantiated by the data in Table 2-38 [51,157,159] for various $(\eta^5\text{-}C_5H_4Y)$ derivatives of $[Mn(\eta^5\text{-}C_5H_5)(CO)_3]$ in that the $(\eta^5\text{-}C_5H_4Y)$ systems having the lowest π^* orbitals have the lowest energy first absorption. However, the first absorption probably has some M → CO π^*CT character as well. The LF transitions are probably obscured by the more intense CT absorption. The band near 250 nm in the benzoyl-cyclopentadienyl complexes depends on the benzoyl substituents in the manner found for the $\pi \to \pi^*$ absorption in substituted benzophenones [160]. Thus, in these complexes the IL assignment is appropriate.

Substitution of CO by a ligand L containing no low-lying π^* levels yields complexes exhibiting a weak, but low-lying, first absorption significantly to the red of the first transition in the tricarbonyls (Figs. 2-30, 2-31; Table 2-37) [157]. The shift is that expected for the substantial decrease in the LF strength for the ligands studied. For L having low-lying π^* levels, the first absorption is fairly intense and very dependent on the substituents in L (Fig. 2-32 and Table 2-37) [157]. Further, the first absorption in such complexes has the typical large solvent dependence found for organometallic MLCT absorptions (Table 2-39). One final note of interest concerning these MLCT bands is that there is invariably a two-component band system (Fig. 2-32) which is better resolved in the Re compared to the Mn complexes. This correlates nicely with the photoelectron spectroscopy in these systems [158] which shows ionizations from the M orbitals with more well-resolved splittings in Re due to the larger spin–orbit coupling. The band width in the ionizations from the "t_{2g}" orbitals is comparable to that found in the internal "ionizations" associated with the MLCT absorptions.

TABLE 2-39

Solvent Dependence of
First Absorption Band System in
$[M(\eta^5\text{-}C_5H_5)(CO)_2(\text{Pyridine})]^{a,b}$

	Absorbance max, nm (ε)	
Solvent	M = Mn	M = Re
Isooctane	412 (5100)	393 (5000)
C_6H_6	388 (4970)	
$(C_2H_5)_2O$	388 (4910)	370
$CHCl_3$	380 (4670)	
THF	380 (4970)	360
CH_2Cl_2	378 (5850)	
C_2H_5OH	378 (5150)	
Acetone	370 (5980)	
CH_3CN	365 (5600)	345

[a] Reprinted with permission from Giordano and Wrighton [157], *Inorg. Chem.* **16**, 160 (1977). Copyright by the American Chemical Society.

3. Metal–Metal Bonded Complexes

A number of dinuclear metal–metal bonded complexes involving Mn and Re have been studied, and indeed, the cornerstone concerns $[Mn_2(CO)_{10}]$ [161]. The, now familiar, intense near-UV absorption in this d^7–d^7 system was shown to be polarized along the Mn–Mn axis as demanded for the $\sigma_b \to \sigma^*$ excitation [61,161], and the orbital scheme in Fig. 2-15 was first proposed for this system. Some key spectra are shown in Figs. 2-34–2-37 [162], and spectral data are summarized in Table 2-40 [163–165].

The simple substituted derivatives of $[M_2(CO)_{10}]$ exhibit electronic spectral features in accord with transitions which terminate in the M—M σ^* orbital. Thus, for $[M_2(CO)_nL_{10-n}]$ ($n = 1, 2$; L = P- or As-donor ligand), the spectrum consists of a low-lying $d\pi \to \sigma^*$ absorption with a more intense $\sigma_b \to \sigma^*$. For M = Mn, the absorptions are typically red-shifted as CO groups are replaced with L; but, for M = Re, the $[Re_2(CO)_8(PPh_3)_2]$ actually exhibits a higher energy $\sigma_b \to \sigma^*$ than in $[Re_2(CO)_{10}]$ itself. Heterodinuclear complexes containing the $[M(CO)_5]$ moiety bonded to another 17-e^- center also exhibit $d\pi \to \sigma^*$ and $\sigma_b \to \sigma^*$ absorptions (Fig. 2-37).

The complexes $[M_2(CO)_8L]$ (M = Mn, Re; L = 1,10-phenanthroline; 2,2'-biquinoline) are unusual in that the lowest transition is not one which terminates in σ^* [165]. The absorption spectrum of $[Re_2(CO)_8(1,10\text{-phen-anthroline})]$ is shown in Fig. 2-38 and reveals the presence of an extraordinarily low-lying absorption band. The lowest band shifts considerably with solvent, atypical of absorptions associated with M—M bonded complexes. The assignment of the lowest band is $\sigma_b \to L\pi^*$; i.e., it is a form of MLCT absorption where the orbital of origin is delocalized over the two metal atoms. A one-electron diagram comparing the σ bonding in [Cl-Re(CO)_3(1,10-phenanthroline)] and $[(OC)_5Re\text{—}Re(CO)_3(1,10\text{-phenanthro-line})]$ is given in Fig. 2-39. The shift in the position of the $\sigma_b \to L\pi^*$ for L = 1,10-phenanthroline compared to 2,2'-biquinoline is in accord with the assignment, and the solvent shift is like that for other MLCT absorptions in complexes such as $[ReCl(CO)_3(1,10\text{-phenanthroline})]$ [150].

C. LUMINESCENCE STUDIES

1. Ligand-Field Emission

Optical emission has been found from the pure solid $[Re(CO)_5X]$ (X = Cl, Br, I) complexes at low temperature [37]. Corrected emission spectra for the

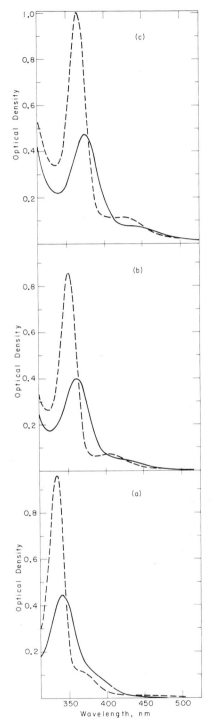

Fig. 2-34. Comparison of 298(——) and 77°K (————) absorption spectra of (a) $[Mn_2(CO)_{10}]$, (b) $[Mn_2(CO)_9PPh_3]$, and (c) $[Mn_2(CO)_8(PPh_3)_2]$ in EPA. The spectra at 77°K are for the same solution as at 298°K, but the low-temperature spectra are not corrected for solvent contraction. Band maxima and molar extinction coefficients are given in Table 2-40. Reprinted with permission from Wrighton and Ginley [162], *J. Am. Chem. Soc.* **97**, 2065 (1975). Copyright by the American Chemical Society.

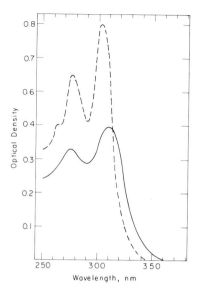

FIG. 2-35. Comparison of 298(——) and 77°K (----) absorption spectra of $[Re_2(CO)_{10}]$ in EPA. Spectral changes are not corrected for solvent contraction upon cooling. Spectral maxima and absorptivities are given in Table 2-40. Reprinted with permission from Wrighton and Ginley [162], *J. Am. Chem. Soc.* **97**, 2065 (1975). Copyright by the American Chemical Society.

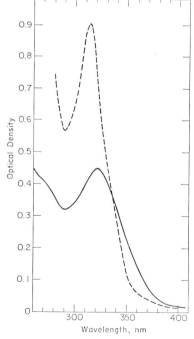

FIG. 2-36. Comparison of 298(——) and 77°K (——) absorption spectra of $[MnRe(CO)_{10}]$ in EPA. Reprinted with permission from Wrighton and Ginley [162], *J. Am. Chem. Soc.* **97**, 2065 (1975). Copyright by the American Chemical Society.

109

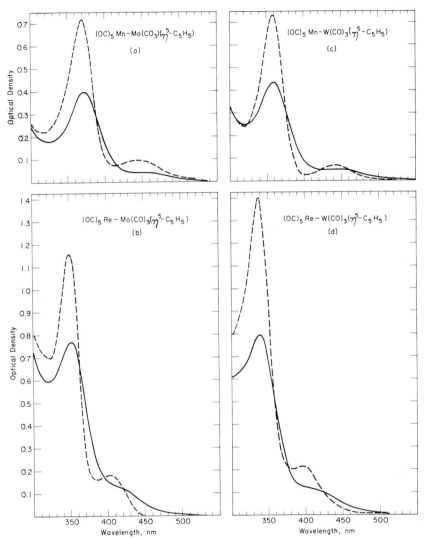

FIG. 2-37. Electronic absorption spectra of $[M(\eta^5\text{-}C_5H_5)(CO)_3M'(CO)_5]$ in EPA at 298 (——) and 77°K (----). Spectral changes upon cooling from 298 to 77°K shown here are not corrected for solvent contraction. Band positions and absorptivities are given in Table 2-15. Reprinted with permission from Ginley and Wrighton [65], *J. Am. Chem. Soc.* **97**, 4908 (1975). Copyright by the American Chemical Society.

TABLE 2-40

Electronic Absorption Spectral Features of Mn, Tc, and Re Metal–Metal Bonded Complexes

Complex	Solvent $(T, °K)$	Bands, cm^{-1} (ε)	Assignment	Reference
$[Mn_2(CO)_{10}]$	3-PIP (77)[a]	26,700 (2900)	$d\pi \rightarrow \sigma^*$	[61]
	3-PIP (77)	29,740 (33,700)	$\sigma_b \rightarrow \sigma^*$	
	3-PIP (77)	37,600 (8200)	$d\pi \rightarrow CO\pi^*$	
	CH$_3$CN (300)	49,100 (84,400)	$M \rightarrow CO\pi^*$	
$[Tc_2(CO)_{10}]$	3-PIP (77)	32,400 (26,600)	$\sigma_b \rightarrow \sigma^*$	[61]
	3-PIP (77)	35,500 (12,700)	$\sigma_b \rightarrow CO\pi^*$	
	3-PIP (77)	38,200 (11,200)	$d\pi \rightarrow CO\pi^*$	
	CH$_3$CN (300)	51,500 (104,000)	$M \rightarrow CO\pi^*$	
$[Re_2(CO)_{10}]$	3-PIP (77)	32,800 (24,000)	$\sigma_b \rightarrow \sigma^*$	[61]
	3-PIP (77)	36,000 (18,000)	$\sigma_b \rightarrow CO\pi^*$	
	3-PIP (77)	38,100 (12,500)	$d\pi \rightarrow CO\pi^*$	
	CH$_3$CN (300)	> 52,500 (> 100,000)	$M \rightarrow \pi^*$	
$[MnRe(CO)_{10}]$	3-PIP (77)	31,950 (20,100)	$\sigma_b \rightarrow \sigma^*$	[61]
	3-PIP (77)	36,000 (11,400)	$\sigma_b \rightarrow \pi^*$	
	3-PIP (77)	37,700 (9900)	$d\pi \rightarrow \pi^*$	
	CH$_3$CN (300)	51,700 (86,600)	$M \rightarrow \pi^*$	
$[Mn_2(CO)_9PPh_3]$	3-PIP (77)	24,150 (2600)	$d\pi \rightarrow \sigma^*$	[61]
		28,400 (45,800)	$\sigma_b \rightarrow \sigma^*$	
$[Mn_2(CO)_8(PPh_3)_2]$	CH$_3$CN (300)	22,500 (3000)	$d\pi \rightarrow \sigma^*$	[61]
		26,800 (31,800)	$\sigma_b \rightarrow \sigma^*$	
$[Mn_2(CO)_8(P(OPh)_3)_2]$	CH$_3$CN (300)	24,700 (4770)	$d\pi \rightarrow \sigma^*$	[163]
		28,300 (29,200)	$\sigma_b \rightarrow \sigma^*$	
$[Mn_2(CO)_8(P(OEt)_3)_2]$	CH$_3$CN (300)	23,800	$d\pi \rightarrow \rho^*$	[163]
		27,800	$\sigma_b \rightarrow \sigma^*$	

(continued)

TABLE 2-40 (continued)

Complex	Solvent (T, °K)	Bands, cm⁻¹ (ε)	Assignment	Reference
[Mn₂(CO)₈(P(m-CH₃C₆H₄)₃)₂]	CH₃CN (300)	23,100 27,000	$d\pi \to \sigma^*$ $\sigma_b \to \sigma^*$	[163]
[Mn₂(CO)₈(AsPh₃)₂]	CH₃CN (300)	22,300 26,900	$d\pi \to \sigma^*$ $\sigma_b \to \sigma^*$	[163]
[Mn₂(CO)₈(PPh₂Me)₂]	CH₃CN (300)	23,300 (3850) 27,500 (31,700)	$d\pi \to \sigma^*$ $\sigma_b \to \sigma^*$	[163]
[Mn₂(CO)₈(PPhEt₂)₂]	CH₃CN (300)	23,600 (3490) 27,800 (27,300)	$d\pi \to \sigma^*$ $\sigma_b \to \sigma^*$	[163]
[Mn₂(CO)₈(PEt₃)₂]	CH₃CN (300)	23,800 (3000) 28,300 (22,600)	$d\pi \to \sigma^*$ $\sigma_b \to \sigma^*$	[163]
[MnRe(CO)₉(PPh₃)]	DMF (300)	29,070 (29,600)	$\sigma_b \to \sigma^*$	[164]
[Re₂(CO)₉(PPh₃)]	DMF (300)	31,250 (—)	$\sigma_q \to \sigma^*$	[164]
[Re₂(CO)₈(PPh₃)₂]	DMF (300)	33,300 (40,000)	$\sigma_b \to \sigma^*$	[164]
[Mn(CO)₅Co(CO)₃PPh₃]	Alkane (298)	26,600 (—)	$\sigma_b \to \sigma^*$	[67]
[Mn₂(CO)₈(1,10-phen)]	1:1 CH₂Cl₂/CCl₄ (298)	17,420 (11,800) 26,310 (12,300) sh 28,320 (13,500)	$\sigma_b \to \mathrm{phen}\pi^*$ $d\pi \to \mathrm{phen}\pi^*$ $\sigma_b \to \sigma^*$	[165]
[Re₂(CO)₈(1,10-phen)]	1:1 CH₂Cl₂/CCl₄ (298)	18,940 (7300) 28,570 (5200) sh	$\sigma_b \to \mathrm{phen}\pi^*$ $d\pi \to \mathrm{phen}\pi^*$	[165]
[Re₂(CO)₈(2,2'-biquin)]	1:1 CH₂Cl₂/CCl₄ (298)	14,530 (7100) 22,940 (5160) 27,170 (35,000)	$\sigma_b \to \mathrm{biquin}\pi^*$ $d\pi \to \mathrm{biquin}\pi^*$ $\pi \to \pi^*\mathrm{biquin}$	[165]

a 3-PIP is a 6:1 mixture of isopentane and 3-methylpentane.

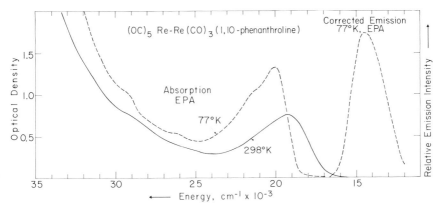

FIG. 2-38.   Electronic absorption and emission of [Re$_2$(CO)$_8$(1,10-phenanthroline)] in EPA. Absorption spectral changes upon cooling have not been corrected for solvent contraction. The absorption spectra were recorded using a 1.0 cm pathlength cell, and the concentration of the complex is 1.1×10^{-4} M. The emission was excited using 514.5 nm light from an argon ion laser. Reprinted with permission from Morse and Wrighton [165], *J. Am. Chem. Soc.* **98**, 3931 (1976). Copyright by the American Chemical Society.

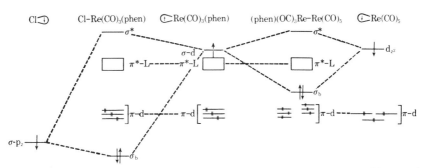

FIG. 2-39.   Correlation of one-electron levels. Reprinted with permission from Morse and Wrighton [165], *J. Am. Chem. Soc.* **98**, 3931 (1976). Copyright by the American Chemical Society.

three complexes are shown in Fig. 2-40. Efficiency and lifetime of the broad, featureless emission depend significantly on temperature (Fig. 2-41), but the spectral distribution of the emitted light was found to be essentially independent of temperature. At the lowest temperatures the absolute quantum efficiency was found to be quite high, while lifetimes remained in the 10^{-5}–10^{-6} sec range. Lack of detectable emission from the Mn analogs under the same conditions points to an important role for spin–orbit coupling and suggests that the lowest (emissive) excited state has considerable triplet character. Thus, as in the [W(CO)$_5$X] complexes, the emission in the isoelectronic

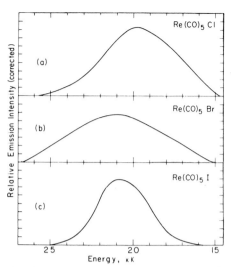

FIG. 2-40. Corrected emission spectra at 25°K of pure powdered [Re(CO)₅X]. Excitation wavelength is 370 nm. Reprinted with permission from Wrighton *et al.* [37], *J. Am. Chem. Soc.* **98**, 1111 (1976). Copyright by the American Chemical Society.

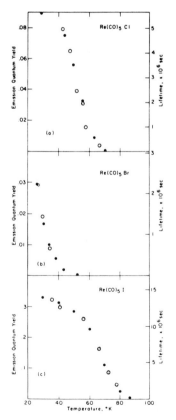

FIG. 2-41. Temperature dependence of emission lifetimes (●) and quantum yields (○) for pure powdered [Re(CO)₅X]. Excitation wavelength is 370 nm. Reprinted with permission from Wrighton *et al.* [37], *J. Am. Chem. Soc.* **98**, 1111 (1976). Copyright by the American Chemical Society.

$[Re(CO)_5X]$ species is assigned as LF, $^3E(e^3b_2{}^2a_1{}^1) \rightarrow {}^1A_1(e^4b_2{}^2)$. As found for the anionic $[M(CO)_6]^-$ (M = Nb, Ta) species (*vide supra*), the temperature-dependence data (Fig. 2-41) demands that Eq. (2-3) hold; i.e., radiative decay constants are temperature-independent while nonradiative decay is, phenomenologically, an activated process.

Table 2-41 summarizes the emission properties of $[Re(CO)_5X]$ at one temperature. Quite interestingly, the dependence on X is not as would be expected. The lowest absorption band (Table 2-33) is in the spectrochemical series ordering I < Br < Cl, while there is no such clear dependence in the emission. However, the emission is considerably red-shifted compared to the first absorption. These facts suggest a highly distorted emissive state which needs to be characterized more fully in order to assess unambiguously its role in photoreactions. One final confusing point here is that the emission study was carried out using the pure powders as the sample. Dilute-solution or doped-crystal studies are needed to ascertain molecular properties.

TABLE 2-41

Emission Data for $[Re(CO)_5X]$ Complexes[a]

X	Half-width of band I,[b] cm^{-1}	Emission maximum,[c] cm^{-1} × 10^{-3}	Emission half-width,[c] cm^{-1}	Φ[c]	$10^6\tau$[c]
Cl	1500 ± 100	19.9 ± 0.2	5200 ± 200	0.09 ± 20%	5.70 ± 10%
Br	1440	21.0	6700	0.02$_8$	2.21
I	1000	20.8	3630	0.33	14.0

[a] Reprinted with permission from Wrighton *et al.* [37], *J. Am. Chem. Soc.* **98**, 1111 (1976). Copyright by the American Chemical Society.

[b] At 77°K in EPA solution; band I is the first absorption band for the complexes; cf. Table 2-33.

[c] At 26°K for pure solid $[Re(CO)_5X]$. Excitation λ 370 nm for Φ.

2.   Charge-Transfer Emission

Complexes *fac*-$[ReX(CO)_3L]$ and *fac*-$[ReX(CO)_3L_2]$ have been found to be emissive in fluid solution at room temperature [150,151]. Except for some isocyanide complexes (to be discussed in Chapter 6), these Re species are the only organometallics known to be emissive under conditions where photochemistry is normally encountered and studied. Certainly it is surprising to find any metal carbonyl which is emissive, owing to the general photosensitivity with respect to CO substitution.

The complexes $[ReCl(CO)_3(1,10\text{-phenanthroline})]$ and related complexes were studied first, and the emission was associated with the Re $\rightarrow$ 1,10-phenanthroline π^* absorption. Some data are given in Table 2-42. Lifetimes

TABLE 2-42

Luminescence Characteristics for $[ReCl(CO)_3L]^{a,b}$

L	Emission max, $cm^{-1} \times 10^{-3}$		Lifetime, sec $\times 10^6$		$\Phi \pm 15\%^c$	$\Phi \pm 15\%^c$
	298°K	77°K	298°K	77°K	298°K	77°K
1,10-Phenanthroline	17.33	18.94	0.3	9.6	0.036	0.33
2,2'-Bipyridine		18.87	0.6	3.8		
5-Methyl-1,10-phenanthroline	17.01	18.83	≤0.65	5.0	0.03_0	0.33
4,7-Diphenyl-1,10-phenanthroline	17.24	18.18	0.4	11.25		
5-Chloro-1,10-phenanthroline	17.12	18.69	≤0.65	6.25		
5-Bromo-1,10-phenanthroline	17.12	18.69	≤0.65	7.6	0.20_0	0.20
5-Nitro-1,10-phenanthroline	d	18.28		11.8		
1,10-phenanthroline-5,6-dione	d	18.45		2.5		0.03_3
2,2'-Biquinoline	d	14.58^d				

^{a} Reprinted with permission from Wrighton and Morse [150], J. Am. Chem. Soc. **96**, 998 (1974). Copyright by the American Chemical Society.

^{b} 77°K measurements in EPA; 298°K measurements in CH_2Cl_2 using a 1P21 PMT detector unless specified otherwise.

^{c} RCA 7102 PMT detector, quantum yields in 298°K solutions measured in benzene solvent.

^{d} Luminescence at 298°K is not detectable from these complexes in solution.

in the 10^{-5}–10^{-6} sec range and lack of emission from the Mn analogs suggest that the emission has considerable triplet → singlet character. Changing 1,10-phenanthroline to 2,2′-biquinoline results in a red-shifted emission, in accord with the shift of the Re → L CT absorption band.

An interesting find concerning the emission properties of the $[ReCl(CO)_3L]$ complexes is that the emission changes markedly with changes in the rigidity of the medium. Generally, rigid media give blue-shifted, more efficient, emission compared to that found in the fluid solutions. The effect is *not* solely a temperature effect, since room-temperature rigid media were used as well (Table 2-43). Further, the blue-shift and intensification are substantial. A complete understanding of this must await the results of time-dependent emission studies, but for now it is important to recognize that the phenomenon is found for all of these emissive Re systems.

At this point the rationale for the easily detectable emission in the $[ReX(CO)_3L]$ complexes resides in the fact that the excited state is one which

TABLE 2-43

Summary of Environmental Effects on Absorption
and Emission Maxima of $[ReCl(CO)_3L]^a$

L	Environment, T, °K	First Absorption max, $cm^{-1} \times 10^{-3}$	Emission max, $cm^{-1} \times 10^{-3}$ ($\tau \times 10^6$ sec)
1,10-Phenanthroline	CH$_2$Cl$_2$, 298	26.53	17.33 (0.3)
	Polyester resin, 298		18.52 (3.67)
	EPA, 77		18.94 (9.6)
5-CH$_3$-1,10-Phenanthroline	Benzene, 298	25.65	17.00 ($\leq$0.65)
	CH$_2$Cl$_2$, 298	26.32	17.01
	CH$_3$OH, 298	27.05	17.00
	Pure solid, 298		18.42
	Polyester resin, 298		18.48 (3.5)
	EPA, 77		18.83 (5.0)
5-Br-1,10-Phenanthroline	Benzene, 298	25.32	17.15 ($\leq$0.65)
	CH$_2$Cl$_2$, 298	25.84	17.12
	CH$_3$OH, 298	26.88	17.04
	Pure solid, 298		17.83
	Polyester resin, 298		18.32 (2.2)
	EPA, 77		18.69 (7.6)
5-Cl-1,10-Phenanthroline	CH$_2$Cl$_2$, 298	25.91	17.12
	Pure solid, 298		17.99
	EPA, 77		18.69 (6.25)

^{a} Reprinted with permission from Wrighton and Morse [150], *J. Am. Chem. Soc.* **96**, 998 (1974). Copyright by the American Chemical Society.

does not undergo unimolecular dissociative-type processes at a fast enough rate to compete with radiative decay. From the quantum yield and lifetime data, the radiative rate constants are not extraordinarily large; it is the nonradiative constants which are small. And here it is important to note that *both* chemical and nonchemical nonradiative decay channels must be blocked in order to observe efficient emission.

Observation [151,166] of fairly efficient emission from $[ReX(CO)_3L_2]$ (L = substituted pyridine) serves to dispel the notion that the rigidity of the chelate ligand is necessary to observe radiative decay in fluid solution. The result of changing the medium rigidity still persists (Table 2-44), and the other features which undergird the CT assignment obtain. Figure 2-42 shows a representative spectral shift in changing from 298°K to 77°K in solution.

One intriguing fact concerning the $[ReX(CO)_3L]$ and $[ReX(CO)_3L_2]$ complexes is that the Re → L CT excited states are lowest in energy even though the donor sphere has a much lower LF strength than in $[Re(CO)_5X]$. The point is that replacing two CO groups by N-donor ligands should substantially lower the LF excited states. Given that the $[Re(CO)_5X]$ species are found to be emissive at ~20,000 cm^{-1} from LF states [37], it is surprising that CT emission at nearly the same energy is found from the N-donor substituted species. An explanation here might be that the lowest LF excited

TABLE 2-44

Emission Properties of *fac*-$[ReX(CO)_3L_2]$ Complexes[a]

X	L	Solvent	T, °K	Emission max, cm$^{-1} \times 10^{-3}$	$\tau \times 10^6$ sec	$\Phi \pm 15\%$
Cl	4,4'-Bipyridine	MeOH	298	16.5	0.45	0.005
Cl	4,4'-B pyridine	EtOH	77	19.0	45	0.35
Br	4,4'-Bipyridine	MeOH	298	16.5	0.40	—
Br	4,4-Bipyridine	EtOH/MeOH (4:1)	77	18.5	40	—
I	4,4'-Bipyridine	MeOH	298	16.2	0.30	—
I	4,4'-Bipyridine	EtOH/MeOH (4:1)	77	18.6	35	—
Cl	4-Phenylpyridine	EtOH/MeOH (4:1)	298	18.6	1.0	0.005
Br	4-Phenylpyridine	EtOH/MeOH (4:1)	298	18.7	0.65	0.004
I	4-Phenylpyridine	EtOH/MeOH (4:1)	298	18.6	0.25	0.003
Cl	3-Benzoylpyridine	C$_6$H$_6$	298	18.3	<0.5	—
Cl	4-Benzoylpyridine	C$_6$H$_6$	298	16.7	<0.5	—
Cl	4-Benzoylpyridine	EPA	77	19.5	39	—

[a] Data from Giordano and Wrighton [151].

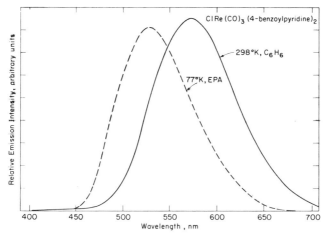

FIG. 2-42.   Total emission spectra in solution excited at 330 nm; concentration is $\sim 10^{-3}$ M. The 298 and 77°K emissions are not shown at the same sensitivity. Reprinted with permission from Giordano *et al.* [166], *J. Am. Chem. Soc.* **100**, 2257 (1978). Copyright by the American Chemical Society.

state remains at nearly the same position, while the upper LF excited states shift downward substantially. This can be understood in terms of the one-electron diagram given in Fig. 2-43.

One additional CT emissive system should be mentioned. The M—M bonded [Re$_2$(CO)$_8$(1,10-phenanthroline)] has been found to be emissive at low temperature [165]. This is noteworthy since it is the only organometallic metal–metal bonded system known to be emissive. No detailed study has been reported, but the $\sim 10^{-4}$ sec lifetime and the overlap of emission and absorption suggest a CT excited state assignment having triplet character.

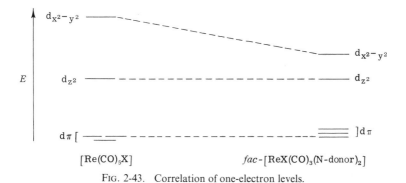

FIG. 2-43.   Correlation of one-electron levels.

3. Intraligand Emissions

As already pointed out, IL excited states are possibly in the same vicinity as LF and CT states. In order to observe emission from an IL state, it is probably necessary that the relaxed IL excited state be lowest in energy. For the fac-[ReX(CO)$_3$(4-phenylpyridine)$_2$] complexes the emission maximum in fluid solution at 298°K is at $\sim 18,700$ cm^{-1} (Table 2-44). This is fairly close to, but lower than, the $\pi-\pi^*$ IL triplet emission of free 4-phenylpyridine or its protonated form. The effect of lowering the temperature in the case of the complex is shown in Fig. 2-44, and some quantitative information is given in Table 2-45 [151]. The usual blue-shift and intensification are found, but the low-temperature lifetime and the vibrational structure on the emission are extraordinary. The low-temperature emission clearly has a large component of $\pi-\pi^*$ IL character; the free ligand emits at a similar energy, and the emission is structured. Some CT character is preserved or the Re extends a large spin–orbit perturbation, since the lifetime of the free ligand at 77°K is in the seconds range whereas the excited complexes are still in the hundreds of microseconds range.

A more unequivocal case of IL emission can be found for fac-[ReX(CO)$_3$(3-benzoylpyridine)$_2$]. These complexes exhibit two emissions at 77°K (Fig. 2-45

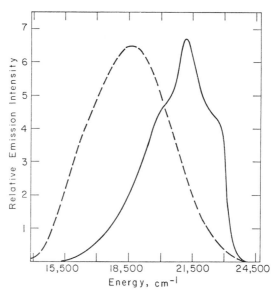

FIG. 2-44.   Emission of fac-[ReCl(CO)$_3$(4-phenylpyridine)$_2$] in EPA solution at 298 (-----) and 77°K (——) not recorded at the same sensitivity. Excitation wavelength is 350 nm. From Giordano and Wrighton [151].

TABLE 2-45

Emission Properties of *fac*-[ReX(CO)$_3$(4-Phenylpyridine)$_2$] at 77°K in EtOH/MeOH (4:1)[a]

X	Emission max, cm^{-1}	$\tau \times 10^6$ sec	$\Phi \pm 15\%$
Cl	20,000 (sh)	450	0.8
	21,300		
	22,600 (sh)		
Br	19,500 (sh)	410	0.8
	21,200		
	22,250 (sh)		
I	19,600 (sh)	260	0.9
	21,100		
	22,250 (sh)		

[a] Data from Giordano and Wrighton [151].

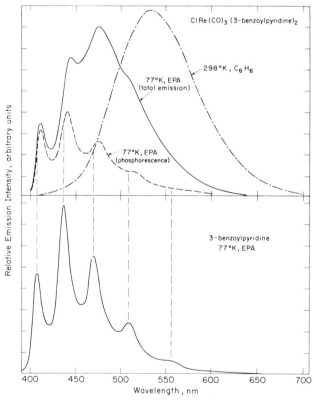

FIG. 2-45. Emission spectra excited at 330 nm for 10^{-3} M solutions. The spectra are not shown at the same sensitivity. All spectra are total emission spectra except that one marked "phosphorescence" which was recorded with the phosphoroscope in place in an Aminco-Bowman emission spectrophotometer. The phosphoroscope allows only the detection of emissions longer than ∼0.1 msec. Reprinted with permission from Giordano *et al.* [166], *J. Am. Chem. Soc.* **100**, 2257 (1978). Copyright by the American Chemical Society.

and Table 2-46). The short-lived emission is structureless and has been assigned as the CT emission. However, structured, long-lived emission is observed, and it is very close to the free ligand $n-\pi^*$ triplet emission. The 4-benzoylpyridine ligand has a very similar $n-\pi^*$ emission (energy, lifetime) as 3-benzoylpyridine, but the Re → L CT excited state is at substantially lower energy for the 4- compared to the 3-benzoylpyridine. Thus, the emission from fac-[ReCl(CO)$_3$(4-benzoylpyridine)$_2$] is CT at any temperature, and it is always the lowest excited state. In the 3-benzoylpyridine complex the CT emission obtains at 298°K, but in rigid media at 77°K the blue-shifted CT and $n-\pi^*$ IL (triplet) states are nearly isoenergetic and both are emissive. Two lifetimes are measurable indicating nonthermally equilibrated CT and IL states. Such is reasonable in view of the very different electronic configuration of the two emissive states and resulting geometric structural differences.

D. PHOTOREACTIONS

Owing to the existence of fluid solution emissive Re complexes, there is developing a rich bimolecular chemistry of these complexes. Additionally, the nonemissive complexes have the usual selection of unimolecular dissociative-type processes. Accordingly, we divide this section into two broad classes of reactions to emphasize a growing importance for bimolecular processes.

1. Bimolecular Reactions

Electronic Energy Transfer. Collisional energy transfer has played an important role in developing an understanding of the behavior of excited molecules. Quenching of an excited molecule by electronic energy transfer depends on whether the donor has enough excitation energy to produce a spectroscopic excited state in the acceptor. Further, the total spin must be conserved. When spin is conserved and the energy transfer is exothermic by a couple of kcal/mol, the quenching rate is often diffusion-controlled.

Excited fac-[ReX(CO)$_3$L] and fac-[ReX(CO)$_3$L$_2$] complexes have been studied with respect to their donor properties toward certain quenchers [150,151]. For example, excited fac-[ReCl(CO)$_3$(1,10-phenanthroline)] can be quenched by O$_2$, anthracene, and $trans$-stilbene [150]. The quenching by anthracene and $trans$-stilbene was shown to obey the Stern–Volmer kinetics, and the quenching constant was determined to be within an order of magnitude of diffusion-controlled for each quencher, with that of the $trans$-stilbene about an order of magnitude lower than that of anthracene. This ordering is in accord with the excitation energy available from the excited Re complex ($\sim$ 50–55 kcal/mol) and the triplet energies of anthracene

TABLE 2-46

Excited-State Properties of fac-[ReX(CO)$_3$(Benzoylpyridine)$_2$] Complexes (X = Cl, Br, I)[a]

Compound	T, °K	Solvent	Total emission[b] band max, cm$^{-1} \times 10^{-3}$ (lifetime, μsec)	Long-lived emission[c] band max, cm$^{-1} \times 10^{-3}$ (lifetime, μsec)
[ReCl(CO)$_3$(3-Benzoylpyridine)$_2$]	298	Benzene	18.3 ($\leq$0.5)	19.47, 21.15, 22.17, 24.57 (1400)
	77	EPA	19.5, 21.2, 22.8, 24.6 (18.0 and 1400)	
[ReCl(CO)$_3$(4-Benzoylpyridine)$_2$]	298	Benzene	16.7 ($\leq$0.5)	
	77	EPA	19.38 (39)	
3-Benzoylpyridine	77	EPA	18.02, 19.80, 21.40, 23.50, 24.30 (5000)	Same as total emission
4-Benzoylpyridine	77	EPA	16.86, 19.19, 20.79, 22.32, 23.98 (3200)	Same as total emission

[a] Reprinted with permission from Giordano et al. [166], J. Am. Chem. Soc. **100**, 2257 (1978). Copyright by the American Chemical Society.

[b] Total emission spectra were obtained from an Aminco-Bowman spectrophotofluorometer; room-temperature spectra are corrected for variation in response of detector. 77°K spectral maxima represent raw data; the detector sensitivity varies by less than a factor of 2 over the wavelength range of interest and band maxima are within 200 cm^{-1}.

[c] Long-lived component of emission spectra obtained from Aminco-Bowman Spectrophotofluorometer using the phosphoroscope attachment. Lifetimes represent the long-lived component of the emission.

(42 kcal/mol) [167] and *trans*-stilbene (50 kcal/mol) [168]. That triplet excitation of the quencher obtains is evidenced by the fact that *trans*- to *cis*-stilbene isomerization accompanies the quenching. Indeed, for several Re complex sensitizers the sensitized isomerization quantum yield is essentially the same as that for the benzophenone (triplet yield is unity) [167] sensitized isomerization [168] (Table 2-47). This is an important result because it shows that the emissive state in the Re complexes has enough triplet character to produce the quencher triplet with unit efficiency. Note that none of the Re complexes has enough excitation energy to produce the stilbene singlet excited state; and, since *trans*- to *cis*-stilbene isomerization represents movement away from the thermodynamic ratio, an excited-state stilbene must be achieved, and the transfer of excitation energy from the Re to the stilbene logically produces the stilbene triplet.

TABLE 2-47

Sensitized *trans*- to *cis*-Stilbene Isomerization[a]

Sensitizer	$\Phi_{t \to c}$[b]	PSS (% *cis*-stilbene)[c]
[ReCl(CO)$_3$(5-CH$_3$-1,10-phen)]	0.62	>97
[ReCl(CO)$_3$(5-Cl-1,10-phen)]	0.65	>97
[ReCl(CO)$_3$(5-Br-1,10-phen)]	0.52	>97
[ReCl(CO)$_3$(4-Ph-py)$_2$]	0.58	66
[ReBr(CO)$_3$(4-Ph-py)$_2$]	0.60	66
[ReI(CO)$_3$(4-Ph-py)$_2$]	0.62	66
Benzophenone[c]	0.60	60

[a] Wrighton and Morse [150] and Giordano and Wrighton [151].
[b] Saltiel *et al.* [168].
[c] Benzene solution, 298°K.

The photostationary state (PSS) composition of *trans*- and *cis*-stilbene is also informative. For the phenanthroline species the ratio is very cis-rich, which is in accord with the fact that the donors do not have quite enough excitation energy to transfer to *cis*-stilbene (triplet energy ~57 kcal/mol) [168]. Consequently, once the *cis*-stilbene is produced, it is essentially inert. But for the bis-4-Ph-py complexes the excitation energy is larger, and transfer to *cis*-stilbene is possible. That is, these complexes are capable of sensitizing the back-reaction to form the trans isomer, yielding a less cis-rich PSS.

2. Excited-State Acid–Base Reactions

Figures 2-24 and 2-25 and Table 2-36 show that the Re → L CT absorption band in *fac*-[ReX(CO)$_3$L$_n$] [L = 2,2'-bipyridine-4,4'-dicarboxylic acid

($n = 1$) or 4,4'-bipyridine ($n = 2$)] depends on whether the site of protonation is protonated. For the protonated species the Re → L CT absorption is at lower energy than in the deprotonated form. Such a spectral shift is in accord with the direction of the CT. But, aside from this spectral information, the fact that the band is at lower energy leads to the conclusion that the excited base form is a stronger base than the ground state. As pointed out previously, such is reasonable, since the negative charge density on L is greater in the Re → L CT excited state than in the ground state. The energetic situation is crudely sketched in Fig. 2-46. Assuming that the spectroscopic energies are a good approximation to the free energy difference between ground and excited states, one predicts a large difference in the equilibrium constant in the excited compared to the ground state. Knowing the ground-state equilibrium constant and the excited-state energies allows calculation of the excited-state equilibrium constant [169]. Such calculations, though, are often only approximate since the spectroscopic energies are often not accurate owing to the absence of well-defined 0–0 transitions.

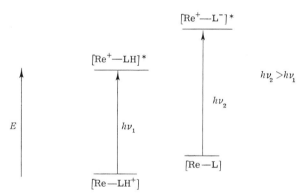

FIG. 2-46. Relative ground and excited-state energies for protonated and deprotonated Re(I) complexes.

In certain situations the ground- and excited-state acid–base equilibrium constants are different enough such that it should be possible to excite one or the other form at a given pH and observe emission from the form not present in the ground state. When both protonated and deprotonated forms are luminescent, it is particularly easy to detect the excited-state reaction, as in Eq. (2-46). Such bimolecular chemistry is probable for two reasons:

$$[Re-L] \xrightarrow{h\nu} [Re^+-L^-]^* \xrightarrow{H^+} [Re^+-LH]^* \qquad (2-46)$$

(1) proton transfer reactions are fast, and (2) there are no examples of an excited-state proton transfer reaction which results in deactivation to the ground state.

TABLE 2-48

Emission and Absorption Maxima of Protonated and
Deprotonated fac-[ReX(CO)$_3$(4,4'-Bipyridine)$_2$]^a

X	Form	Absorption max,^{b} cm^{-1}	Emission max,^{c} cm^{-1}
Cl	Neutral	31,750	19,100
	Protonated	29,410	17,100
Br	Neutral	31,470	18,500
	Protonated	29,240	17,500
I	Neutral	31,250	18,600
	Protonated	29,410	18,000

^{a} Giordano and Wrighton [151].
^{b} MeOH solution at 298°K.
^{c} EtOH/MeOH, 4:1 at 77°K.

For the fac-[ReX(CO)$_3$(4,4'-bipyridine)$_2$] complexes the protonated form does not emit detectably in fluid solution at room temperature. Interestingly, though, the emission at low temperature for both forms can be detected (Table 2-48). Taking the emission maxima as a relative measure of the excited-state energy, we see that there is quite a difference in the energies. But, curiously, the comparable data from the absorption bands suggest an even larger change in excited-state base strength. Recognizing that the absorption is predominantly singlet → singlet in character while the emission is triplet → singlet in character does offer an explanation for the observations: there is a significant dependence on the excited-state multiplicity. This is not an unprecedented result; a number of organic molecules have different base strengths in their singlet and triplet excited states [167,169].

Both fac-[ReX(CO)$_3$(2,2'-bipyridine-4,4'-dicarboxylic acid)] and its conjugate base exhibit detectable emission in fluid solution [170]. Spectrophotometric and luminescence titrations show ground- and excited-state acid dissociation constants which differ by less than 2 pK_a units. However, the shift is in accord with the notion that the base form is a stronger base in the excited state. It has been shown that excitation of a solution at a pH where predominantly the base form absorbs gives emission predominantly from the protonated species, showing that proton transfer can occur without excited-state deactivation.

3.   Excited-State Electron Transfer

Electron transfer properties of excited fac-[ReCl(CO)$_3$L] (L = 1,10-phenanthroline, 4,7-diphenyl-1,10-phenanthroline) have been investigated [171]. Cyclic voltammetric studies show that the ground-state species exhibit somewhat reversible oxidation or reduction, revealing that the radical cation and anion have some lifetime in solution. The ground state potentials were found to be $E^\circ([ReCl(CO)_3L]/[ReCl(CO)_3L]^-) = -1.3$ V vs. SCE and $E^\circ([ReCl(CO)_3L]^+/[ReCl(CO)_3L]) = +1.3$ V vs. SCE. Since the lowest (emissive) excited state of $[ReCl(CO)_3L]$ is ~ 2.3 eV above the ground state, the excited state potentials are estimated to be

$$E^\circ([ReCl(CO)_3L]^*/[ReCl(CO)_3L]^+) = +1.0 \text{ V vs. SCE}$$

and $E^\circ([ReCl(CO)_3L]^+/[ReCl(CO)_3L]^* = -1.0$ V vs. SCE. That is, the excited state is ~ 2.3 V more oxidizing *and* ~ 2.3 V more reducing than the ground state.

That the excited $[ReCl(CO)_3L]$ can be quenched by electron transfer was unequivocally established by detection of the flash-photolysis generated transient quencher product. For the quenchers N,N'-dimethyl-4,4'-bipyridinium (MV^{2+}) and N,N'-dibenzyl-4,4'-bipyridinium (BV^{2+}) the one-electron-reduced MV^+ and BV^+ are detectable as electron transfer products. The MV^{2+} and BV^{2+} both quench excited $[ReCl(CO)_3L]$ at a diffusion-controlled rate. But no net chemical change results, since MV^+ or BV^+ back-react with the $[ReCl(CO)_3L]^+$ species at nearly a diffusion-controlled rate. The sequence is given in Eqs. (2-47)–(2-49).

$$[ReCl(CO)_3L] \xrightarrow{h\nu} [ReCl(CO)_3L]^* \qquad (2\text{-}47)$$

$$[ReCl(CO)_3L]^* + Q^{2+} \xrightarrow{k_q} [ReCl(CO)_3L]^+ + Q^+ \qquad (2\text{-}48)$$

$$Q^+ + [ReCl(CO)_3L]^+ \xrightarrow{k_{br}} [ReCl(CO)_3L] + Q^{2+} \qquad (2\text{-}49)$$

Thus, the net result is the light-catalyzed exchange of electrons between Q^{2+} and $[ReCl(CO)_3L]$. Excited-state reduction of Q^{2+} can be used as the central element in a solar-energy conversion scheme, but the rapid back-reaction severely limits the efficiency and utility of such a process.

In the sequence (2-47)–(2-49) the quenching of the excited Re complex occurs at a diffusion-controlled rate. However, since the quenching is via electron transfer, the value of the quenching constant should depend on the redox properties of the quencher. Table 2-49 shows that a reasonable correlation between k_q and the ease of reduction of the quencher does exist. The point at which k_q drops precipitously is in accord with the excited-state potentials mentioned previously.

One final interesting point concerning the electron transfer properties of $[ReCl(CO)_3L]$ is that the excited-state potentials ($+1.0$; -1.0 V vs. SCE) are straddled by the ground-state potentials ($+1.3$; -1.3). This means that annihilation reaction, Eq. (2-50), can result in one excited $[ReCl(CO)_3L]$ species.

$$[ReCl(CO)_3L]^+ + [ReCl(CO)_3L]^- \rightarrow [ReCl(CO)_3L] + [ReCl(CO)_3L]^* \qquad (2\text{-}50)$$

TABLE 2-49

Quenching Constants for Electron Transfer Quenching of
Excited [ReCl(CO)$_3$(1,10-Phenanthroline)]^a

Quencher	$E_{1/2}$, V vs. SCE	k_q (M^{-1} sec^{-1})
TCNE	+0.24	7.43 × 10^9
[N,N'-Dibenzyl-4,4'-bipyridinium](PF$_6$)$_2$	−0.36	2.67 × 10^9
[N,N'-Dimethyl-4,4'-bipyridinum](PF$_6$)$_2$	−0.45	3.08 × 10^9
[N-Methyl-4-cyanopyridinium]PF$_6$	−0.79	2.25 × 10^9
p-NO$_2$-Benzaldehyde	−0.86	2.58 × 10^9
4,4'-DiNO$_2$-Biphenyl	−1.00	1.94 × 10^9
m-NO$_2$-Benzaldehyde	−1.02	6.40 × 10^8
4-Cl-Nitrobenzene	−1.06	2.36 × 10^8
4-CH$_3$-Nitrobenzene	−1.20	<2 × 10^7

^{a} Reprinted with permission from Luong et al. [171], J. Am. Chem. Soc. 100, 5790 (1978). Copyright by the American Chemical Society.
^{b} Data for degassed CH$_3$CN solutions of 0.1 M [n-Bu$_4$N]ClO$_4$ at 298°K.

Since the excited species is emissive, the annihilation should be accompanied by luminescence characteristic of the excited species. Electrochemical generation of the radical-anion and radical-cation does result in luminescence. Such electrogenerated chemiluminescence has been previously observed for a number of organic substances and for [Ru(2,2'-bipyridine)$_3$]$^{2+}$ and related species [172].

4. Unimolecular Reactions

a. Substitution in [M(CO)$_5$X] *Species.* While the photochemistry of the parent [M(CO)$_6$]$^+$ species has not been reported, a number of reports on the photochemistry of [M(CO)$_5$X] (X = Cl, Br, I, H, Me, etc.) have appeared. As usual in such mononuclear metal carbonyls, CO dissociation dominates the decay processes of the lowest excited states.

Quantum yield data have been reported for substitution reactions of [Re(CO)$_5$X] (X = Cl, Br, I) [Eqs. (2-51)–(2-53); Table 2-50] [37]. Efficient

$$2[Re(CO)_5X] \xrightarrow[CCl_4]{hv} [Re(CO)_4X]_2 + 2CO \qquad (2\text{-}51)$$

$$[Re(CO)_5X] \xrightarrow[pyridine]{hv} cis\text{-}[ReX(CO)_4(pyridine)] + CO \qquad (2\text{-}52)$$

$$[Re(CO)_5X] \xrightarrow[PPh_3]{hv} cis\text{-}[ReX(CO)_4PPh_3] + CO \qquad (2\text{-}53)$$

substitution obtains upon excitation to LF states. There does appear to be a wavelength dependence, as often found, but a detailed analysis is not possible without CO-labeling experiments. Qualitative observations point

TABLE 2-50

Reaction Quantum Yields for $[Re(CO)_5X]^{a,b}$

X	Reaction No.[c]	$\Phi_{366\ nm}\ (\pm 10\%)^d$	$\Phi_{313\ nm}\ (\pm 10\%)^e$
Cl	(2-51)	0.06	0.44
Br	(2-51)	0.18	0.65
I	(2-51)	0.08	0.35
Cl	(2-52)	0.20	0.76
Br	(2-52)	0.34	0.58
I	(2-52)	0.10	0.67
Cl	(2-53)	0.21	—
Br	(2-53)	0.31	—
I	(2-53)	0.07	—

[a] Reprinted with permission from Wrighton *et al.* [37], *J. Am. Chem. Soc.* **98**, 1111 (1976). Copyright by the American Chemical Society.

[b] Reactions carried out in CCl_4 and measured by monitoring decline of near-IR CO overtone of $[Re(CO)_5X]$.

[c] Reaction number (2-51) corresponds to formation of $[Re(CO)_4X]_2$; number (2-52) to $[ReX(CO)_4(py)]$, and number (2-53) to formation of $[ReX(CO)_4PPh_3]$.

[d] 1.0×10^{-7} einstein/min.

[e] 6.65×10^{-8} einstein/min.

to the conclusion that the Mn analogs are also photosubstitution-labile [173]. There have been no claims of photosubstitution of the X ligand for X = Cl, Br, I. Lack of lability is consistent with the fact that these are π-donor ligands and their π bonding is strengthened in the lowest (LF) states.

The photoreactions of a number of $[Mn(CO)_5X]$ (X = strong σ-bonded ligand) complexes have been reported. Among these studies is the recent low-temperature photolysis of $[Mn(CO)_5H]$ which undergoes loss of CO to form a five-coordinate species of trigonal-bipyramidal structure (Eq. 2-54) [174]. The regeneration of the $[Mn(CO)_5H]$ by lower-energy photolysis may

$$[Mn(CO)_5H] \underset{285\ nm}{\overset{228\ nm}{\underset{15\ K,\ Ar}{\rightleftharpoons}}} [Mn(CO)_4H] + CO \qquad (2\text{-}54)$$

be due to localized softening of the environment, allowing thermal recombination of the coordinatively unsaturated intermediate and the ligand. The direct observation of the five-coordinate species does serve to establish the dissociative nature of the CO photosubstitution. Additionally, the implication of the trigonal–bipyramidal structure for $[Mn(CO)_4H]$ leads to the expectation that the stereochemical significance of photoproducts $[Mn(CO)_4(L)H]$ will be clouded. Interestingly, the photoinduced ^{13}CO incorporation into $[Mn(CO)_5Br]$ revealed no difference in the rate of axial vs. equatorial substitution, while the thermal reaction proceeded to give axial substitution at a rate equal to 0.74 times the equatorial rate [175]. These experiments provide support for different intermediates in the thermal and photosubstitution reactions; but, aside from this difference in reactivity, no information is available regarding other properties of the intermediates with the exception that they are fluxional and five-coordinate [175].

Irradiation of $[R_3CMn(CO)_5]$ (R = H, D, F) at 17°K in an argon matrix has been shown to produce a five-coordinate acyl derivative (Eq. 2-55) [176].

$$[R_3CMn(CO)_5] \xrightarrow[\text{17°K, Ar}]{hv} [R_3COMn(CO)_4] \qquad (2\text{-}55)$$

$$R = H, D, F$$

The role of the light in this reaction is not clear: Is the primary process rupture of the Mn—CR_3 or the Mn—CO bond, or does light induce migration of CR_3?

The photolysis [177] of $[Mn(\eta^1\text{-}C_3H_5)(CO)_5]$ represents one of the earliest reports of a reorganization of bonding between the metal and the hydrocarbon group induced by photodissociation of CO (Eq. 2-56). The

$$[Mn(\eta^1\text{-}C_3H_5)(CO)_5] \xrightarrow{hv} [Mn(\eta^3\text{-}C_3H_5)(CO)_4] + CO \qquad (2\text{-}56)$$

decarbonylation reaction (2-57) proceeds in 10.5% yield at $-68°$ [178], but

$$\left[\begin{array}{c} \overset{O}{\underset{\|}{\diagup}} \\ \diagdown C-Mn(CO)_5 \end{array} \right] \xrightarrow[-68°C]{hv} \left[\begin{array}{c} \\ Mn(CO)_3 \end{array} \right] + 3\,CO \qquad (2\text{-}57)$$

the primary photoprocess is not obvious. Interestingly, the room-temperature photolysis yields only $[Mn_2(CO)_{10}]$ and bitropyl $(C_{14}H_{24})$, implicating $[Mn(CO)_5]$ radical intermediates.

Finally, with respect to the six-coordinate $[M(CO)_5X]$ complexes, highly substituted derivatives of $[Mn(CO)_5H]$ are formed via irradiation in the presence of PF_3 (Eq. 2-58) [179]. The PF_3 is, as usual in these cases, a strong

$$[Mn(CO)_5H] \xrightarrow[PF_3]{hv} [Mn(CO)_n(PF_3)_{5-n}H]$$
$$n = 4, 3, 2, 1, 0 \qquad (2\text{-}58)$$

π-acceptor ligand capable of stabilizing low oxidation states of the central metal.

It is worth noting that luminescent complexes such as fac-$[ReX(CO)_3L_2]$ (X = Cl, Br, I; L = phenylpyridine, 4,4′-bipyridine) are essentially photosubstitution-inert [151]. This fact is in accord with nonlabile MLCT excited states; consequently, the photosubstitution lability of the mononuclear species in this section can be attributed to low-lying LF excited states having substantial σ-antibonding character.

   b.   Substitution   in $[M(\eta^5\text{-}C_5H_5)(CO)_nL_{3-n}]$ Species.   As with the $[Cr(\eta^6\text{-arene})(CO)_3]$ system, the isoelectronic $[M(\eta^5\text{-}C_5H_5)(CO)_3]$ (M = Mn, Re) photochemistry is dominated by CO photosubstitution. Most work has involved M = Mn, and a number of substituted derivatives have been prepared. Some examples are given in Table 2-51 [107, 122, 180–188]. The quantum yield for reaction (2-59) was claimed to be 1.0 for L = acetone and

$$[Mn(\eta^5\text{-}C_5H_5)(CO)_3] \xrightarrow[L]{hv} [Mn(\eta^5\text{-}C_5H_5)(CO)_2L] + CO \qquad (2\text{-}59)$$

PhCCPh [189]. A more recent study [157] shows that the quantum yield is somewhat less, but the fact is that the dissociative loss of CO is quite efficient. This is particularly interesting since it has been claimed that $[Mn(\eta^5\text{-}C_5H_5)(CO)_3]$ can be recovered unchanged from PPh$_3$-containing solutions refluxed at 200°C. Excitation obviously leads to a large change in substitution lability. Though less has been done with it, $[Re(\eta^5\text{-}C_5H_5)(CO)_3]$ is also known to give fairly efficient simple dissociative loss of CO upon photoexcitation [157, 190]. As Table 2-51 suggests, virtually any entering group can be used in reaction (2-59). Further loss of CO from $[Mn(\eta^5\text{-}C_5H_5)(CO)_2L]$ has been observed in several cases notably for L = good π-acceptor ligand as is evident from the examples given in Table 2-51. All CO groups have been displaced in the formation of $[Mn(\eta^5\text{-}C_5H_5)(\eta^6\text{-}C_6H_6)]$ [188].
   The coordinatively unsaturated intermediate from $[Mn(\eta^5\text{-}C_5H_5)(CO)_3]$ is susceptible to oxidative addition, like its $[Cr(\eta^6\text{-arene})(CO)_3]$ analog

TABLE 2-51

Photosubstitution Reactions of $[Mn(\eta^5\text{-}C_5H_5)(CO)_3]$

Entering group, L	Product(s)	Reference
Tetrahydrofuran	$[Mn(\eta^5\text{-}C_5H_5)(CO)_2L]$	[180]
$(Ph_2PCH_2CH_2)_3N$	$[Mn(\eta^5\text{-}C_5H_5)(CO)L]$	[181]
$(CH_3)_2NPF_2$	$[Mn(\eta^5\text{-}C_5H_5)(CO)L_2]$	[122]
$C_5H_{10}NPF_2$	$[Mn(\eta^5\text{-}C_5H_5)(CO)L_2]$	[122]
2,3-Diazabicyclo[2.2.2]hept-2-ene	$[Mn(\eta^5\text{-}C_5H_5)(CO)_2L]$	[107]
Pyridine	$[Mn(\eta^5\text{-}C_5H_5)(CO)_2L]$	[182]
Piperidine	$[Mn(\eta^5\text{-}C_5H_5)(CO)_2L]$	[183]
SO$_2$	$[Mn(\eta^5\text{-}C_5H_5)(CO)_2L]$	[184]
PPh$_3$	$[Mn(\eta^5\text{-}C_5H_5)(CO)_2L]$	[185]
	$[Mn(\eta^5\text{-}C_5H_5)(CO)L_2]$	
1,3-Butadiene	$[Mn(\eta^5\text{-}C_5H_5)(CO)L]$	[186]
Ethylene (other alkenes)	$[Mn(\eta^5\text{-}C_5H_5)(CO)_2L]$	[187]
Benzene	$[Mn(\eta^5\text{-}C_5H_5)(L)]$	[188]

(Eq. 2-60) [116]. The resulting product is a distorted square–pyramid and formally seven-coordinate.

$$[Mn(\eta^5\text{-}C_5H_5)(CO)_3] \xrightarrow[Cl_3SiH]{h\nu} (OC)_2Mn\overset{\displaystyle \big}{\underset{H}{-}}SiCl_3 \tag{2-60}$$

Several examples of dinuclear complexes formed via photolysis in the presence of a bidentate ligand have been reported [191–194]. One typical example is shown in Eq. (2-61). Photolysis of the dinuclear complex can

$$[Mn(\eta^5\text{-}C_5H_5)(CO)_3] \xrightarrow[Ph_2PCH_2CH_2PPh_2]{h\nu}$$

$$[Mn(\eta^5\text{-}C_5H_5)(CO)_2P(Ph_2)CH_2CH_2(Ph)_2PMn(\eta^5\text{-}C_5H_5)(CO)_2] \tag{2-61}$$

result in the formation of the mononuclear $[Mn(\eta^5\text{-}C_5H_5)(CO)L]$ [191].

TABLE 2-52

Quantum Yields for $[Mn(\eta^5\text{-}C_5H_5)(CO)_2X] + Y$ to
$[Mn(\eta^5\text{-}C_5H_5)(CO)_2Y] + X$ Ligand Substitution[a,b]

X	Y	Irradiation, λ, nm	$\Phi \pm (0.15\Phi)$
CO	py	313	0.65
Piperidine	4-Acetyl-py	436	0.14
	1-Pentene	436	0.16
py	4-Acetyl-py	405	0.35
	1-Pentene	405	0.37
3,5-DiCl-py	Piperidine	436	0.42
	1-Pentene	436	0.40
	3,4-DiMe-py	436	0.38
4-Acetyl-py	1-Pentene	550	0.25
	Piperidine	550	0.24
	3,4-DiMe-py	550	0.25
	4-Me-py	550	0.23
	py	550	0.25
4-Formyl-py	3,4-DiMe-py	550	0.25
	4-Me-py	550	0.25
	py	550	0.25
	3,4-DiMe-py	633	0.25
	4-Me-py	633	0.25
	py	633	0.25

[a] Reprinted with permission from Giordano and Wrighton [157], *Inorg. Chem.* **16**, 160 (1977). Copyright by the American Chemical Society.

[b] All photoreactions in isooctane solutions of 0.25 *M* Y at 25°C; py = pyridine; Me = methyl.

No one has claimed that substitution of the η^5-C_5H_5 ring is a primary photoprocess. As with $[Cr(\eta^6$-arene)$(CO)_3]$, it is probable that the six-electron donor system is not labilized by a one-electron excitation to the same degree as are the two-electron donor CO groups. Finally, the only derivative of a substituted cyclopentadienyl that has been studied, $[Mn(\eta^5$-$C_5H_4CH_3)(CO)_3]$, behaves like the parent species [116, 182, 191, 195].

A detailed study of the photosubstitution of $[M(\eta^5$-$C_5H_5)(CO)_2X]$ (M = Mn, Re; X = CO, THF, amine, pyridine, or substituted pyridine) has been reported [157], and for all complexes studied the only photoreaction found was (2-62). Some quantum yield data are given in Tables 2-52 and 2-53.

$$[M(\eta^5\text{-}C_5H_5)(CO)_2X] \xrightarrow[Y]{h\nu} [M(\eta^5\text{-}C_5H_5)(CO)_2Y] + X \qquad (2\text{-}62)$$

Treating these complexes as pseudo-C_{4v} species (η^5-$C_5H_5 \equiv$ three $2e^-$ carbon donors), we see that the photochemistry can be related to that for an

TABLE 2-53

Quantum Yields for $[Re(\eta^5$-$C_5H_5)(CO)_2X] + Y$ to $[Re(\eta^5$-$C_5H_5)(CO)_2Y] + X$ Ligand Substitution[a]

X	Y	Irradiation, λ, nm	$\Phi \pm (0.15\Phi)$
CO	py	313	0.30
NH_3	py	313	0.34
	3,5-DiCl-py	313	0.28
4-Me-py	4-Acetyl-py	436	0.28
4-Me-py	1-Pentene	436	0.30
py	4-Acetyl-py	436	0.09
	1-Pentene	436	0.11
3-Br-py	1-Pentene	436	0.013
	3,4-DiMe-py	436	0.013
	4-Me-py	436	0.013
3,5-DiCl-py	1-Pentene	436	0.005
	3,4-DiMe-py	436	0.005
	4-Me-py	436	0.005
3-Acetyl-py	1-Pentene	436	0.005
3-Benzoyl-py	1-Pentene	436	0.005
4-Acetyl-py	3,4-DiMe-py	436 or 550	$<10^{-4}$
	4-Me-py	436 or 550	$<10^{-4}$
	py	436 or 550	$<10^{-4}$
4-Benzoyl-py	3,4-DiMe	436 or 550	$<10^{-4}$

[a] Reprinted with permission from Giordano and Wrighton [157], *Inorg. Chem.* **16**, 160 (1977). Copyright by the American Chemical Society.

[b] All photoreactions in isooctane solution of 0.25 M Y at 25°C; py = pyridine; Me = methyl.

analogous $[W(CO)_5X]$ series of complexes. Thus, the loss of X is explicable in terms of LF lowest excited states which are σ-antibonding with respect to the M—X bond. For M = Mn all complexes studied were found to be photosensitive with quantum yields for loss of X in the range 0.14–0.65. It is particularly noteworthy that the $[Mn(\eta^5\text{-}C_5H_5)(CO)_2(4\text{-formylpyridine})]$ is photosensitive, even at 550 nm and despite the fact that the absorption spectrum is dominated by the Mn → 4-formylpyridine CT band. However, the first absorption in $[Mn(\eta^5\text{-}C_5H_5)(CO)_2\text{piperidine}]$ tails very substantially into the low-energy visible region. Presumably, this is a spin-allowed LF absorption which is not only present in the piperidine species but also has a spin-forbidden (triplet) component at lower energy. The point is that LF reaction obtains despite efficient, direct population of the Mn → X CT excited state. Decay via the excited LF state is still the reasonable explanation since the CT state should not labilize X. If anything, CO lability might be expected owing to the formal Mn(I) → Mn(II) oxidation accompanying the CT excitation.

The Re series is very different from the Mn series in that the photosubstitution of X is not uniformly efficient. For the complexes where the Re → XCT excited state is lowest in energy, the photosubstitution yield is very low. The difference between Mn and Re rests in the fact that the LF states for Re are much higher in energy and it is possible to push the Re → X CT state well below the σ-antibonding LF states. The situation is summarized in Fig. 2-47 like that in Fig. 2-20 for $[W(CO)_5X]$.

   c. Photochemistry of $[Mn(CO)_4NO]$. As with $[Fe(CO)_5]$, photo-induced loss of CO from $[Mn(CO)_4NO]$ leads to a dinuclear carbonyl (Eq. 2-63) [196]. Low-temperature photolysis in an argon matrix has revealed formation of a coordinatively unsaturated species (Eq. 2-64), which

$$2[Mn(CO)_4NO] \xrightarrow{\ h\nu\ } Mn_2(CO)_7(NO)_2 + CO \qquad (2\text{-}63)$$

$$[Mn(CO)_4NO] \underset{\Delta,\ 30°K}{\overset{h\nu}{\underset{\longleftarrow}{\xrightarrow{15°K,\ Ar}}}} [Mn(CO)_3NO] + CO$$

$$\qquad\qquad\qquad\qquad\qquad 15°K,\ Ar \Big\downarrow h\nu \qquad\qquad (2\text{-}64)$$

$$[Mn(CO)_2NO] + CO$$

itself is photosensitive [197]. The $[Mn(CO)_3NO]$ intermediate is presumably involved in the room-temperature photochemistry [Eqs. (2-63) or (2-65)] to produce $Mn(CO)_3L(NO)$ complexes.

$$[Mn(CO)_4(NO)] \xrightarrow[\ L\ ]{h\nu} [Mn(CO)_3(L)NO] + CO$$

$$\qquad\qquad\qquad\qquad\qquad\qquad\qquad (2\text{-}65)$$

$$L = PPh_3,\ AsPh_3,\ P(n\text{-}Bu)_3$$

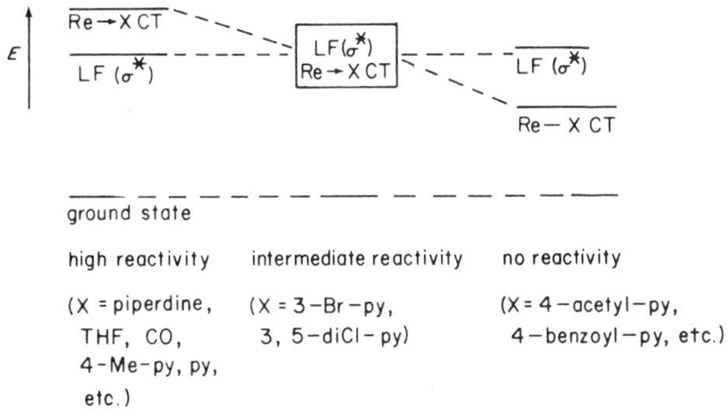

FIG. 2-47. $[Re(\eta^5\text{-}C_5H_5)(CO)_2X]$ photoreactivity. Reprinted with permission from Giordano and Wrighton [157], *Inorg. Chem.* **16**, 160 (1977). Copyright by the American Chemical Society.

Quantum yields for reaction (2-65) were less than unity and found to depend both on the nature of entering group L and, for PPh_3, its concentration [198]. These two facts were interpreted as indicative of an associative contribution to the dissociative path for substitution. Such a result may be reasonable if the nature of the Mn—NO interaction is changed photochemically such as a change from NO = three-electron donor to NO = one-electron donor.

d. Photoactivation of the M—X Bond in $[Mn(CO)_5X]$. A number of reports now exist revealing that the Mn—X bond can be cleaved by photolysis of $[Mn(CO)_5X]$ for a wide variety of σ-bonded X groups including R_3Sn, R_3Ge, and alkyl. Examples have already been presented of a photoassisted carbonylation (reaction 2-55) and decarbonylation (reaction 2-57), but in these examples it is not clear that cleavage of the Mn—X bond is the primary excited-state decay path. The homolytic scission of $[Mn(CO)_5X]$ to yield $[Mn(CO)_5]$ and X might be expected if high-energy excitation is used to ensure achievement of the $\sigma_b \rightarrow \sigma_z^*$ (LMCT) excitation.

Existing experimental reports lead to conflicting conclusions regarding the dominant primary photoreaction. The insertion reaction (2-66) is reported

$$[CH_3Mn(CO)_5] \xrightarrow[hv,\ 1\text{–}2\ \text{atm}]{F_2C=CF_2} [CH_3CF_2CF_2Mn(CO)_5] \qquad (2\text{-}66)$$

to go in essentially quantitative yield [199], while simple substitution in reaction (2-67) gives good yields as well [200]. Additionally, reaction (2-68)

$$[C_6F_5Mn(CO)_5] \xrightarrow[\text{pyridine}]{hv} [C_6F_5Mn(CO)_4(\text{pyridine})] \qquad (2\text{-}67)$$

$$[HC_2F_4Mn(CO)_5] \xrightarrow[\text{PF}_3]{hv} [HC_2F_4Mn(CO)_4PF_3] \qquad (2\text{-}68)$$

goes smoothly, yielding simple substitution products [179].

However, aside from the insertion reaction in (2-66), other systems give varying ratios of insertion and what appear to be free-radical coupling products. For example, consider the photoreactions of $[(CH_3)_3GeMn(CO)_5]$ in the presence of C_2F_4 and CF_3CFCF_2 as in reactions (2-69) and (2-70),

$$[(CH_3)_3GeMn(CO)_5] \xrightarrow[F_2C=CF_2]{hv} (CH_3)_3GeF + [CH_3GeCF_2CF_2Mn(CO)_5] + [Mn_2(CO)_{10}]$$
$$(2\text{-}69)$$

$$[(CH_3)_3GeMn(CO)_5] \xrightarrow[CF_3CF=CF_2]{hv} [Mn_2(CO)_{10}] + [CF_3CFCFMn(CO)_5]$$
$$\text{50°C, pentane} \qquad\qquad (2\text{-}70)$$

respectively [201]. The analogous $[(CH_3)_3SnMn(CO)_5]$ undergoes no decomposition at 130°C; but, upon photolysis under conditions as for the $(CH_3)_3Ge$ compound, rich chemistry is obtained [202] as in reaction (2-71).

$$[(CH_3)_3SnMn(CO)_5] \xrightarrow[F_2C=CF_2]{hv} [C_5F_9Mn(CO)_5] + [(CH_3)_3SnCF_2CF_2Mn(CO)_5]$$
$$+ [CF_2=CFCOMn(CO)_5] + [CF_2=CFMn(CO)_4]_2 \quad (2\text{-}71)$$

It is therefore clear that photolysis can activate the Mn—X bond in $[Mn(CO_5X]$ complexes, and substantial activation of hydro- and fluorocarbons can be obtained [201–203]. However, answers to questions regarding structure–reactivity relationships are not yet available.

    e. Photoinduced Cleavage of Transition Metal–Metal Bonds. The parent carbonyls of Mn and Re, $[M_2(CO)_{10}]$, may be viewed as special cases of $[M(CO)_5X]$ where X = $[M(CO)_5]$. As has already been pointed out, these complexes exhibit an intense electronic absorption band corresponding to a $\sigma_b \rightarrow \sigma^*$ transition associated with the metal–metal bond. And, like the other metal–metal bonded complexes described, these undergo efficient photoinduced cleavage of the metal–metal bond. A number of qualitative reports exist suggesting that simple photosubstitution of $[M_2(CO)_{10}]$ can occur (Table 2-54) [204–213], but it is now known that such products actually *do not* result from dissociative loss of CO to generate $[M_2(CO)_9]$. Results of a number of qualitative photochemical studies suggested that metal–metal bond cleavage in dinuclear Mn and Re complexes is a component of the excited state chemistry (Table 2-55) [214–221].

    In 1973, quantitative data were reported [222] for the first time on a metal–metal bonded complex, and reaction (2-72) was said to occur with a

$$[Re_2(CO)_{10}] \xrightarrow[CCl_4]{hv} 2[Re(CO)_5Cl] \qquad\qquad (2\text{-}72)$$

quantum yield of 0.60 for $[Re_2(CO)_{10}]$ disappearance upon 313 nm excitation. The high quantum yield and quantitative chemical yield of what appears to be the result of radical trap reactions indicates that CO loss is not as

TABLE 2-54

Photosubstitution Products of $[M_2(CO)_{10}]$

M	Entering group, L	Product(s)	Reference
Mn	PPh_3	$[Mn_2(CO)_9L]$	[204]
		$[Mn_2(CO)_8L_2]$	[204–207]
	$P(OC_6H_5)_3$		
	$P(C_2H_5)_3$		
	$P(C_6H_4F)_3$	$[Mn_2(CO)_8L_2]$	[204]
	$As(C_6H_5)_3$		
Re	PF_3	$[Re_2(CO)_{10-n}L_n]$	[208]
Mn	PF_3	$[Mn_2(CO)_{10-n}L_n]$	[208]
		($n = 1,2,3$)	
Re	$CH_3P(C_6H_5)_2$	$[Re_2(CO)_{10-n}L_n]$	[209]
		($n = 1,2,3$)	
Mn	^{13}CO	$[Mn_2(CO)_n(^{13}CO_{10-n})]$	[210]
Mn	$CH_2{=}CHCN$	$[Mn_2(CO)_9L]$	[211]
		(equatorial)	
	C_6H_5CN		
	CH_3CN		
Re, Mn	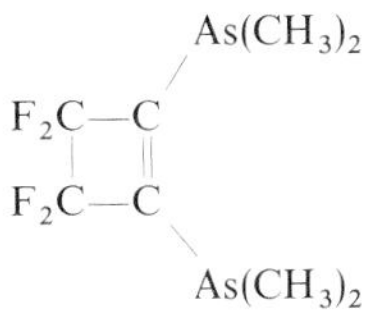 $F_2C{-}C\overset{As(CH_3)_2}{\underset{As(CH_3)_2}{}}$ $F_2C{-}C$	$[M_2(CO)_8L]$ (bridging L)	[212,213]

TABLE 2-55

Photoreactions of $[M_2(CO)_{10}]$ Involving Rupture of the M—M Bonds

Starting M—M bonded species	Added reagent	Product(s)	Reference
$[Mn_2(CO)_8]$	Chlorinated solvent	$[MnCl(CO)_3(phen)]$ + $[(CO)_5MnCl]$	[214]
$[Re_2(CO)_{10}]$	$Fe(CO)_5$	$[(CO)_5ReFe(CO)_4Re(CO)_5]$	[215]
$[Mn_2(CO)_{10}]$	$Fe(CO)_5$	$[(CO)_5MnFe(CO)_4Mn(CO)_5]$	
$[Tc_2(CO)_{10}]$	$Fe(CO)_5$	$[(CO)_5TcFe(CO)_4Tc(CO)_5]$	
$[Re_2(CO)_{10}]$ + $[Mn_2(CO)_{10}]$	—	$[ReMn(CO)_{10}]$	[216]
$[Mn_2(CO)_{10}]$	HBr	$[Mn(CO)_5Br]$	[217]
$[Mn_2(CO)_{10}]$	CCl_4	$[Mn(CO)_5Cl]$	[218]
$[Re_2(CO)_{10}]$	I_2	$[Re(CO)_5I]$	[219]
$[(CO)_5MnRe(CO)_3(phen)]$	—	$[Mn_2(CO)_{10}]$ + $[Re(CO)_3(phen)]_2$	[220,221]

important as Re—Re cleavage. Table 2-56 [162] summarizes some quantum yield and stoichiometry data for this and other systems. Generally, the quantum yields are high and the stoichiometry is clean. Irradiation of $[M_2(CO)_{10}]$ (M = Mn, Re) and $[MnRe(CO)_{10}]$ in alkane solutions containing I_2 gives quantitative chemical yields of $[M(CO)_5I]$, but the quantum yields are viscosity-dependent. For example, the disappearance quantum yield for $[Re_2(CO)_{10}]$ is 0.64 in isopentane and only 0.30 in Nujol [162]. This viscosity dependence is in accord with a mechanism which involves the photogeneration of free $[M(CO)_5]$ radicals.

TABLE 2-56

Quantum Yield Data for Dinuclear Mn and Re Carbonyl Complexes
in the Presence of $CCl_4{}^a$

Reactant	Product	λ_{irrdn}, nm	$\Phi_{dis} \pm 10\%{}^b$	$\Phi_{formn} \pm 10\%{}^c$
$[Mn_2(CO)_{10}]$	$[Mn(CO)_5Cl]$	366	0.41	0.72
		313	0.48	1.02
$[Mn_2(CO)_9PPh_3]$	$[Mn(CO)_5Cl]$	366	0.45	0.36
	$[Mn(CO)_4PPh_3Cl]$			0.40
$[Mn_2(CO)_8(PPh_3)_2]$	$[Mn(CO)_4PPh_3Cl]$	366	0.70	0.48
$[Re_2(CO)_{10}]$	$[Re(CO)_5Cl]$	313	0.60	1.20
$[MnRe(CO)_{10}]$	$[Mn(CO)_5Cl]$	366	0.42	0.46
	$[Re(CO)_5Cl]$			0.43

^{a} Reprinted with permission from Wrighton and Ginley [162], *J. Am. Chem. Soc.* **97**, 2065 (1975). Copyright by the American Chemical Society.
^{b} Disappearance quantum yield for reactant.
^{c} Formation quantum yield for product.

The cross-coupling of 17-valence-electron radicals as in reaction (2-73) [162] in an inert solvent is the result which is compelling with respect to the

$$[Re_2(CO)_{10}] + [Mn_2(CO)_{10}] \underset{\text{alkane solvent}}{\overset{hv}{\rightleftharpoons}} 2[MnRe(CO)_{10}] \qquad (2\text{-}73)$$

conclusion that the primary photoprocess leads to the production of radicals which can couple. The chemistry indicated in Eq. (2-73) is quite clean, consistent with little other than metal–metal bond cleavage as the primary excited state reaction. In connection with such cross-coupling reactions, it is important to note that the coupling of $[Mn(CO)_5]$ (Eq. 2-74) occurs at a

$$2[Mn(CO)_5] \longrightarrow [Mn_2(CO)_{10}] \qquad (2\text{-}74)$$

diffusion-controlled rate as measured in flash photolysis experiments [223]. The photogenerated $[M(CO)_5]$ radicals can be spin-trapped [133] and presumably react with NO to produce $[M(CO)_4NO]$ complexes [224]. The interesting cluster $[Re(CO)_3OH]_4$ results from irradiation of $[Re_2(CO)_{10}]$ in the presence of H_2O [225].

In most of the examples studied there is relatively little wavelength dependence of the quantum yield for metal–metal bond cleavage, particularly in the region where the $\sigma_b \to \sigma^*$ and the $d\pi \to \sigma^*$ would be distinguished. Thus, it appears that either of these one-electron transitions is active, in accord with electrochemical experiments where it is known that the reduction of $[M_2(CO)_{10}]$ is irreversible. Presumably, reduction of $[M_2(CO)_{10}]$ to produce $[M_2(CO)_{10}]^{\doteq}$ involves putting an electron in σ^*, which is very destabilizing.

Oxidation of $[M_2(CO)_{10}]$ to produce $[M_2(CO)_{10}]^{\dotplus}$ is also irreversible presumably owing to the fact that an electron is removed from σ_b to produce the radical-cation. Photochemical studies [165] of $[M_2(CO)_8L]$ (M = Mn, L = 1,10-phenanthroline, 2,2′-biquinoline) provide a test of whether removal of electron density from σ_b leads to extreme metal–metal bond lability. As discussed previously, the lowest excited state in the $[M_2(CO)_8L]$ species is the (M—M) $\sigma_b \to \pi^*L$ CT. Table 2-57 summarizes the photoreactivity of these complexes in halogen donor solvents. Each complex is photosensitive with a quantum yield which is essentially independent of excitation wavelength. The products are as indicated in reaction (2-75), consistent with symmetrical metal–metal bond cleavage. Flash excitation of $[M_2(CO)_8L]$

$$[M_2(CO)_8L] \xrightarrow[CH_2Cl_2/CCl_4]{h\nu} [M(CO)_5Cl] + [MCl(CO)_3L] \qquad (2\text{-}75)$$

in an inert solvent gives respectable yields of the metal radical coupling products. Thus, the $\sigma_b \to \pi^*L$ CT state is reactive enough to give fairly efficient chemistry within the excited-state lifetime.

Having labored to establish that metal–metal bond cleavage is the only primary excited-state reaction in the dinuclear Mn and Re complexes, it is now appropriate to discuss the mechanism of the photochemical synthesis

TABLE 2-57

Quantum Yields for Reaction of $[M_2(CO)_8(L)]$ Complexes[a,b]

Complex	$\Phi_{313\,nm}$[b]	$\Phi_{366\,nm}$	$\Phi_{436\,nm}$	$\Phi_{550\,nm}$	$\Phi_{633\,nm}$
$[Mn_2(CO)_8(phen)]$		0.85	1.10	0.93	
$[Re_2(CO)_8(phen)]$	0.15	0.22	0.19	0.13	
$[Re_2(CO)_8(biquin)]$		0.032	0.02	0.04	0.02

[a] Reprinted with permission from Morse and Wrighton [165], *J. Am. Chem. Soc.* **98**, 3931 (1976). Copyright by the American Chemical Society.

[b] Reactions carried out at 25°C in CH_2Cl_2/CCl_4 ($\frac{1}{2}$ by volume). The photoproducts in every case are stoichiometric yields of $[M(CO)_5Cl]$ and $[MCl(CO)_3L]$.

[c] All Φs are ±20%.

of simple substitution products. Wrighton and Ginley [2,162], first offered a mechanism involving thermal substitution at the radical stage as in Eq. (2-76). Their mechanism was based on the observation that $[Mn_2(CO)_8(PPh_3)_2]$

$$[Mn_2(CO)_{10}] \overset{hv}{\underset{}{\rightleftharpoons}} 2[Mn(CO)_5] \overset{PPh_3}{\underset{\Delta}{\longrightarrow}} 2[Mn(CO)_4PPh_3] + 2CO$$

$$[Mn_2(CO)_9PPh_3] \qquad [Mn_2(CO)_8(PPh_3)_2]$$

$$(2\text{-}76)$$

was the major *primary* photoproduct from irradiation of $[Mn_2(CO)_{10}]$ in the presence of PPh_3. The monosubstituted $[Mn_2(CO)_9PPh_3]$ was the minor ($\sim 5\%$) component of the two primary products. If the usual mechanism of CO photosubstitution were operating, $[Mn_2(CO)_9]$ would be the intermediate and $[Mn_2(CO)_9PPh_3]$ would be the principal product at short irradiation times. Work by Brown and co-workers has elegantly established that a number of 17-valence-electron radicals are substitution-labile [226].

$[Mn(CO)_5]$ and $[Re(CO)_5]$ have been ordered in terms of reactivity toward halogen atom donors [131]. Thus far, $[Re(CO)_5]$ has been determined to be the most reactive photogenerated, 17-valence-electron radical followed by $[Mn(CO)_5]$, *vide supra*. These radicals are about three orders of magnitude more reactive than $[Fe(\eta^5\text{-}C_5H_5)(CO)_2]$ toward CCl_4 or 1-iodopentane. The ordering of radical reactivity has depended on the consistency of the photochemistry and electronic structure of a number of heterodinuclear metal–metal bonded complexes where $[M(CO)_5]$ (M = Mn, Re) are involved [cf. Eq. (2-28); Tables 2-15 and 2-30; Fig. 2-37]. In addition to the dinuclear complexes which can be prepared photochemically, it is worth noting that irradiation of $[M_2(CO)_{10}]$ and $[Fe(CO)_5]$ yields $[(OC)_5MFe(CO)_4M(CO)_5]$ species [227]. The detailed mechanism is not clear but probably involves reaction of photogenerated $[Fe(CO)_4]$ and $[M(CO)_5]$.

Photoinduced metal–metal bond cleavage also obtains for $[H_3Re_3(CO)_{12}]$ which was shown [228] to convert smoothly and stoichiometrically to $[H_2Re_2(CO)_8]$ upon photolysis (Eq. 2-77). The 366 and 313 nm quantum yields are of the order of 0.1 in degassed solution. The quantum yields are independent of light intensity, indicating that the declusterification reaction occurs via a one-photon process. However, the quantum yield is significantly

$$2[H_3Re_3(CO)_{12}] \overset{hv}{\longrightarrow} 3[H_2Re_2(CO)_8] \qquad (2\text{-}77)$$

diminished by CO; for example, the 366 nm quantum yield measured under a CO atmosphere was smaller by a factor of five. Mechanisms for final product formation can be written which begin either with the cleavage of a Re–Re bond or with the loss of CO subsequent to the absorption of light, and the nature of the lowest excited states is such that either result could be explained. Such reactions merit further mechanistic attention in order to assess the relative importance of the two differing primary photoreactions.

    f. Photochemistry of a Metal Carbonyl Anion: $[Mn(CO)_5]^-$. The anions $[M(CO)_5]^-$ (M = Mn, Re) are d^8 species and, like $[Fe(CO)_5]$, adopt a D_{3h} structure in solution. Irradiation of $[Mn(CO)_5]^-$ has been found to lead to dissociative loss of CO with a high quantum yield [229]. In the presence of P-donor ligands, irradiation of $[Mn(CO)_5]^-$ leads to $[Mn(CO)_4L]^-$ products [229]. For L = PPh_3 the $[Mn(CO)_4L]^-$ species is not photosensitive with respect to CO loss, unlike $[Fe(CO)_4PPh_3]$ [230]. However, $[Mn(CO)_4P(OMe)_3]^-$ does appear to be photosensitive with respect to CO loss. Irradiation of $[Mn(CO)_4PPh_3]^-$ does result in substitution of PPh_3 with a quantum yield in the range of 0.02.

    The interesting find concerning the photochemistry of $[Mn(CO)_5]^-$ is that oxidative addition of certain cations obtains as in Eq. (2-78a). For

$$[EPh_4]^+ + [Mn(CO)_5]^- \xrightarrow[\text{THF}]{hv} cis\text{-}[PhMn(CO)_4EPh_3] + CO \qquad (2\text{-}78a)$$

$$E = As, P$$

$[PPh_3Me]^+$ only one of the possible products is found (Eq. 2-78b). The

$$[PPh_3Me]^+ + [Mn(CO)_5]^- \xrightarrow[\text{THF}]{hv} cis\text{-}[PhMn(CO)_4PPh_2Me] + CO \qquad (2\text{-}78b)$$

role of ion pairing must be quite great, since changing the solvent from THF to CH_3CN leads to a precipitous decline in the efficiency of such processes. Photoinduced oxidative addition of neutral oxidative addition substrates, $HSiPh_3$ and $Ph_3SnSnPh_3$, results in anionic products [229].

    g. Intraligand Photochemistry. The photochemistry of *fac*-$[ReX(CO)_3$ (*trans*-styrylpyridine)] involves trans → cis isomerization of the coordinated ligand via IL lowest excited states [162]. These species are treated in more detail in Chapter 3.

V. IRON, RUTHENIUM, AND OSMIUM CARBONYLS

A. Geometric Structure

    The commonly available carbonyls of iron include $[Fe(CO)_5]$ (VII) [231], $[Fe_2(CO)_9]$ (VIII) [232,233], and $[Fe_3(CO)_{12}]$ (IX) [234,235]. In

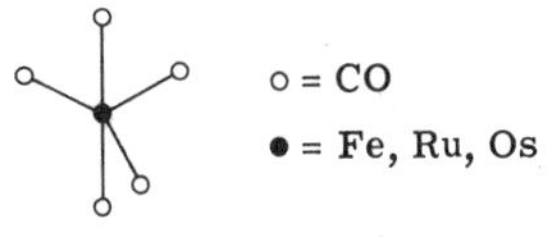

o = CO
● = Fe, Ru, Os

(VII)

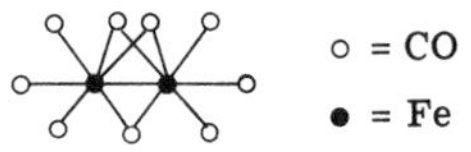

o = CO
● = Fe

(VIII)

o = CO
● = Fe

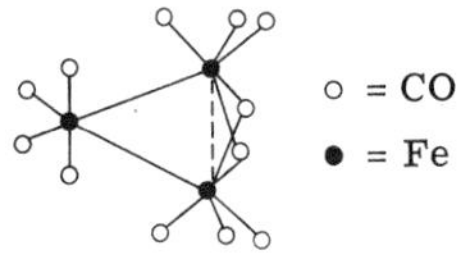

(IX)

each case the central metal is formally in a zero oxidation state. The mono-nuclear and dinuclear complexes for Ru and Os are known, but the most common complexes containing only the metal and CO are $[M_3(CO)_{12}]$ which have no bridging carbonyls as shown in (X) [236–239]. A large number of $[M(CO)_nL_{5-n}]$ complexes of Fe, Ru, Os are known.

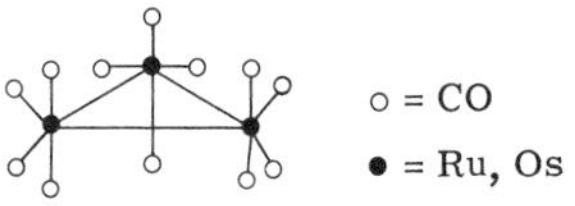

o = CO
● = Ru, Os

(X)

Cyclopentadienyl complexes containing Fe, Ru, Os, and CO are known and exist as dimeric species with a metal–metal bond. Again, the first-row Fe complex has bridging CO groups (XI) [240], while the second-row Ru has both the bridged and nonbridged forms and the Os complex has no bridging CO groups (XII) [240–243]. These complexes may be viewed as

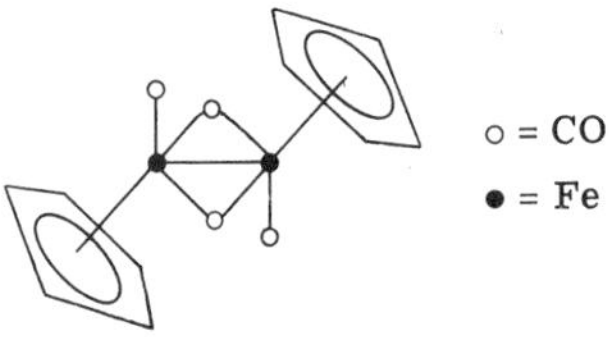

o = CO
● = Fe

o = CO
● = Ru, Os

(XI) 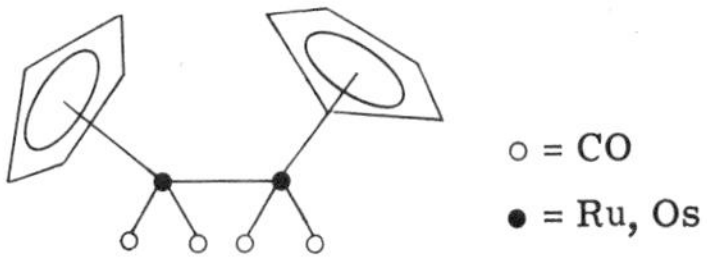   (XII)

having the central metal in a +1 oxidation state, but the complexes are diamagnetic by virtue of the direct M–M bond. Numerous $[Fe(\eta^5\text{-}C_5H_5)(CO)_n(L_{2-n})]$ complexes are known.

Higher oxidation state Fe, Ru, and Os carbonyl complexes do exist but have been of relatively little interest to photochemists thus far. Most mononuclear species of higher oxidation states seem to be six-coordinate complexes.

B. ELECTRONIC STRUCTURE

Even though the photochemistry of $[Fe(CO)_5]$ and its derivatives has been pursued vigorously for about 15 years, little effort has been devoted to a detailed study of electronic spectra. The spectrum of $[Fe(CO)_5]$ itself exhibits a shoulder in the vicinity of 40,000 cm^{-1} and another at 35,500 cm^{-1} but otherwise seems featureless [51,244]. Some UV–visible absorption data for Fe, Ru, and Os carbonyls are set out in Table 2-58 [51,244–253].

For the $[M(CO)_nL_{5-n}]$ complexes one might expect both CT and LF absorption. The d-orbital ordering in D_{3h} symmetry is shown in Fig. 2-48, and the first absorption maximum in $[Fe(CO)_5]$ has been identified as the $d_{xy}, d_{x^2-y^2} \to d_{z^2}$ transition [244]. Much of the intense high-energy absorption in $[M(CO)_nL_{5-n}]$ is logically associated with M $\to \pi^*CO$ and MLCT absorption.

No detailed spectral studies on the $[Fe(\eta^5-C_5H_5)(CO)_2X]$ species have been reported, but the available data [253] are consistent with LF lowest excited states. Likewise the lowest excited state in most $[Fe(CO)_3diene]$ species are probably LF states.

Metal–metal bonded complexes have received some attention; and, as in other dinuclear complexes which have direct metal–metal bonds, the lowest excited states involve the metal–metal σ_b and σ^* orbitals. A $\sigma_b \to \sigma^*$ absorption has been identified for $[Fe_2(\eta^5-C_5H_5)_2(CO)_4]$ at $\sim 29,000$ cm^{-1} [251]. This dinuclear complex is $>99\%$ bridged in solution, and it would be of interest to evaluate the $\sigma_b \to \sigma^*$ position for a nonbridged form, since the relative $\sigma_b \to \sigma^*$ positions have been correlated with M–M dissociation energies. The heterodinuclear $[Fe(\eta^5-C_5H_5)(CO)_2M(\eta^5-C_5H_5)(CO)_3]$ (M = Mo, W) are nonbridged in solution and exhibit $\sigma_b \to \sigma^*$ transitions [66] (Fig. 2-49) which are lower in energy than found for $[M_2(\eta^5-C_5H_5)_2(CO)_6]$ or $[Fe_2(\eta^5-C_5H_5)_2(CO)_4]$. Such information has been used to calculate [66] the $\sigma_b \to \sigma^*$ position for the nonbridged $[Fe_2(\eta^5-C_5H_5)_2(CO)_4]$ as $\sim 24,000$ cm^{-1}, or about 5,000 cm^{-1} lower than observed for the bridged species. Consistent with this result, the $\sigma_b \to \sigma^*$ absorptions for *both* the bridged and nonbridged forms of $[Ru_2(\eta^5-C_5H_5)_2(CO)_4]$ are observable [254]: the bridged $\sigma_b \to \sigma^*$ is at $\sim 38,000$ cm^{-1} and the nonbridged at $\sim 30,500$ cm^{-1} (Fig. 2-50). The spectrum for $[Ru_2(\eta^5-C_5H_5)_2(CO)_4]$ is very temperature-dependent, in accord with a bridged/nonbridged distribution of $\sim 50/50$ at 298°K and completely bridged at 77°K.

TABLE 2-58

UV–Visible Absorption Data for Iron, Ruthenium, and Osmium Carbonyls

Complex	Bands, cm^{-1} (ε)	Reference
$[Fe(CO)_5]$	~35,500 (3800)	[51,244]
$[Fe_3(CO)_{12}]$	~40,000 (40,000) 16,580 (2900) 22,270 sh (2380) 31,750 sh (12,400) 36,360 sh (17,700) ~52,000 (>70,000)	[245]
$[Ru_3(CO)_{12}]$	25,320 (7700) 37,310 sh (27,000) 41,840 (35,500) ~49,360 sh (48,000)	[245]
$[Os_3(CO)_{12}]$	25,970 sh (3700) 30,400 (9300) 34,720 sh (8500) 40,980 (26,000)	[245]
$[Fe(CO)_4]^{2-}$	33,000 (—)	[246]
$[Fe_2(CO)_8]^{2-}$	28,800 (8920)	[246]
$[Fe_3(CO)_{11}]^{2-}$	20,600 (3160)	[246]
$[Fe_4(CO)_{13}]^{2-}$	20,000 (4470)	[246]

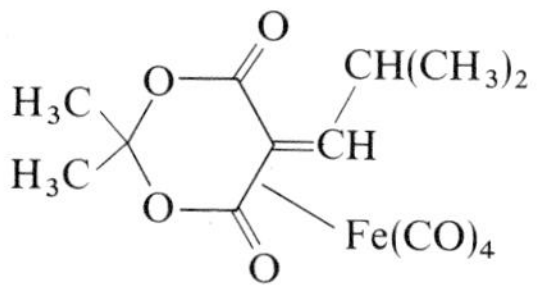

Structure (H$_3$C)$_2$C dioxdione ring with CH(CH$_3$)$_2$, =CH, Fe(CO)$_4$

29,000 (1600)
41,000 (12,000)   [247]

Structure (H$_3$C)$_2$C dioxdione ring with CH(CH$_3$)$_2$, =CH, Fe(CO)$_3$

24,000 (520)
30,700 (2400)   [247]

Structure: bicyclic =CH$_2$, =O, Fe(CO)$_3$

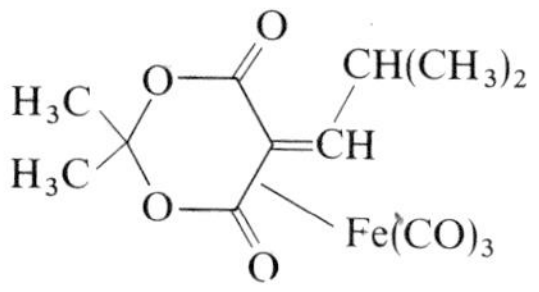

23,500 (550)
29,700 (1980)
42,200 (11,900)   [248]

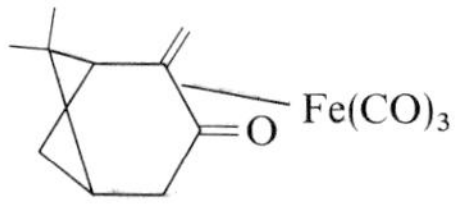

Structure: Ph, Ph, O, Fe(CO)$_3$

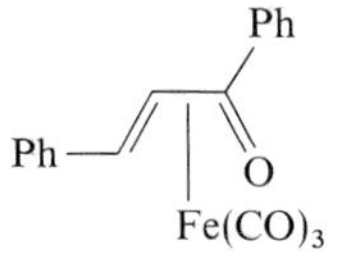

29,500 (5500)
32,500 (8600)
38,800 (22,400)   [248]

Structure: Ph, Ph, O, Fe(CO)$_4$

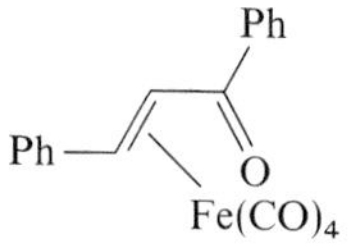

29,000 (4400)
33,200 (13,200)
40,000 (22,800)   [248]

(*continued*)

TABLE 2-58   (*continued*)

Complex	Bands, cm^{-1} (ε)	Reference
$[Fe(CO)_3Br]_2$	21,700 (900)	[249]
	28,300 (7600)	
$[Fe(CO)_3I]_2$	19,500 (700)	[249]
	27,400 (3200)	
$[Fe(1,3\text{-Butadiene})(CO)_3]$	36,000 (2460)	[51]
	$\sim 47,300$ (23,000)	
(benzoquinone)—Fe(CO)$_3$	28,400 (2820)	[250]
	33,800 (2760)	
	37,900 (5900)	
(cyclopentadienone)—Fe(CO)$_3$	29,400 (1780)	[250]
	33,600 (5000)	
$[Fe_2(\eta^5\text{-}C_5H_5)_2(CO)_4]$	$\sim 17,250$ sh (270)	[251]
	$\sim 19,800$ sh (430)	
	$\sim 24,000$ sh (1400)	
	29,000 (8240)	
$[Fe_2(\eta^5\text{-}C_5H_5)_2(CO)_3(P(OCH_3)_3)]$	17,900 (520)	[251]
	29,200 (4150)	
$[Ru_2(\eta^5\text{-}C_5H_5)_2(CO)_4]$	23,000 sh (600)	[251]
	30,300 (6300)	
	37,600 (8800)	
$[Fe_4(\eta^5\text{-}C_5H_5)_4(CO)_4]$	12,900 (2900)	[252]
	25,380 (18,300)	
$[Fe(\eta^5\text{-}C_5H_5)(CO)_2Cl]$	25,770 sh (565)	[253]
	29,760 (935)	
$[Fe(\eta^5\text{-}C_5H_5)(CO)_2Br]$	25,910 sh (700)	[253]
	28,570 (1028)	
$[Fe(\eta^5\text{-}C_5H_5)(CO)_2I]$	29,240 (2090)	[253]
$[Fe(\eta^5\text{-}C_5H_5)(CO)_2NCS]$	23,920 (795)	[253]
	29,410 (1590)	
	37,700 sh (4650)	
$[Fe(\eta^5\text{-}C_5H_5)(CO)_2SCN]$	19,050 (1094)	[253]
	28,985 (1604)	
	35,970 (7420)	
$[Fe(\eta^5\text{-}C_5H_5)(CO)(PPh_3)Br]$	16,287 (160)	[253]
	22,523 (775)	
	27,780 sh (785)	
$[Fe(\eta^5\text{-}C_5H_5)(CO)(PPh_3)I]$	16,155 (168)	[253]
	22,727 (784)	
	30,675 sh (2510)	
$[Fe(\eta^5\text{-}C_5H_5)(CO)(PPh_3)NCS]$	18,182 (362)	[253]
	22,989 (948)	
	32,468 (2430)	

FIG. 2-48.   d-Orbital diagram for d^8, D_{3h} complexes.

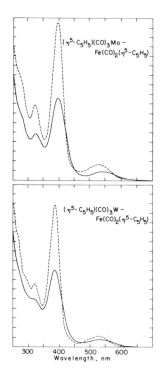

FIG. 2-49. Electronic absorption spectra in EPA solution at 298 (——) and 77°K (----). See Table 2-15 for absorptivities and band positions. Reprinted with permission from Abrahamson and Wrighton [66], *Inorg. Chem.* **17**, 1003 (1978). Copyright by the American Chemical Society.

Absorption spectra for $[M_3(CO)_{12}]$ (M = Fe, Ru, Os) in solution are shown in Fig. 2-51 [245]. The lowest energy absorption maximum is ordered as Os > Ru > Fe, perhaps reflecting the greater strength of M—M bonds in the heavier atom systems. The transitions logically involve M—M orbitals and terminate in antibonding (M—M) levels.

Other polynuclear Fe complexes have been examined. The series $[Fe(CO)_4]^{2-}$, $[Fe_2(CO)_8]^{2-}$, $[Fe_3(CO)_{11}]^{2-}$, $[Fe_4(CO)_{13}]^{2-}$ is particularly interesting and shows qualitatively that the position of the lowest absorption band depends on the number of metal atoms in the cluster (Table 2-58). The spectrum of the tetranuclear $[Fe_4(\eta^5\text{-}C_5H_5)_4(CO)_4]$ complex (Fig. 2-52) [252] also shows a very low energy absorption compared to $[Fe_2(\eta^5\text{-}C_5H_5)_2(CO)_2]$. Quite interestingly, the tetranuclear species is oxidizable and

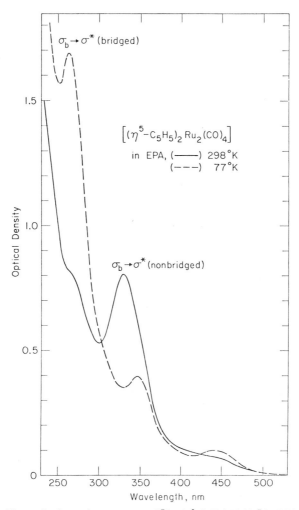

FIG. 2-50. Electronic absorption spectrum of $[Ru_2(\eta^5\text{-}C_5H_5)_2(CO)_4]$ in EPA at 298 (——) and 77°K (————) uncorrected for solvent contraction [254].

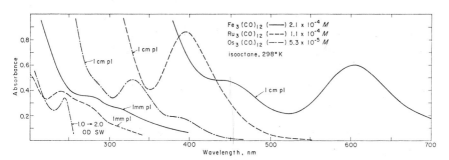

FIG. 2-51. Optical absorption spectra of $[M_3(CO)_{12}]$ complexes at 298°K in isooctane solution. Reprinted with permission from Austin *et al.* [245], *In* Advances in Chemistry Series No. 168 (M.S. Wrighton, ed.), p. 193 (1978).

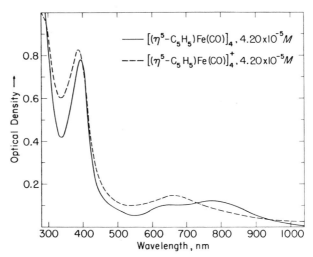

FIG. 2-52. Optical absorption spectra of Fe_4 clusters in CH_3CN at $25°C$ in a 1.00 cm pathlength cell. The cation is as the $PF_6{}^-$ salt. Reprinted with permission from Bock and Wrighton [252], *Inorg. Chem.* **16**, 1309 (1977). Copyright by the American Chemical Society.

reducible by one electron [255], but there are only relatively small electronic spectral changes accompanying this redox chemistry. This fact is consistent with a spectrum dominated by the Fe_4 core.

C. PHOTOREACTIONS

1. Substitution Reactions

a. $[M(CO)_nL_{5-n}]$ *Complexes.* It now appears that all the $[M(CO)_5]$ ($M = Fe, Ru, Os$) complexes are photosensitive with respect to substitution of CO. Dissociative decay of $[M(CO)_5]^*$ to yield $[M(CO)_4]$ is a certainty. The photochemical formation of $[Fe_2(CO)_9]$ from irradiation of $[Fe(CO)_5]$ is the earliest reported [256] photoreaction of a metal carbonyl; presumably $[Fe_2(CO)_9]$ forms via the reaction of the photogenerated $[Fe(CO)_4]$ species with a second $[Fe(CO)_5]$ molecule. Likewise, irradiation of $[Ru(CO)_5]$ or $[Os(CO)_5]$ in the absence of nucleophiles results in the formation of di- and trinuclear complexes [257–259].

In the presence of a wide variety of entering groups $[M(CO)_n(L)_{5-n}]$ complexes are formed from $[M(CO)_5]$ ($M = Fe, Ru, Os$), and some representative examples are set out in Table 2-59 [257,260–269]. The examples

TABLE 2-59

Representative Photosubstitution Chemistry of $[M(CO)_5]$ (M = Fe, Ru, Os)

Starting complex	Entering group, L	Product(s)	Reference
$[Fe(CO)_5]$	CF_2CFCl	$[Fe(CO)_4L]$	[260]
$[Fe(CO)_3[(MeO)_3P]_2]$	Fluoro olefin	$[Fe(CO)_2[(MeO)_3P]_2L]$	[261]
$[Ru(CO)_3[(MeO)_3P]_2]$	Fluoro olefin	$[Ru(CO)_2[(MeO)_3P]L_2]$	[261]
$[Ru(CO)_5]$	PPh_3	$[Ru(CO)_4L]$	[257]
$[Os(CO)_5]$	PPh_3	$[Os(CO)_4L]$	[257]
$[Fe(CO)_5]$	$F_2C{-}CF_2$ $(CH_3)_2AsC{=}CAs(CH_3)_2$	$[Fe(CO)_4L]$	[262]
$[Fe(CO)_5]$	$F_2C{-}CF_2$ $(Ph)_2PC{=}CP(Ph)_2$	$[Fe(CO)_3L]$	[262]
$[(1,3\text{-Butadiene})Fe(CO)_3]$	^{13}CO	$[(1,3\text{-Butadiene}) Fe(CO)_n(L)_{3-n}]$ $(n = 2,1,0)$	[263]
$[(1,3\text{-Butadiene})Fe(CO)_3]$	PF_3	$[(1,3\text{-Butadiene})Fe(CO)L_2]$ $(n = 2,1,0)$	[263]
$[Fe(CO)_2(PF_3)_3]$	PF_3	$[Fe(PF_3)_5]$	[265]
$[Fe(CO)_5]$	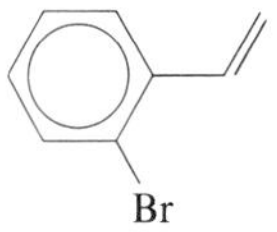 (2-bromostyrene)	$[Fe(CO)_4L]$ $[Fe(CO)_3]_2L$	[266]
$[Fe(CO)_5]$	$AsPh_3$ PPh_3	$[Fe(CO)_4L]$ $[Fe(CO)_3L_2]$	[267]
$[Fe(CO)_5]$	Pyridine Piperidine Acetonitrile $CH_2{=}CH{-}CN$ CN^- SCN^-	$[Fe(CO)_4L]$ (axial substitution)	[268] [269]

cited clearly demonstrate the fact that CO lability in $[M(CO)_nL_{5-n}]$ and its derivatives is achieved upon photolysis. Further, the work of Clark and co-workers [263–265] provides the now-familiar result that ligands most like CO will be incorporated to the highest degree. Various isotopically labeled CO groups have been incorporated into $[Fe(CO)_5]$ by photosubstitution [270].

The photochemistry of $[Fe(CO)_5]$ in low-temperature matrices has been elegantly elucidated, and a number of $[Fe(CO)_n]$ species have been identified under such conditions [271]. Recently, it has been claimed that the C_{2v} $[Fe(CO)_4]$ is paramagnetic when photogenerated in a matrix at low temperature [272,273]. Selective CO-laser-induced chemistry of $[Fe(CO)_4]$ [273,274] has proved useful in elucidating some structural and dynamic aspects of the species.

Photoinduced oxidative-addition reactions of $[Fe(CO)_5]$ via $[Fe(CO)_4]$ have been reported. Chemistry represented by Eq. (2-79) [116] and (2-80) [275] is typical.

$$[Fe(CO)_5] \xrightarrow[R_3SiH]{hv} cis\text{-}[HFe(CO)_4SiR_3] + CO \qquad (2\text{-}79)$$

$$R=Cl, Ph$$

$$[Fe(CO)_5] \xrightarrow[H_2CCHCH_2X]{hv} [(\eta^3\text{-}C_3H_3)Fe(CO)_3X] \qquad (2\text{-}80)$$

b.   Photochemistry of $[Fe(\eta^5\text{-}C_5H_5)(CO)_2X]$ *and* $[Fe(\eta^4\text{-}C_4H_4)(CO)_3]$. Both $[Fe(\eta^5\text{-}C_5H_5)(CO)_2X]$ and $[Fe(\eta^4\text{-}C_4H_4)(CO)_3]$ are photosensitive with respect to CO loss, and the primary photoprocess in each is dissociative loss of CO. There are several reports of simple photosubstitution of $[Fe(\eta^4\text{-}C_4H_4)(CO)_3]$ [276–279] and likewise for $[Fe(\eta^5\text{-}C_5H_5)(CO)_2X]$; some representative examples for $[Fe(\eta^5\text{-}C_5H_5)(CO)_2X]$ are given in Table 2-60 [118,122,181,280,281]. Table 2-61 [253] gives some quantum yield data for the photosubstitution of CO in $[Fe(\eta^5\text{-}C_5H_5)(CO)_2X]$

TABLE 2-60

Photosubstitution Reactions of $[Fe(\eta^5\text{-}C_5H_5)(CO)_2X]$

Starting material	Entering group, L	Product(s)	Reference
$[Fe(\eta^5\text{-}C_5H_5)(CO)_2I]$	Et_2NPF_2	$[Fe(\eta^5\text{-}C_5H_5)(CO)LI]$	[122]
		$[Fe(\eta^5\text{-}C_5H_5)(L)_2I]$	
	$C_5H_{10}NPF_2$	$[Fe(\eta^5\text{-}C_5H_5)(L)_2I]$	
	$[Ph_2PCH_2CH_2PPhCH_2]_2$	$[Fe(\eta^5\text{-}C_5H_5)L]^+I^-$	[181]
$[Fe(\eta^5\text{-}C_5H_5)(CO)_2Br]$	$C_5H_{10}NPF_2$	$[Fe(\eta^5\text{-}C_5H_5)(L)_2I]$	[122]
	$(Ph_2PCH_2CH_2)_2PPh$	$[Fe(\eta^5\text{-}C_5H_5)L]^+Br^-$	[118]
$[Fe(\eta^5\text{-}C_5H_5)(CO)_2Cl]$	$Ph_2PCH_2CH_2PPh_3$	$[Fe(\eta^5\text{-}C_5H_5)LCl]$	[118]
	$(Ph_2PCH_2CH_2)_2PPh$	$[Fe(\eta^5\text{-}C_5H_5)L]^+Cl^-$	[118]
$[Fe(\eta^5\text{-}C_5H_5)(CO)_2Sn(CH_3)_2]$	$SbPh_3$	$[Fe(\eta^5\text{-}C_5H_5)(CO)LSnPh_3]$	[280]
		$[Fe(\eta^5\text{-}C_5H_5)L_2SnPh_3]$	
$[Fe(\eta^5\text{-}C_5H_5)(CO)_2SnPh_3]$	$(CH_3)_2PPh$	$[Fe(\eta^5\text{-}C_5H_5)L_2SnPh_3]$	[280]
$[Fe_2(\eta^5\text{-}C_5H_5)_2(CO)_4]$	$Ph_2PC\equiv CPPh_2$	$[Fe_2(\eta^5\text{-}C_5H_5)(CO)_3L]$	[281]
	$P(OMe)_3$	$[Fe_2(\eta^5\text{-}C_5H_5)_2(CO)_3L]$	[254]

TABLE 2-61

Quantum Yields for Conversion of
$[Fe(\eta^5\text{-}C_5H_5)(CO)_2X]$ to
$[Fe(\eta^5\text{-}C_5H_5)(CO)(PPh_3)X]^a$

X	Irradiation, λ	Φ ($\pm 10\%$)^b
Br	366	0.087
	436	0.12
I	366	0.071
	436	0.071

^{a} Reprinted with permission from Alway and Barnett [253], In "Inorganic and Organometallic Photochemistry," Advances in Chemistry Series No. 168 (M.S. Wrighton, ed.), p. 120 (1978).
^{b} C_6H_6 solution, room temperature.

(X = Br, I). The photochemistry probably originates from the LF states as in the analogous $[Cr(\eta^6\text{-arene})(CO)_3]$ and $[Mn(\eta^5\text{-}C_5H_5)(CO)_3]$ species. There is no convincing evidence that either $\eta^4\text{-}C_4H_4$ or $\eta^5\text{-}C_5H_5$ is photolabile.

Interesting consequences of CO dissociation from $[Fe(\eta^5\text{-}C_5H_5)(CO)_2X]$ include intramolecular rearrangements as shown in reactions (2-81) [282]

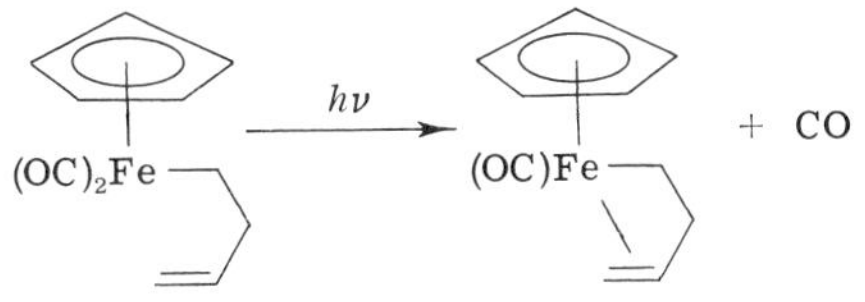

$$\text{(2-81)}$$

and (2-82) [283]. Additionally, it is known that the coordinately unsaturated

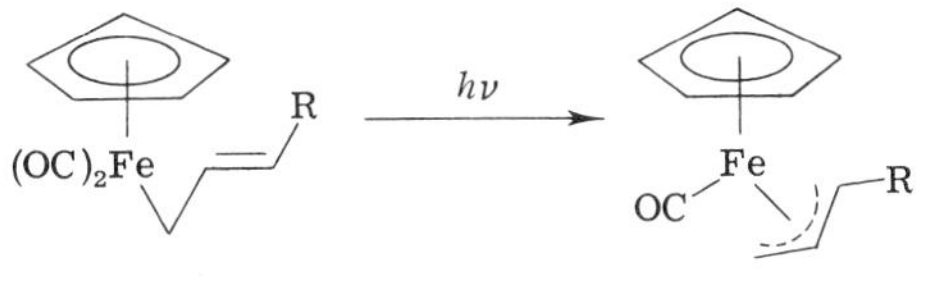

$$\text{(2-82)}$$

metal is susceptible to oxidative addition as shown in reaction (2-83) [116].

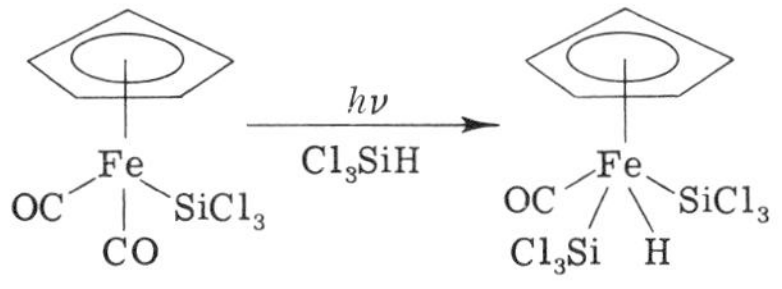

$$\text{(2-83)}$$

2. Photoreactions of Polynuclear Iron, Ruthenium, and Osmium Carbonyls

Excited-state reactions of polynuclear Fe, Ru, and Os carbonyls include both substitution and declusterification. Some recent low-temperature work with $[Fe_2(CO)_9]$ reveals in this case that loss of CO occurs to yield $[Fe_2(CO)_8]$ upon photolysis [284] according to Eq. (2-84). When the same

$$[Fe_2(CO)_9] \xrightarrow[\text{Ar, 20 K}]{h\nu} [Fe_2(CO)_8] + CO \tag{2-84}$$

reaction was carried out in a nitrogen matrix, evidence was obtained for the nitrogen complex $[Fe_2(CO)_8N_2]$. On the other hand, numerous mononuclear $[Fe(CO)_nL_{5-n}]$ complexes can be formed via the photolysis of $[Fe_2(CO)_9]$ in the presence of an entering group L. It is plausible that the mechanism of formation of the mononuclear species is first the generation of $[Fe_2(CO)_8]$ which scavenges L and subsequently decomposes, yielding $[Fe(CO)_4L]$ and a second $[Fe(CO)_4]$ which reacts with L. The bridging CO groups in $[Fe_2(CO)_9]$ may preclude efficient cleavage of the Fe—Fe bond by $\sigma_b \rightarrow \sigma^*$ excitation.

The $M_3(CO)_{12}$ (M = Fe, Ru, Os) clusters have all received some attention, and a variety of confusing observations have been made. In Table 2-62 [245,285–290] we present a number of examples of photodeclusterification involving conversion of the trinuclear carbonyls to dinuclear and mononuclear species. The primary photoprocess is very likely that represented by Eq. (2-85) [245]. The resulting fragment probably undergoes unimo-

$$\triangle \underset{\Delta}{\overset{h\nu}{\rightleftarrows}} \bigwedge \tag{2-85}$$

lecular thermal processes to give final products. The disappearance quantum yields are fairly low (<0.05) and probably reflect efficient re-formation of the M—M bond to regenerate $[M_3(CO)_{12}]$. Though a number of simple substitution products can be photochemically produced by irradiation of $[M_3(CO)_{12}]$ in the presence of ligands [291–293], the primary photoproduct is very likely that given in Eq. (2-85).

Photochemistry of dinuclear complexes containing the 17-valence-electron $[Fe(\eta^5\text{-}C_5H_5)(CO)_2]$ moiety attached to other metal radicals involves M—M bond cleavage [66,136]. Irradiation of $[Ru_2(\eta^5\text{-}C_5H_5)_2(CO)_4]$ also gives efficient, clean, cleavage of the M—M bond [254], despite the presence of bridging CO groups.

The tetranuclear species $[Fe_4(\eta^5\text{-}C_5H_5)_4(CO)_4]$ is inert with respect to M—M bond cleavage or CO loss. This species does undergo photooxidation in solutions of halocarbons via $Fe_4 \rightarrow$ solvent CT excited states as discussed in Chapter 5 [252].

TABLE 2-62

Photodeclusterification of $[M_3(CO)_{12}]$ Complexes

Starting carbonyl	Entering group	Product(s)	Reference
$[Os_3(CO)_{12}]$	R_3SiH $R = CH_3, C_2H_5$	$[R_3SiOs(CO)_4H]$ $[(R_3Si)_2Os(CO)_4]$ $[R_3SiOs(CO)_4]_2$ $[(CH_3)_3SnOs(CO)_4H]$ $[(CH_3)_3GeOs(CO)_4H]$	[285–287]
	1,5-Cyclooctadiene	$[(1,3\text{-Cyclooctadiene})Os(CO)_3]$ $[(1,5\text{-Cyclooctadiene})Os(CO)_3]$	[288]
	1,3-Cyclooctadiene	$[(1,3\text{-Cyclooctadiene})Os(CO)_3]$	
$[Ru_3(CO)_{12}]$	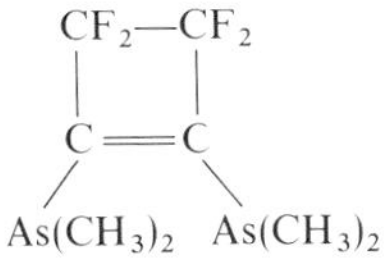 $\equiv$ ffars	$[Ru_2(CO)_6(\text{ffars})]$	[289]
	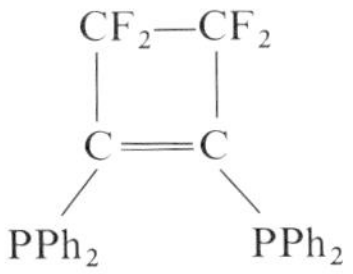 $\equiv$ ffos	$[Ru_2(CO)_6(\text{ffos})]$ $[Ru(CO)_3(\text{ffos})]$	
	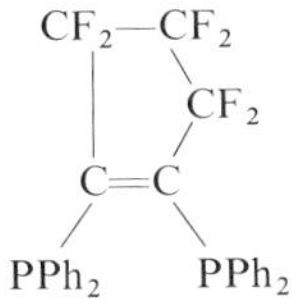 $\equiv$ f$_6$fos	$[Ru_2(CO)_6(\text{f}_6\text{fos})]$ $[Ru(CO)_3(\text{f}_6\text{fos})]$	
	$(CH_3)_3SiH$ $(CH_3)_3GeH$ Cl_3GeH $(CH_3)_5Si_2H$	$[((CH_3)_3Si)_2Ru(CO)_4]$ $[(CH_3)_3GeRu(CO)_4]_2$ $cis\text{-}[(Cl_3Ge)_2Ru(CO)_4]$ $[(CH_3)_3SiRu(CO)_3\text{-}$ $\mu\text{-}Si(CH_3)_2]_2$	[285–287]
	PPh_3	$[Ru(CO)_3(PPh_3)_2]$ $[Ru(CO)_4(PPh_3)]$	[290]
	1-Pentene	$[Ru(CO)_4(\text{Alkene})]$	[245]
	CO	$[Ru(CO)_5]$	[290]
$[Fe_3(CO)_{12}]$	CO	$[Fe(CO)_5]$	[245]
	1-Pentene	$[Fe(CO)_4(1\text{-Pentene})]$	[245]

VI. COBALT, RHODIUM, AND IRIDIUM CARBONYLS

A. GEOMETRIC AND ELECTRONIC STRUCTURES

Dicobaltoctacarbonyl has the structure in (XIII) according to an x-ray determination [294] but in solution exists in equilibrium with (XIV) [295]. The central metal is zerovalent and is formally d^9, but the complex is dia-

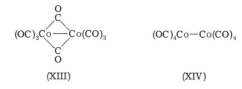

(XIII)   (XIV)

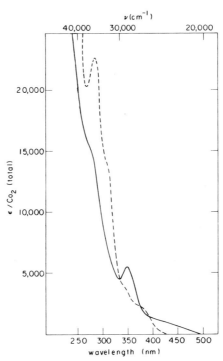

FIG. 2-53. Electronic spectra of $[Co_2(CO)_8]$ in 2-methylpentane solutions: at $298°K$ (——); after cooling to $50°K$ (– – – –). The spectral changes are not corrected for solvent contraction. Reprinted with permission from Abrahamson et al. [67], Inorg. Chem. **16**, 1554 (1977). Copyright by the American Chemical Society.

TABLE 2-63

Electronic Spectral Features of Dinuclear Co—Co Bonded Molecules[a]

Complex	v, cm^{-1}	ε, M^{-1} cm^{-1b}	Assignment
$[Co_2(CO)_8]$	23,470[c]	980	$d\pi \to \sigma^*$ (nonbridged)
	28,570[c]	5500	$\sigma_b \to \sigma^*$ (nonbridged)
	35,460[c]	14,400	$\sigma_b \to \sigma^*$ (bridged)
	26,460[d]	2200	$d\pi \to \sigma^*$ (bridged)
	(29,000)[d]	(3700)	e
	(32,100)[d]	(13,500)	e
	35,460[d]	22,600	$\sigma_b \to \sigma^*$ (bridged)
$[Co_2(CO)_6(PPh_3)_2]^f$	22,200 (sh)	5300	$d\pi \to \sigma^*$
	25,450	23,800	$\sigma_b \to \sigma^*$
$[Co_2(CO)_6(P(OMe)_3)_2]^g$	20,800[h]	h	$d\pi \to \sigma^*$
	27,780[g]	24,000[g]	$\sigma_b \to \sigma^*$

[a] Reprinted with permission from Abrahamson *et al.* [67], *Inorg. Chem.* **16**, 1554 (1977). Copyright by the American Chemical Society.
[b] In 2-methylpentane; ε/Co_2 (total) values are reported.
[c] 298°K.
[d] 50°K; uncorrected for solvent contraction.
[e] Not assigned.
[f] In dichloromethane at 298°K.
[g] In isooctane at 298°K.
[h] Shoulder, observable only in low-temperature spectrum; cf. Fig. 2-54.

magnetic because of the Co—Co bond. As in the $[Ru_2(\eta^5\text{-}C_5H_5)_2(CO)_4]$ described previously, the $[Co_2(CO)_8]$ exhibits a $\sigma_b \to \sigma^*$ excitation which is highly dependent on the temperature (Fig. 2-53) [67]. The $\sigma_b \to \sigma^*$ excitation shifts ~ 7000 cm^{-1} in going from the nonbridged to the bridged form (Table 2-63) [67]. By way of contrast, simple nonbridged dinuclear Co complexes exhibit the usual $\sigma_b \to \sigma^*$ (Table 2-63 and Fig. 2-54) [67]. The one-electron orbital diagram appropriate to $[Co_2(CO)_6L_2]$ is shown in Fig. 2-55. Heterodinuclear metal–metal bonded species involving $[Co(CO)_nL_{4-n}]$ fragments exhibit the usual $\sigma_b \to \sigma^*$ absorption (Table 2-15 and Fig. 2-56). The binary $[Co_4(CO)_{12}]$ exists and has a tetrahedrane-like Co_4 core.

Tricobalt species exhibit intense visible absorptions which can be associated with the Co_3 core (Fig. 2-57 and Table 2-64) [296]. As in the trinuclear Fe, Ru, and Os complexes, these absorptions can be associated with transitions terminating in antibonding (M—M) orbitals.

The $[M(\eta^5\text{-}C_5H_5)(CO)_2]$ (M = Co, Rh, Ir) species are complexes of the M in a d^8 configuration where the metal is in a formal +1 oxidation state. The electronic spectra of the complexes have not been assigned, but the colors of the complexes are red, orange, and yellow for the Co, Rh, and Ir,

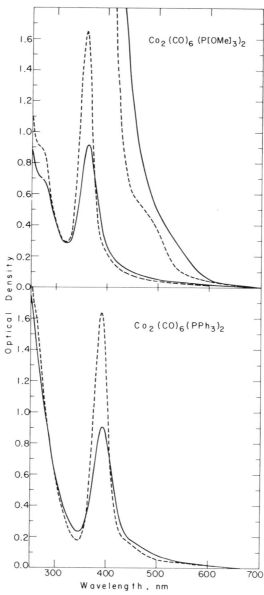

FIG. 2-54. Electronic spectra of $[Co_2(CO)_6(P(OMe)_3)_2]$ in EPA solution and $[Co_2(CO)_6 (PPh_3)_2]$ in 2-methyltetrahydrofuran solution: at 298°K (——), after cooling to 77°K (----). The high optical density in the $[Co_2(CO)_6(P(OMe)_3)_2]$ spectrum is for a higher concentration. Changes in the spectra upon cooling are not corrected for solvent contraction. Reprinted with permission from Abrahamson *et al.* [67], *Inorg. Chem.* **16**, 1554 (1977). Copyright by the American Chemical Society.

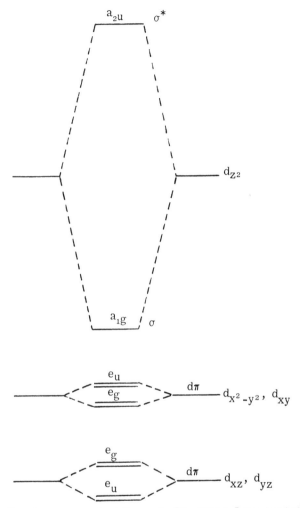

FIG. 2-55.   One-electron energy level diagram for $[Co_2(CO)_6L_2]$. Reprinted with permission from Abrahamson *et al.* [67], *Inorg. Chem.* **16**, 1554 (1977). Copyright by the American Chemical Society.

respectively, consistent with the expected increase of LF splittings going down the group.

Finally, of interest here is the four-coordinate $[Co(CO)_3NO]$ complex which may be viewed as isoelectronic with $[Ni(CO)_4]$ having a d^{10} configuration. Electron diffraction studies [297] indicate a tetrahedral arrangement of the four ligands surrounding the central metal.

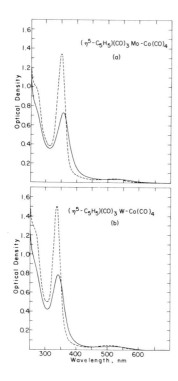

FIG. 2-56.   Electronic absorption spectra in EPA solution at 298 (——) and 77°K (————). Reprinted with permission from Abrahamson and Wrighton [66], *Inorg. Chem.* **17**, 1003 (1978). Copyright by the American Chemical Society.

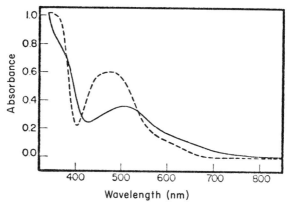

FIG. 2-57.   Electronic absorption spectra of [CH$_3$CCo$_3$(CO)$_9$] at 300 (——) and 77°K (————) in EPA solution. Reprinted with permission from Geoffroy and Epstein [296], *Inorg. Chem.* **16**, 2795 (1977). Copyright by the American Chemical Society.

TABLE 2-64

Electronic Spectra of Trinuclear Co Complexes[a]

Cluster	300°K[b]		77°K[c]	
	λ_{max}, nm	ε_{max}	λ_{max}, nm	ε_{max}
$[HCCo_3(CO)_9]$	505	1590	470	2240
	370 sh	3740	360	3780
	290 sh	13,500		
$[CH_3CCo_3(CO)_9]$	510	1840	475	2430
			448	2320
	370 sh	3680	360	3750
	295 sh	13,600		
$[CH_3CCo_3(CO)_8PPh_3]$	502[c]	2490	485	3830
			466	3770
	400 sh	5320	390	8080
	310 sh	16,170		
$[HFeCo_3(CO)_{12}]$	528	3980	591[e]	
			525	5150
	380 sh	6940	377	9510
	324 sh	11,000	324	13,270
	280 sh	12,400		
$[HFeCo_3(CO)_{10}(PPh_3)_2]$	680 sh[d]	2450	660	4070
	592	4590	580	7790
	381	12,140	380	17,370
	308	16,350	305	16,290
$Na[Co_3(CO)_{10}]$	505 sh			
	371 sh			

[a] Reprinted with permission from Geoffroy and Epstein [296], *Inorg. Chem.* **16**, 2795 (1977). Copyright by the American Chemical Society.

[b] 2,2,4-Trimethylpentane solution.

[c] Diethyl ether/isopentane/ethyl alcohol (5:5:2) solution.

[d] CH_2Cl_2 solution.

[e] Diethyl ether/isopentane (1:1) solution.

B. PHOTOCHEMISTRY OF COBALT, RHODIUM, AND IRIDIUM CARBONYLS

1. Substitution Reactions of Mononuclear Species

Examples of photosubstitution reactions of cobalt carbonyls are presented in Table 2-65 [111,298–301]. The entering groups in each case are all good π-acceptor ligands, and with PF_3 all of the CO groups can be replaced in either $[Co(CO)_3NO]$ [299] or $[RCo(CO)_4]$ [300]. The examples in Table

TABLE 2-65

Photosubstitution Reactions of Mononuclear Cobalt Carbonyls

Starting carbonyl	Entering group, L	Product(s)	Reference
$[Co(CO)_3(NO)]$	PPh_3	$[Co(CO)_2(NO)L]$	[298]
	PF_3	$[Co(CO)_{3-n}(NO)(L)_n]$ ($n = 1, 2, 3$)	[299]
	*CO	$[Co(*CO)_3(NO)]$	[111]
$[RCo(CO)_4]$ R = alkyl	PF_3	$[RCo(CO)_{4-n}(L)_n]$ ($n = 1, 2, 3, 4$)	[300]
$[Co(\eta^5\text{-}C_5H_5)(CO)_2]$	Cyclooctatetraene	$[CoL(\eta^5\text{-}C_5H_5)]$	[301]

2-65 include systems of d^8, d^9, and d^{10} configuration. Though studies with the second- and third-row systems are not common, there are some results which suggest that similar photoprocesses obtain. The $[Rh(\eta^5\text{-}C_5H_5)(CO)_2]$ complex apparently undergoes loss of CO upon photolysis, giving reactions (2-86) [302] and (2-87) [303]. The dinuclear Co analog of the photoproduct

$$[Rh(\eta^5\text{-}C_5H_5)(CO)_2] \rightarrow [Rh_2(\eta^5\text{-}C_5H_5)_2(CO)_3] + CO \qquad (2\text{-}86)$$

$$[Rh(\eta^5\text{-}C_5H_5)(CO)_2] \xrightarrow[R_3SiH]{hv} [Rh(\eta^5\text{-}C_5H_5)(CO)(H)(SiR_3)] + CO \qquad (2\text{-}87)$$

in reaction (2-86) has also been prepared photochemically [304]. The oxidative addition to the coordinatively unsaturated intermediate should generally yield a fairly stable complex, since the low-spin d^6 configuration will be achieved. An interesting example of this is indicated in Eq. (2-88) where the Co(I) changes to the substitution-inert Co(III) state [305].

$$\qquad \xrightarrow{hv} \qquad + \ 2CO \ + \ CO_2 \qquad (2\text{-}88)$$

2. Photoreaction of Di- and Tricobalt Clusters

Since $[M(\eta^5\text{-}C_5H_5)(CO)_3Co(CO)_4]$ can be prepared by the photoreaction [66,67] indicated in Eq. (2-89) (M = Mo, W), the $[Co_2(CO)_8]$ species presumably undergoes Co—Co bond cleavage subsequent to electronic

$$[M_2(\eta^5\text{-}C_5H_5)_2(CO)_6] + [Co_2(CO)_8] \underset{}{\overset{hv}{\rightleftharpoons}} [M(\eta^5\text{-}C_5H_5)(CO)_3Co(CO)_4] \qquad (2\text{-}89)$$

$$M = Mo, W$$

excitation. However, low-temperature photolysis does give some indication that $[Co_2(CO)_7]$ can be generated photochemically [306]. Heterodinuclear complexes involving the $[Co(CO)_4]$ fragment do seem to give high quantum yields for M—Co cleavage [66]. Net photosubstitution [307] of CO in $[Co_2(CO)_8]$ probably occurs via substitution-labile $[Co(CO)_4]$ radicals [308].

Tricobalt species having the Co_3C-tetrahedrane core are photosensitive under H_2 and give rise to reactions according to Eq. (2-90) [296].

$$[HCCo_3(CO)_9] \begin{array}{c} \xrightarrow[\text{H}_2]{hv} CH_4 + [Co_4(CO)_{12}] \\ \downarrow_{hv, \text{ CO}} \\ \xrightarrow[\text{H}_2/\text{CO}]{hv} CH_4 + [Co_2(CO)_8] \end{array} \qquad (2\text{-}90)$$

The quantum yield of $\sim 10^{-2}$ is similar to that for the trinuclear Fe, Ru, and Os carbonyls. Interestingly, the Co_3C-based systems are photoinert in the absence of H_2. It is noteworthy that the Co_4-based species $[Co_4(CO)_{12}]$ undergoes photodeclusterification when irradiated under CO. All of these reactions probably involve the cleavage of a Co—Co bond as the primary excited-state reaction. The intermediate "diradical" is then scavenged and proceeds to give net chemistry or simply back-reacts to starting material.

VII.   NICKEL CARBONYL

Nickel carbonyl $[Ni(CO)_4]$ has a d^{10} electronic configuration and is of tetrahedral geometry [309]. The absorption spectrum [26] is banded, exhibiting shoulders at 42,270 and 44,500 cm^{-1} and a maximum at 48,550 cm^{-1} ($\varepsilon \approx 10^5$). These transitions have been assigned as Ni $\rightarrow$ COπ^* transitions of the sort $t_2 \rightarrow$ COπ^* or $e \rightarrow$ COπ^* [26]. Owing to the considerable evidence mounting *against* such excited states being substitution-labile, we offer the explanation that the $[Ni(CO)_4]$ has absorptions which have d $\rightarrow$ s(σ^*) character. This explanation seems possible since the observed absorptions are at fairly high energy.

Consistent with an excited state with considerable substitution lability, photolysis of $[Ni(CO)_4]$ in the gas phase [310], solution [311], or the solid state [312] results in CO dissociation (Eq. 2-91). The quantum yield of

$$[Ni(CO)_4] \xrightarrow{hv} [Ni(CO)_3] + CO \qquad (2\text{-}91)$$

formation of CO is somewhat dependent on the excitation wavelength in
n-hexane, increasing from 0.22 at 366 nm to 0.50 near 240 nm. At 15°K, the
$[Ni(CO)_3]$ intermediate has C_{3v} symmetry from IR measurements [312].
Elucidation of the photoprocesses in substituted nickel carbonyls has not
been reported. However, photoassisted reactions of olefins have been carried out using derivatives of $[Ni(CO)_4]$ [313].

REFERENCES

1. F. A. Cotton and G. Wilkinson, "Advanced Inorganic Chemistry," 3rd ed., pp. 683–721.
    Wiley (Interscience), New York, 1972; I. Wender and P. Pino, "Organic Synthesis via
    Metal Carbonyls," Vol. 1, pp. 1–273. Wiley (Interscience), New York, 1968; E. W. Abel
    and F. G. A. Stone, *Q. Rev., Chem. Soc.* **24**, 498 (1970), and references therein.
2. M. Wrighton, *Chem. Rev.* **74**, 401 (1974); *Top. Curr. Chem.* **65**, 37 (1976).
3. A. Vogler, in "Concepts of Inorganic Photochemistry" (A. W. Adamson and P. D.
    Fleischauer, eds.), Chapter 6. Wiley, New York, 1975.
4. V. Balzani and V. Carassiti, eds., "Photochemistry of Coordination Compounds,"
    pp. 323–348. Academic Press, New York, 1970.
5. E. Koerner von Gustorf and F.-W. Grevels, *Fortschr. Chem. Forsch.* **13**, 366 (1969).
6. W. Strohmeier, *Angew. Chem.* **76**, 873 (1964); *Fortschr. Chem. Forsch.* **10**, 306 (1968).
7. C. R. Bock and E. A. Koerner von Gustorf, *Adv. Photochem.* **10**, 221 (1977).
8. J. J. Turner, *Angew. Chem., Int. Ed. Engl.* **14**, 304 (1975).
9. J. E. Ellis, Ph.D. Thesis, Massachusetts Institute of Technology, Cambridge (1971).
10. N. A. Beach and H. B. Gray, *J. Am. Chem. Soc.* **90**, 5713 (1968).
11. M. S. Wrighton, D. I. Handeli, and D. L. Morse, *Inorg. Chem.* **15**, 434 (1976).
12. M. S. Wrighton, unpublished results.
13. G. Natta, R. Ercoli, F. Calderazzo, A. Alberola, P. Corradini, and G. Allegra, *Atti
    Accad. Naz. Lincei, Cl. Sci. Fis., Mat. Nat., Rend.* [8] **27**, 107 (1959).
14. R. Ercoli, F. Calderazzo, and A. Alberola, *J. Am. Chem. Soc.* **82**, 2966 (1960); H. Haas
    and R. K. Sheline, *ibid.* **88**, 3219 (1966); D. W. Pratt and R. J. Myers, *ibid.* **89**, 6470 (1967);
    R. L. Pruett and J. E. Wyman, *Chem. Ind. (London)* p. 119 (1960); Calderazzo, R. Cini,
    P. Corradini, R. Ercoli, and G. Nata, *ibid.* p. 500; D. G. Schmidling, *J. Mol. Struct.* **24**,
    1 (1975); R. E. Sullivan, M. S. Lupin, and R. W. Kiser, *Chem. Commun.* p. 655 (1969);
    S. Evans, J. C. Green, A. F. Orchard, T. Saito, and D. W. Turner, *Chem. Phys. Lett.* **4**,
    361 (1969).
15. T. C. DeVore and H. F. Franzen, *Inorg. Chem.* **15**, 1318 (1976).
16. T. A. Ford, H. Huber, W. Klotzbucher, M. Moskovits, and G. A. Ozin, *Inorg. Chem.* **15**,
    1666 (1976).
17. H. J. Keller, P. Laubereau, and D. Nothe, *Z. Naturforsch., Teil B* **24**, 257 (1969); J. C.
    Bernier and O. Kahn, *Chem. Phys. Lett.* **19**, 414 (1973).
18. A. Davison and J. E. Ellis, *J. Organomet. Chem.* **31**, 239 (1971).
19. P. S. Braterman and A. Fullarton, *J. Organomet. Chem.* **31**, C27 (1971).
20. A. N. Nesmeyanov, *Adv. Organomet. Chem.* **10**, 56 (1972); E. O. Fischer, E. Louis, and
    R. J. J. Schneider, *Angew. Chem., Int. Ed Engl.* **7**, 136 (1968); E. O. Fischer and R. J. J.
    Schneider, *Angew. Chem.* **79**, 537 (1967); R. Tsumura and N. Hagihara, *Bull. Chem. Soc.
    Jpn.* **38**, 1901 (1965).

21. L. O. Brockway, R. V. G. Ewens, and M. W. Lister *Trans. Faraday Soc.* **34**, 1350 (1938); F. A. Cotton and C. S. Kraihanzel, *J. Am. Chem. Soc.* **84**, 4432 (1962).
22. W. Hieber, W. Beck, and G. Braun, *Angew. Chem.* **72**, 795 (1960).
23. L. B. Handy, J. K. Ruff, and L. F. Dahl, *J. Am. Chem. Soc.* **92**, 7312 (1970).
24. F. A. Cotton and G. Wilkinson, "Advanced Inorganic Chemistry," 3rd ed., pp. 968–969. Wiley (Interscience), New York, 1972.
25. H. B. Gray and N. A. Beach, *J. Am. Chem. Soc.* **85**, 2922 (1963).
26. A. F. Schreiner and T. L. Brown, *J. Am. Chem. Soc.* **90**, 3366 (1968).
27. B. N. Figgis, "Introduction to Ligand Fields," p. 158. Wiley, New York, 1966.
28. M. Wrighton, H. B. Gray, and G. S. Hammond, *Mol. Photochem.* **5**, 165 (1973).
29. J. K. Burdett, M. A. Graham, R. N. Perutz, M. Poliakoff, A. J. Rest, J. J. Rurner, and R. F. Turner, *J. Am. Chem. Soc.* **97**, 4805 (1975); R. N. Perutz and J. J. Turner, *Inorg. Chem.* **14**, 262 (1975); M. A. Graham, A. J. Rest, and J. J. Turner, *J. Organomet. Chem.* **24**, C54 (1970); M. A. Graham, R. N. Perutz, M. Poliakoff, and J. J. Turner, *ibid.* **34**, C34 (1972); M. A. Graham, M. Poliakoff, and J. J. Turner, *J. Chem. Soc. A* p. 2939 (1971).
30. J. D. Black and P. S. Braterman, *J. Am. Chem. Soc.* **97**, 2908 (1975).
31. E. P. Kündig and G. A. Ozin, *J. Am. Chem. Soc.* **96**, 3820 (1974).
32. R. N. Perutz and J. J. Turner, *J. Am. Chem. Soc.* **97**, 4791 (1975).
32a. J. K. Burdett, R. N. Perutz, M. Poliakoff, and J. J. Turner, *Chem. Commun.* p. 157 (1975).
33. M. Wrighton, G. S. Hammond, and H. B. Gray, *J. A. Chem. Soc.* **93**, 4336 (1971).
34. K. Mitteilung, *Z. Phys. Chem. (Frankfurt am Main)* **27**, 439 (1961).
35. P. S. Braterman and A. P. Walker, *Discuss. Faraday Soc.* **47**, 121 (1969).
36. M. Wrighton, *Inorg. Chem.* **13**, 905 (1974).
37. M. S. Wrighton, D. L. Morse, H. B. Gray, and D. K. Ottesen, *J. Am. Chem. Soc.* **98**, 1111 (1976).
38. M. S. Wrighton, H. B. Abrahamson, and D. L. Morse, *J. Am. Chem. Soc.* **98**, 4105 (1976).
39. P. Ford, De, F. P. Rudd, R. Gaunder, and H. Taube, *J. Am. Chem. Soc.* **90**, 1187 (1968).
40. M. Y. Darensbourg and D. J. Darensbourg, *Inorg. Chem.* **9**, 32 (1970).
41. C. P. Casey and T. J. Burkhardt, *J. Am. Chem. Soc.* **95**, 5833 (1973).
42. M. Wrighton, G. S. Hammond, and H. B. Gray, *J. Organomet. Chem.* **70**, 283 (1974).
43. M. Herberhold, *Angew. Chem., Int. Ed. Engl.* **7**, 305 (1968).
44. M. Wrighton, G. S. Hammond, and H. B. Gray, *Mol. Photochem.* **5**, 179 (1973).
45. H. Saito, J. Fujita, and K. Saito, *Bull. Chem. Soc. Jpn.* **41**, 359 (1968).
46. H. Saito, J. Fujita, and K. Saito, *Bull. Chem. Soc. Jpn.* **41**, 863 (1968).
47. Y. Kaizi, I. Fujita, and H. Kobayashi, *Z. Phys. Chem. (Frankfurt am Main)* **79**, 298 (1972).
48. M. S. Wrighton and D. L. Morse, *J. Organomet. Chem.* **97**, 405 (1975); H. B. Abrahamson and M. S. Wrighton, *Inorg. Chem.*, **17**, 3385 (1978).
49. L. Pdungsap and M. S. Wrighton, *J. Organomet. Chem.* **127**, 337 (1977).
50. D. G. Carroll and S. P. McGlynn, *Inorg. Chem.* **7**, 1285 (1968).
51. R. T. Lundquist and M. Cais, *J. Org. Chem.* **27**, 1167 (1962).
52. S. Fredericks, L. Pdungsap, and M. S. Wrighton, to be submitted for publication.
53. M. S. Wrighton and J. L. Haverty, *Z. Naturforsch., Teil B* **30**, 254 (1975).
54. L. Lang, ed., "Absorption Spectra in the UV and Visible Region," Vol. 9, pp. 13–16. Academic Press, New York, 1967.
55. M. S. Wrighton and D. S. Ginley, *J. Am. Chem. Soc.* **97**, 4246 (1975).
56. A. R. Burkett, T. J. Meyer, and D. G. Whitten, *J. Organomet. Chem.* **67**, 67 (1974).
57. J. L. Hughey, IV and T. J. Meyer, *Inorg. Chem.* **14**, 947 (1975).
58. H. Brunner and W. A. Hermann, *J. Organomet. Chem.* **74**, 423 (1974).

59. H. tomDieck and I. W. Renk, *Angew. Chem., Int. Ed. Engl.* **9**, 793 (1970).
60. A. T. T. Hsieh and B. O. West, *J. Organomet. Chem.* **78**, C40 (1974).
61. R. A. Levenson and H. B. Gray, *J. Am. Chem. Soc.* **97**, 6042 (1975).
62. D. C. Harris and H. B. Gray, *J. Am. Chem. Soc.* **97**, 3073 (1975).
63. J. R. Johnson, R. J. Ziegler, and W. M. Risen, Jr., *Inorg. Chem.* **12**, 2349 (1973).
64. D. S. Ginley, C. R. Bock, and M. S. Wrighton, *Inorg. Chim. Acta* **23**, 85 (1977).
65. D. S. Ginley and M. S. Wrighton, *J. Am. Chem. Soc.* **97**, 4908 (1975).
66. H. B. Abrahamson and M. S. Wrighton, *Inorg. Chem.* **17**, 1003 (1978).
67. H. B. Abrahamson, C. C. Frazier, D. S. Ginley, H. B. Gray, J. Lilienthal, D. R. Tyler, and M. S. Wrighton, *Inorg. Chem.* **16**, 1554 (1977).
68. D. Dewit, J. P. Fawcett, A. J. Poe, and M. V. Twigg, *Coord. Chem. Rev.* **8**, 81 (1972); J. P. Fawcett, A. J. Poe, and M. V. Twigg, *Chem. Commun.* p. 267 (1973).
69. R. Klingler, W. Butler, and M. D. Curtis, *J. Am. Chem. Soc.* **97**, 3535 (1975).
70. R. D. Adams, D. M. Collins, and F. A. Cotton, *Inorg. Chem.* **13**, 1086 (1974).
71. W. I. Bailey, Jr., F. A. Cotton, J. D. Jamerson, and J. R. Kobb, *J. Organomet. Chem.* **121**, C23 (1976); W. I. Bailey, Jr., D. M. Collins, and F. A. Cotton, *ibid.* **135**, C53 (1977).
72. D. S. Ginley, C. R. Bock, M. S. Wrighton, B. Fischer, D. L. Tipton, and R. Bau, *J. Organomet. Chem.* **157**, 41 (1978).
73. M. Wrighton, G. S. Hammond, and H. B. Gray, *Inorg. Chem.* **11**, 3122 (1972).
74. P. D. Fleischauer and P. Fleischauer, *Chem. Rev.* **70**, 199 (1970).
75. A. G. Massey and L. E. Orgel, *Nature (London)* **191**, 1387 (1961).
76. I. W. Stolz, G. R. Dobson, and R. K. Sheline, *J. Am. Chem. Soc.* **84**, 3589 (1962); **85**, 1013 (1963).
77. W. Strohmeier and K. Gerlach, *Chem. Ber.* **94**, 398 (1961).
78. J. Nasielski and A. Colas, *J. Organomet. Chem.* **101**, 215 (1975).
79. R. N. Perutz and J. J. Turner, *J. Am. Chem. Soc.* **97**, 4800 (1975).
80. J. Nasielski, P. Kirsch, and L. Wilputte-Steinert, *J. Organomet. Chem.* **29**, 269 (1971).
81. J. M. Kelly and A. Morris, *Rev. Latinoam. Quim.* **2**, 163 (1972); J. M. Kelly, H. Hermann, and E. A. Koerner von Gustorf, *IUPAC Symp. Photochem. 1972* Paper No. 34, p. 119 of the Manuscripts of Contributed Papers (1972); J. M. Kelly, H. Hermann, and E. A. Koerner von Gustorf, *Chem. Commun.* p. 105 (1973).
82. J. A. McIntyre, *J. Phys. Chem.* **74**, 2403 (1970).
83. J. M. Kelly, D. V. Bent, H. Hermann, D. Schulte-Frohlinde, and E. Koerner von Gustorf, *J. Organomet. Chem.* **69**, 259 (1974).
84. A. Vogler, *Z. Naturforsch., Teil B* **25**, 1069 (1970).
85. W. Strohmeier, D. von Hobe, G. Schonauer, and H. Laporte, *Z. Naturforsch., Teil B* **17**, 502 (1962).
86. W. Strohmeier and D. von Hobe, *Chem. Ber.* **94**, 2031 (1961).
87. W. Strohmeier and D. von Hobe, *Z. Phys. Chem. (Frankfurt am Main)* **34**, 393 (1962).
88. D. J. Darensbourg, M. Y. Darensbourg, R. J. Dennenberg, *J. Am. Chem. Soc.* **93**, 2807 (1971); D. J. Darensbourg and M. A. Murphy, *Inorg. Chem.* **17**, 884 (1978).
89. G. Schwenzer, M. Y. Darensbourg, and D. J. Darensbourg, *Inorg. Chem.* **11**, 1967 (1972).
90. G. Malouf and P. C. Ford, *J. Am. Chem. Soc.* **96**, 601 (1974).
91. A. J. Rest and J. R. Sodeau, *Chem. Commun.* p. 696 (1975).
92. J. D. Black and P. S. Braterman, *J. Organomet. Chem* **63**, C19 (1973).
93. M. Poliakoff, *Inorg. Chem.* **15**, 2022 and 2892 (1976).
94. R. Matheiu and R. Poilblanc, *Inorg. Chem.* **11**, 1858 (1972).
95. R. Matheiu, M. Lenzi, and R. Poilblanc, *Inorg. Chem.* **9**, 2030 (1970).
96. R. J. Clark and P. I. Hoberman, *Inorg. Chem.* **4**, 1771 (1965).

97. I. W. Stolz, G. R. Dobson, and R. K. Sheline, *Inorg. Chem.* **2**, 1264 (1963).
98. D. J. Darensbourg and H. H. Nelson, III, *J. Am. Chem. Soc.* **96**, 6511 (1974).
99. G. Platbrood and L. Wilputte-Steinert, *J. Organomet. Chem.* **85**, 199 (1975).
100. D. J. Darensbourg, H. H. Nelson, III, and M. A. Murphy, *J. Am. Chem. Soc.* **99**, 896 (1977).
101. W. P. Anderson, W. G. Blenderman, and K. A. Drews, *J. Organomet. Chem.* **42**, 139 (1972).
102. H. Werner, K. Deckelmann, and U. Schonenberger, *Helv. Chim. Acta* **53**, 2002 (1970); F. A. Cotton and F. Zingales, *J. Am. Chem. Soc.* **83**, 351 (1961); E. W. Abel, M. A. Bennett, and G. Wilkinson, *J. Chem. Soc.* p. 2323 (1959).
103. W. Strohmeier and H. Hellmann, *Z. Naturforsch., Teil B* **18**, 769 (1963); *Chem. Ber.* **96**, 2859 (1963); **97**, 1877 (1964); **98**, 1598 (1965); W. Strohmeier, G. Popp, and J. F. Guttenberger, *ibid.* **99**, 165 (1966); W. Strohmeier, J. F. Guttenberger, and F. J. Muller, *Z. Naturforsch., Teil B* **22** 1091 (1967); J. F. Guttenberger and W. Strohmeier, *Chem. Ber.* **100**, 2807 (1967).
104. W. Strohmeier and D. von Hobe, *Z. Naturforsch., Teil B* **18**, 981 (1963).
105. J. Nasielski and O. Denisoff, *J. Organomet. Chem.* **102**, 65 (1975).
106. A. N. Nesmeyanov, D. N. Kursanov, V. N. Setkina, V. D. Vil'chevaskaya, N. K. Baranetskaya, A. I. Krylova, and L. A. Gluskchenko, *Dokl. Akad. Nauk SSSR* **199**, 1336 (1971).
107. M. Herberhold and W. Golla, *J. Organomet. Chem.* **26**, C27 (1971).
108. B. V. Lokshin, V. I. Zdanovich, N. K. Baranetskaya, V. N. Setkina, and D. N. Kursanov, *J. Organomet. Chem.* **37**, 331 (1972).
109. R. J. Angelici and L. Busetto, *Inorg. Chem.* **7**, 1935 (1968).
110. E. O. Fischer and P. Kuzel, *Z. Naturforsch., Teil B* **16**, 475 (1961).
111. W. Strohmeier and D. von Hobe, *Z. Naturforsch., Teil B* **18**, 770 (1963).
112. W. Strohmeier and F.-J. Muller, *Chem. Ber.* **102**, 3608 (1969).
113. G. Natta, R. Ercoli, F. Calderazzo, and E. Santambrozio, *Chim. Ind. (Milan)* **40**, 1003 (1958).
114. D. A. Brown, D. Cunningham, and W. K. Glass, *Chem. Commun.* p. 306 (1966).
115. D. A. Brown, W. K. Glass, and B. Kumar, *Chem. Commun.* p. 736 (1967).
116. W. Jetz and W. A. G. Graham, *Inorg. Chem.* **10**, 4 (1971).
117. R. B. King, P. N. Kapoor, and R. N. Kapoor, *Inorg. Chem.* **10**, 1841 (1971).
118. R. B. King, L. W. Houk, and K. H. Pannell, *Inorg. Chem.* **8**, 1042 (1969).
119. R. J. Haines, R. S. Nyholm, and M. H. B. Stiddard, *J. Chem. Soc. A* p. 94 (1967).
120. L. H. Ali, A. Cox, and T. J. Kemp, *J. Chem. Soc., Dalton Trans.* p. 1475 (1973).
121. K. W. Barnett and P. M. Treichel, *Inorg. Chem.* **6**, 294 (1967).
122. R. B. King, W. C. Zipperer, and M. Ishaq, *Inorg. Chem.* **11**, 1361 (1972).
123. K. L. Tang Wong, J. L. Thomas, and H. H. Brintzinger, *J. Am. Chem. Soc.* **96**, 3694 (1974).
124. A. T. McPhail, G. R. Knox, C. G. Robertson, and G. A. Sim, *J. Chem. Soc. A* p. 205 (1971).
125. M. L. H. Green and A. N. Stear, *J. Organomet. Chem.* **1**, 230 (1964); R. B. King and R. N. Kapoor, *Inorg. Chem.* **8**, 2535 (1969); R. B. King and M. B. Bisnette, *ibid.* **4**, 486 (1965); R. B. King and A. Fronzaglia, *J. Am. Chem. Soc.* **88**, 709 (1966).
126. H. Alt. *J. Organomet. Chem.* **124**, 167 (1977).
127. A. J. Gingele, A. Harris, A. J. Rest, and R. N. Turner, *J. Organomet. Chem.* **121**, 205 (1976).
128. R. J. Haines, R. S. Nyholm, and M. G. B. Stiddard, *J. Chem. Soc. A* p. 43 (1968).
129. J. L. Hughey, IV, C. R. Bock, and T. J. Meyer, *J. Am. Chem. Soc.* **97**, 4440 (1975).
130. R. M. Laine and P. C. Ford, *Inorg. Chem.* **16**, 388 (1977).

131. H. B. Abrahamson and M. S. Wrighton, *J. Am. Chem. Soc.* **99**, 5510 (1977).

132. A. M. Rosan and J. W. Faller, *Synth. React. Inorg. Met.-Org. Chem.* **6**, 357 (1976).

133. A. Hudson, M. F. Lappert, and B. K. Nicholson, *J. Chem. Sci., Dalton Trans.* p. 551 (1977).

134. D. M. Allen, A. Cox, T. J. Kemp, Q. Sultana, and R. B. Pitts, *J. Chem. Soc., Dalton Trans.* p. 1189 (1976).

135. R. B. King and K. H. Pannell, *Inorg. Chem.* **7**, 2356 (1968).

136. C. Ginnotti and G. Merle, *J. Organomet. Chem.* **105**, 97 (1976).

137. D. S. Ginley and M. S. Wrighton, *J. Am. Chem. Soc.* **97**, 3533 (1975).

138. D. G. Alway and K. W. Barnett, *J. Organomet. Chem.* **99**, C52 (1975).

139. K. Ofele, E. Roos, and M. Herberhold, *Z. Naturforsch., Teil B* **31**, 1070 (1976); K. Ofele and M. Herberhold, *ibid.* **28**, 306(1973).

140. H. D. Gafney, J. L. Reed, and F. Basolo, *J. Am. Chem. Soc.* **95**, 7998 (1973); J. L. Reed, H. D. Gafney, and F. Basolo, *ibid.* **96**, 1363 (1974).

141. D. G. Whitten and M. T. McCall, *J. Am. Chem. Soc.* **91**, 5097 (1969).

142. P. R. Zarnegar, C. R. Bock, and D. G. Whitten, *J. Am. Chem. Soc.* **95**, 4367 (1973); P. Zarnegar and D. G. Whitten, *ibid.* **93**, 3776 (1971).

143. L. F. Dahl and R. E. Rundle, *Acta Crystallogr.* **16**, 419 (1963); M. F. Bailey and L. F. Dahl, *Inorg. Chem.* **4**, 1140 (1965); L. F. Dahl, E. Ishishi, and R. E. Rundle, *J. Chem. Phys.* **26**, 1750 (1957).

144. L. F. Dahl and C. H. Wei, *Acta Crystallogr.* **16**, 611 (1963).

145. R. A. N. McClean, *J. Chem. Soc., Dalton Trans.* p. 1568 (1974).

146. G. B. Blakney and W. F. Allen, *Inorg. Chem.* **10**, 2763 (1971).

147. H. B. Gray, E. Billig, A. Wojcicki, and M. Farona, *Can. J. Chem.* **41**, 1281 (1963).

148. M. S. Wrighton, unpublished observations.

149. C. H. Bamford, J. W. Burley, and M. Coldbeck, *J. Chem. Soc., Dalton Trans.* p. 1846 (1972).

150. M. Wrighton and D. L. Morse, *J. Am. Chem. Soc.* **96**, 998 (1974).

151. P. J. Giordano and M. S. Wrighton, *J. Am. Chem. Soc.* **101**, 2888 (1979).

152. M. S. Wrighton, D. L. Morse, and L. Pdungsap, *J. Am. Chem. Soc.* **97**, 2073 (1975).

153. P. Ford, De, F. P. Rudd, R. Gaunder, and H. Taube, *J. Am. Chem. Soc.* **90**, 1187 (1968).

154. D. K. Lavallee and E. B. Fleischer, *J. Am. Chem. Soc.* **94**, 2583 (1972).

155. P. J. Giordano, C. R. Bock, M. S. Wrighton, L. V. Interrante, and R. F. X. Williams, *J. Am. Chem. Soc.* **99**, 3187 (1977).

156. P. J. Giordano, C. R. Bock, and M. S. Wrighton, *J. Am. Chem. Soc.* **100**, 6960 (1978).

157. P. J. Giordano and M. S. Wrighton, *Inorg. Chem.* **16**, 160 (1977).

158. D. L. Lichtenberger and R. F. Fenske, *J. Am. Chem. Soc.* **98**, 50 (1976); D. L. Lichtenberger, D. Sellmann, and R. F. Fenske, *J. Organomet. Chem.* **117**, 253 (1976).

159. N. J. Gogan and C.-K. Chu, *J. Organomet. Chem.* **93**, 363 (1975).

160. L. Doub and J. M. Vanderbelt, *J. Am. Chem. Soc.* **77**, 4535 (1955).

161. R. A. Levenson, H. B. Gray, and G. P. Ceasar, *J. Am. Chem. Soc.* **92**, 3653 (1970).

162. M. S. Wrighton and D. S. Ginley, *J. Am. Chem. Soc.* **97**, 2065 (1975).

163. L. S. Brenner and A. L. Balch, *J. Organomet. Chem.* **134**, 121 (1977).

164. P. Lemoine and M. Gross, *J. Organomet. Chem.* **133**, 193 (1977).

165. D. L. Morse and M. S. Wrighton, *J. Am. Chem. Soc.* **98**, 3931 (1976); see also J. C. Luong, R. A. Faltynak, and M. S. Wrighton, *ibid.* **101**, 1597 (1979).

166. P. J. Giordano, S. M. Fredericks, M. S. Wrighton, and D. L. Morse, *J. Am. Chem. Soc.* **100**, 2257 (1978).

167. N. J. Turro, "Molecular Photochemistry," Benjamin, New York, 1967.

168. J. Saltiel, J. D'Agostino, E. D. Megarity, L. Metts, K. R. Neuberger, M. Wrighton, and O. C. Zafirion, *Org. Photochem.* **3**, 1 (1973).

169. J. F. Ireland and P. A. H. Wyatt, *Adv. Phys. Org. Chem.* **12**, 131 (1976), and references therein.
170. P. J. Giordano, S. M. Fredericks, and M. S. Wrighton, submitted for publication.
171. J. C. Luong, L. Nadjo, and M. S. Wrighton, *J. Am. Chem. Soc.* **100**, 5790 (1978).
172. N. E. Tokel and A. J. Bard, *J. Am. Chem. Soc.* **94**, 2862 (1972); N. E. Tokel-Takvoryan, R. E. Hemingway and A. J. Bard, *ibid.* **95**, 6582 (1973).
173. M. S. Wrighton, unpublished results.
174. A. J. Rest and J. J. Turner, *Chem. Commun.* p. 375 (1969).
175. A Berry and T. L. Brown, *Inorg. Chem.* **11**, 1165 (1972).
176. J. F. Ogilvie, *Chem. Commun.* p. 323 (1970).
177. M. L. H. Green and P. L. I. Nagy, *Adv. Organomet. Chem.* **2**, 325 (1964).
178. T. H. Whitesides and R. A. Budnick, *Chem. Commun.* p. 1514 (1971).
179. W. J. Miles, Jr. and R. J. Clark, *Inorg. Chem.* **7**, 1801 (1968).
180. J. K. Ruff, *Inorg. Chem.* **10**, 409 (1971).
181. R. B. King, R. N. Kapoor, M. S. Saran, and P. N. Kapoor, *Inorg. Chem.* **10**, 1851 (1971).
182. W. Strohmeier and K. Gerlach, *Z. Naturforsch., Teil B* **15**, 675 (1960).
183. W. Strohmeier, J. F. Guttenberger, and H. Hellmann, *Z. Naturforsch., Teil B* **19**, 353 (1964).
184. W. Strohmeier and J. F. Guttenberger, *Chem. Ber.* **97**, 1871 (1964).
185. W. Strohmeier and C. Barbeau, *Z. Naturforsch., Teil B* **17**, 848 (1962).
186. M. L. Ziegler and R. K. Sheline, *Inorg. Chem.* **4**, 1230 (1965).
187. R. J. Angelici and W. Loewen, *Inorg. Chem.* **6**, 682 (1967).
188. E. O. Fischer and M. Herberhold, *Experientia, Suppl.* **9**, 259 (1964).
189. W. Strohmeier, C. Barbeau, and D. von Hobe, *Chem. Ber.* **96**, 3254 (1963); W. Strohmeier and D. von Hobe, *Z. Phys. Chem. (Frankfurt am Main)* **34**, 393 (1962).
190. A. S. Foust, J. K. Hoyano, and W. A. G. Graham, *J. Organomet. Chem.* **32**, C65 (1971).
191. R. S. Nyhold, S. S. Sandhu, and M. H. B. Stiddard, *J. Chem. Soc.* p. 5916 (1963).
192. W. Strohmeier and J. F. Guttenberger, *Chem. Ber.* **96**, 2112 (1963).
193. R. B. King and M. S. Saran, *Inorg. Chem.* **10**, 1861 (1971).
194. E. O. Fischer and M. Herberhold, *Z. Naturforsch., Teil B* **16**, 841 (1961).
195. W. Strohmeier and J. F. Guttenberger, *Z. Naturforsch., Teil B* **18**, 80 (1963).
196. P. M. Treichel, E. Pitcher, R. B. King, and F. G. A. Stone, *J. Am. Chem. Soc.* **83**, 2593 (1961).
197. A. J. Rest, *Chem. Commun.* p. 345 (1970).
198. D. P. Keeton and F. Basolo, *Inorg. Chim. Acta* **6**, 33 (1972).
199. J. B. Wilford, P. M. Treichel, and F. G. A. Stone, *Proc. Chem. Soc., London* p. 218 (1963).
200. A. J. Oliver and W. A. G. Graham, *Inorg. Chem.* **9**, 2578 (1970).
201. H. C. Clark, J. D. Cotton, and J. H. Tsai, *Inorg. Chem.* **5**, 1582 (1966).
202. H. C. Clark and J. H. Tsai, *Inorg. Chem.* **5**, 1407 (1966).
203. J. B. Wilford, P. M. Treichel, and F. G. A. Stone, *J. Organomet. Chem.* **2**, 119 (1964).
204. A. G. Osborne and M. H. B. Stiddard, *J. Chem. Soc.* p. 634 (1964).
205. J. R. Miller and D. H. Myers, *Inorg. Chim. Acta* **5**, 215 (1970).
206. J. Lewis, R. S. Nyhold, A. G. Osborne, S. S. Sandhu, and M. H. B. Stiddard, *Chem. Ind. (London)* p. 1398 (1963).
207. C. Barbeau, *Can. J. Chem.* **45**, 161 (1967).
208. R. J. Clark, J. P. Hargaden, H. Hass, and R. K. Sheline, *Inorg. Chem.* **7**, 673 (1968).
209. J. T. Moelwyn-Hughes, A. W. B. Garner, and N. Gordon, *J. Organomet. Chem.* **26**, 373 (1971).
210. G. Bor, *Chem. Commun.* p. 641 (1969).
211. M. L. Ziegler, H. Haas, and R. K. Sheline, *Chem. Ber.* **98**, 2454 (1965).
212. J. P. Crow, W. R. Cullen, and F. L. Hou, *Inorg. Chem.* **11**, 2125 (1972).

213. J. P. Crow, W. R. Cullen, F. L. Hou, L. Y. Y. Chan, and F. W. B. Einstein, *Chem. Commun.* p. 1229 (1971).

214. W. Hieber and W. Schropp, Jr., *Z. Naturforsch., Teil B* **15**, 271 (1960).

215. M. W. Lindauer, G. O. Evans, and R. K. Sheline, *Inorg. Chem.* **7**, 1249 (1968).

216. G. O. Evans and R. K. Sheline, *J. Inorg. Nucl. Chem.* **30**, 2862 (1968).

217. J. C. Kwok, Ph.D. Thesis, University of Liverpool (1971).

218. C. M. Bamford and J. Paprotny, *Chem. Commun.* p. 140 (1971).

219. K. Moedritzer, *Synth. Inorg. Met.-Org. Chem.* **1**, 63 (1971).

220. T. Kruck, M. Hofler, and M. Noack, *Chem. Ber.* **99**, 1153 (1966).

221. T. Kruck and M. Hofler, *Angew. Chem.* **76**, 786 (1964).

222. M. Wrighton and D. Bredesen, *J. Organomet. Chem.* **50**, C35 (1973).

223. J. L. Hughey, IV, C. P. Anderson, and T. J. Meyer, *J. Organomet. Chem.* **125**, C49 (1977).

224. M. Herberhold and A. Razavi, *J. Organomet. Chem.* **67**, 81 (1974).

225. M. Herberhold and G. Süss, *Angew. Chem., Int. Ed. Engl.* **14**, 700 (1975).

226. B. H. Byers and T. L. Brown, *J. Am. Chem. Soc.* **97**, 947 (1975); **98**, 3160 (1976); **99**, 2527 (1977).

227. M. W. Lindauer, G. O. Evans, and R. K. Sheline, *Inorg. Chem.* **7**, 1249 (1968).

228. R. A. Epstein, T. R. Gaffney, G. L. Geoffroy, W. L. Gladfelter and R. S. Henderson, *J. Am. Chem. Soc.* **101**, 3847 (1979).

229. R. A. Faltynek and M. S. Wrighton, *J. Am. Chem. Soc.* **100**, 2701 (1978).

230. R. D. Sanner, R. G. Austin, M. S. Wrighton, W. D. Honnick, and C. V. Pittman, Jr., *Inorg. Chem.* **18**, 928 (1979).

231. A. W. Hanson, *Acta Crystallogr.* **15**, 930 (1962); J. Donohue and A. Caron, *ibid.* **17**, 663 (1964).

232. R. Brill, *Z. Kristallogr., Kristallgeom., Kristallphys., Kristallchem.* **65**, 85 (1927); H. M. Powell and V. G. Ewen, *J. Chem. Soc.* p. 286 (1939).

233. R. K. Sheline and K. S. Pitzer, *J. Am. Chem. Soc.* **72**, 1107 (1950).

234. C. H. Wei and L. F. Dahl, *J. Am. Chem. Soc.* **88**, 1821 (1966); **91**, 1351 (1969).

235. K. Noack, *Helv. Chim. Acta* **45**, 1847 (1962); F. A. Cotton and G. Wilkinson, *J. Am. Chem. Soc.* **79**, 752 (1957).

236. E. R. Corey and L. F. Dahl, *Inorg. Chem.* **1**, 521 (1962).

237. E. R. Corey and L. F. Dahl, *J. Am. Chem. Soc.* **83**, 2202 (1961).

238. D. K. Huggins, N. Flitcroft, and H. D. Kaesz, *Inorg. Chem.* **4**, 166 (1965).

239. D. L. Smith, *J. Chem. Phys.* **42**, 1460 (1965).

240. O. S. Mills, *Acta Crystallogr.* **11**, 620 (1958).

241. R. D. Fischer, A. Vogler, and K. Noack, *J. Organomet. Chem.* **7**, 135 (1967).

242. F. A. Cotton and G. Yagupsky, *Inorg. Chem.* **6**, 15 (1967).

243. A. R. Manning, *J. Chem. Soc. A* p. 1319 (1968).

244. M. Dartiguenave, Y. Dartiquenave, and H. B. Gray, *Bull. Soc. Chim. Fr.* **12**, 4223 (1969).

245. R. G. Austin, R. S. Paonessa, P. J. Giordano, and M. S. Wrighton, *Adv. Chem. Ser.* **168**, 189 (1978); J. L. Graff, R. D. Sanner, and M. S. Wrighton, *J. Am. Chem. Soc.* **101**, 273 (1979); D. R. Tyler, R. A. Levenson, and H. B. Gray, *J. Am. Chem. Soc.* **100**, 7888 (1978).

246. W. Hieber and H. Beutner, *Z. Naturforsch., Teil B* **17**, 211 (1962).

247. E. Koerner von Gustorf, O. Jaenicke, and O. E. Polansky, *Z. Naturforsch., Teil B* **27**, 575 (1972).

248. E. Koerner von Gustorf, F.-W. Grevels, C. Kruger, G. Olbrich, F. Mark, D. Schulz, and R. Wagner, *Z. Naturforsch., Teil B* **27**, 392 (1972).

249. E. Koerner von Gustorf, J. C. Hogan, and R. Wagner, *Z. Naturforsch., Teil B* **27**, 140 (1972).

250. G. N. Schrauzer and G. Kratel, *J. Organomet. Chem.* **2**, 336 (1964).

251. D. C. Harris, and H. B. Gray, *Inorg. Chem.* **14**, 1215 (1975).
252. C. R. Bock and M. S. Wrighton, *Inorg. Chem.* **16**, 1309 (1977).
253. D. G. Alway and K. W. Barnett, *Adv. Chem. Ser.* **168**, 115 (1978).
254. H. B. Abrahamson, M. C. Palazzotto, C. L. Reichel, and M. S. Wrighton, *J. Am. Chem. Soc.*, **101**, 4123 (1979).
255. J. A. Ferguson and T. J. Meyer, *Chem. Commun.* p. 623 (1971); *J. Am. Chem. Soc.* **94**, 3409 (1972).
256. J. Dewar and H. O. Jones, *Proc. R. Soc. London, Ser. A* **76**, 558 (1905).
257. F. Calderazzo and F. L.'Eplattenier, *Inorg. Chem.* **6**, 1220 (1967).
258. J. R. Moss and W. A. G. Graham, *Chem. Commun.* p. 835 (1970).
259. J. R. Moss and W. A. G. Graham, *J. Chem. Soc., Dalton Trans.* p. 95 (1977).
260. R. Fields, M. M. Germain, R. N. Haszeldine, and P. W. Wiggans, *J. Chem. Soc A* pp. 1964 and 1969 (1970).
261. R. Burt, M. Cooke, and M. Green, *J. Chem. Soc. A* pp. 2975 and 2981 (1970).
262. W. R. Cullen, D. A. Harbourne, B. V. Liengme, and J. R. Sams, *Inorg. Chem.* **8**, 1464 (1969).
263. J. D. Warren and R. J. Clark, *Inorg. Chem.* **9**, 373 (1970).
264. J. D. Warren, M. A. Busch, and R. J. Clark, *Inorg. Chem.* **11**, 452 (1972).
265. R. J. Clark, *Inorg. Chem.* **3**, 1395 (1964).
266. R. Victor, R. Ben-Shoshan, and S. Sarel, *Chem. Commun.* p. 1241 (1971).
267. J. Lewis, R. S. Nyholm, S. S. Sandhu, and M. H. B. Stiddard, *J. Chem. Soc.* p. 2825 (1964).
268. E. H. Schubert and R. K. Sheline, *Inorg. Chem.* **5**, 1071 (1966).
269. J. K. Ruff, *Inorg. Chem.* **8**, 86 (1969).
270. G. Bor, *Inorg. Chim. Acta* **3**, 191 (1969); K. Noack and M. Ruch, *J. Organomet. Chem.* **17**, 309 (1969); D. F. Keely and R. E. Johnson, *J. Inorg. Nucl. Chem.* **11**, 33 (1959).
271. M. Poliakoff and J. J. Turner, *J. Chem. Soc., Dalton Trans.* p. 1351 (1973); p. 2276 (1974).
272. T. J. Barton, B. Davies, R. Gritner, M. Poliakoff, and A. J. Thomson, cited in Davies *et al.* [273].
273. B. Davies, A. McNeish, M. Poliakoff, and J. J. Turner, *J. Am. Chem. Soc.* **99**, 7573 (1977).
274. A. McNeish, M. Poliakoff, K. P. Smith, and J. J. Turner, *Chem. Commun.* p. 859 (1976).
275. R. F. Heck and C. R. Boss, *J. Am. Chem. Soc.* **86**, 2580 (1964).
276. P. Reeves, J. Henery, and R. Pettit, *J. Am. Chem. Soc.* **91**, 5888 (1969); J. S. Ward and R. Pettit, *ibid.* **93**, 262 (1971).
277. A. Bond and M. Green, *Chem. Commun.* p. 12 (1971).
278. A. Bond and M. Green, *J. Chem. Soc., Dalton Trans.* p. 763 (1972).
279. F. M. Chaudhari and P. L. Pauson, *J. Organomet. Chem.* **5**, 73 (1966).
280. W. R. Cullen, J. R. Sams, and J. A. J. Thompson, *Inorg. Chem.* **10**, 843 (1971).
281. A. J. Carty, A. Efraty, T. W. Ng, and T. Birchall, *Inorg. Chem.* **9**, 1263 (1970).
282. M. L. H. Green and M. J. Smith, *J. Chem. Soc. A* p. 3220 (1971).
283. M. L. H. Green and P. L. I. Nagy, *J. Chem. Soc.* p. 189 (1963).
284. M. Poliakoff and J. J. Turner, *J. Chem. Soc. A* p. 2403 (1971).
285. S. A. R. Knox and F. G. A. Stone, *J. Chem. Soc. A* p. 3147 (1970); 2559 (1969).
286. A. Brockes, S. A. R. Knox, and F. G. A. Stone, *J. Chem. Soc. A* 3469 (1971).
287. S. A. R. Knox and F. G. A. Stone, *J. Chem. Soc. A* p. 2874 (1971).
288. F. A. Cotton, A. J. Deeming, P. L. Josty, S. S. Ullah, A. J. P. Domingos, B. F. G. Johnson, and J. Lewis, *J. Am. Chem. Soc.* **93**, 4624 (1971).
289. W. R. Cullen and D. A. Harbourne, *Inorg. Chem.* **9**, 1839 (1970).
290. B. F. G. Johnson, J. Lewis, and M. V. Twigg, *J. Organomet. Chem.* **67**, C75 (1974).
291. W. R. Cullen, D. A. Harbourne, B. V. Liengme, and J. R. Sams, *Inorg. Chem.* **9**, 702 (1970).

292. P. J. Roberts and J. Trotter, *J. Chem. Soc. A* p. 1479 (1971).
293. W. R. Cullen, D. A. Harbourne, B. V. Liengme, and J. R. Sams, *J. Am. Chem. Soc.* **90**, 3293 (1968).
294. G. G. Summer, H. P. Klug, and L. E. Alexander, *Acta Crystallogr.* **17**, 732 (1964).
295. K. Noack, *Spectrochim. Acta* **19**, 1925 (1963); G. Bor, *ibid.* p. 2065.
296. G. L. Geoffroy and R. A. Epstein, *Inorg. Chem.* **16**, 2795 (1977).
297. L. O. Brockway and J. S. Anderson, *Trans. Faraday Soc.* **33**, 1233 (1937).
298. I. H. Sabherwal and A. Burg. *Chem. Commun.* p. 853 (1969).
299. R. J. Clark, *Inorg. Chem.* **6**, 299 (1967).
300. C. A. Udovich and R. J. Clark, *Inorg. Chem.* **8**, 938 (1969).
301. A. Nakamura and N. Hagihara, *Bull. Chem. Soc. Jpn.* **33**, 425 (1960).
302. E. O. Fischer and K. Bittler, *Z. Naturforsch., Teil B* **16**, 835 (1961).
303. A. J. Oliver and W. A. G. Graham, *Inorg. Chem.* **10**, 1 (1971).
304. K. P. C. Vollhardt, J. E. Bercaw, and R. G. Bergman, *J. Organomet. Chem.* **97**, 283 (1975).
305. M. Rosenblum, B. North, D. Wells, and W. P. Giering, *J. Am. Chem. Soc.* **94**, 1239 (1972).
306. R. L. Sweany and T. L. Brown, *Inorg. Chem.* **16**, 421 (1977).
307. E. O. Fischer, P. Kuzel, and H. P. Fritz, *Z. Naturforsch., Teil B* **16**, 138 (1961).
308. M. Absi-Halabi and T. L. Brown, *J. Am. Chem. Soc.* **99**, 2982 (1977).
309. J. Ladell, B. Post, and I. Fankuchen, *Acta Crystallogr.* **5**, 795 (1952).
310. A. B. Callear, *Proc. R. Soc. London, Ser. A* **265**, 71 (1961).
311. A. P. Garratt and H. W. Thompson, *J. Chem. Soc.* p. 1817 (1934).
312. A. J. Rest and J. J. Turner, *Chem. Commun.* p. 1026 (1969).
313. B. Hill, K. Math, D. Pillsbury, G. Voecks, and W. Jennings, *Mol. Photochem.* **5**, 195 (1973).

Olefin Complexes

3

I. INTRODUCTION

Numerous metal–olefin complexes have been prepared and characterized, and photochemical studies of several have been conducted. These studies have concentrated to a large extent on investigations of diene complexes because of their greater thermal inertness and ease of characterization. In many of these studies, formation of olefin complexes was induced *in situ* either by photochemical or thermal means and their presence determined by spectroscopic measurements. This method often introduces ambiguities since it becomes difficult to know the nature and concentration of all light-absorbing species. Several interesting and potentially useful photochemical transformations of olefins have been reported, particularly the photocatalyzed hydrogenation [1–9] and hydrosilation [10] of 1,3-dienes and the photoinduced valence isomerization of norbornadiene to quadricyclane [11,11a]. The latter has been proposed as an energy conversion and storage system. Although not strictly in the class of olefin complexes, several com-

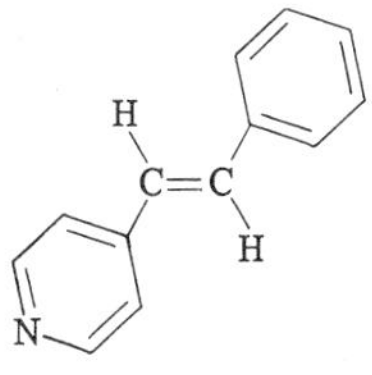

(XV)

plexes of 4-styrylpyridine (XV) are briefly discussed in this chapter because their photochemistry involves cis–trans isomerization of the coordinated ligand.

Many photochemical reactions of metal–olefin carbonyl complexes involve primary photochemical processes which were discussed in Chapter 2. For example, photolysis of $[Fe(CO)_3(\eta^4\text{-}1,3\text{-butadiene})]$ in the presence of haloalkenes leads to products in which the haloalkane inserts into the iron–butadiene bond, but the available evidence indicates that the transformation proceeds through initial CO loss (Eq. 3-1) [12].

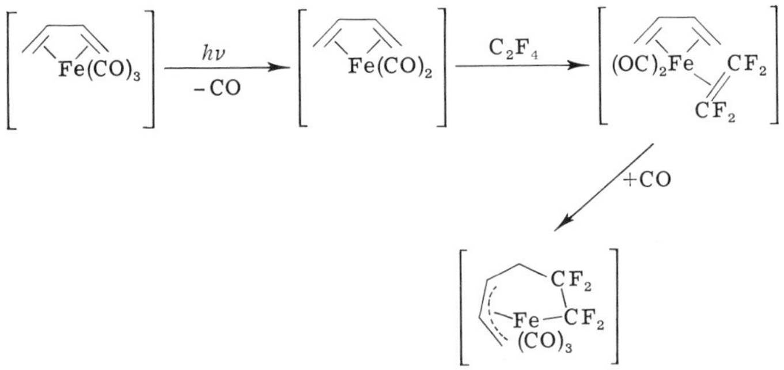

$$
\left[\begin{array}{c} \diagup\!\!\!\diagdown\!\!\!\diagup\!\!\!\diagdown \\ Fe(CO)_3 \end{array}\right] \xrightarrow[-CO]{h\nu} \left[\begin{array}{c} \diagup\!\!\!\diagdown\!\!\!\diagup\!\!\!\diagdown \\ Fe(CO)_2 \end{array}\right] \xrightarrow{C_2F_4} \left[\begin{array}{c} \diagdown\!\!\!\diagup\!\!\!\diagdown\!\!\!\diagup \\ (OC)_2Fe \diagdown\!\!\!\diagup CF_2 \\ CF_2 \end{array}\right]
$$

$$
\Bigg\downarrow +CO
$$
$$
\left[\begin{array}{c} CF_2 \\ \diagdown\!\!\!\diagup Fe{-}CF_2 \\ (CO)_3 \end{array}\right]
$$

(3-1)

II.   NIOBIUM COMPLEXES

$[Nb(\eta^5\text{-}C_5H_5)(CO)(PhC_2Ph)_2]$

Irradiation of $[Nb(\eta^5\text{-}C_5H_5)(CO)(PhC_2Ph)_2]$ in boiling acetonitrile solution, freshly distilled over P_2O_5, produced the unusual complex (XVI) [13].

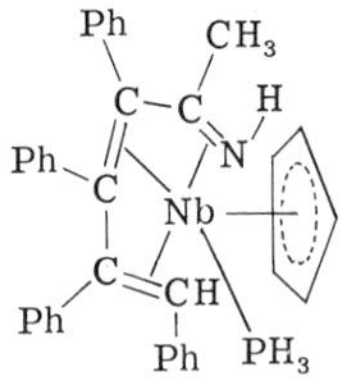

$$
\begin{array}{c}
Ph \quad CH_3 \\
C{-}C \quad H \\
Ph{-}C \diagup \quad \diagdown N \diagup \\
\quad C \quad Nb{-} \\
Ph \diagup C{=}CH \\
\quad Ph \quad PH_3
\end{array}
$$

(XVI)

No mechanistic details were presented although it was noted that the PH_3 ligand must have derived from P_2O_5. Insertion of the solvent (acetonitrile) obtains, but it is not clear which (if any) of the steps are photoactivated.

III. CHROMIUM, MOLYBDENUM, AND TUNGSTEN COMPLEXES

A. $[Cr(1,3\text{-}Diene)(CO)_4]$ AND $[Cr(Norbornadiene)(CO)_4]$

Wilputte-Steinert and co-workers [1] reported in 1971 that irradiation of $[Cr(CO)_6]$ under H_2 would effect the hydrogenation of 2,3-dimethyl-butadiene and 1,3-cyclohexadiene. The hydrogenation proceeds smoothly at room temperature and 1 atm H_2 to produce in good yield 2,3-dimethyl-2-butene and cyclohexene, respectively. This is in marked contrast to the high temperatures and pressures required for the same reaction without light using $[Cr(\eta^6\text{-}C_6H_6)(CO)_3]$ as the catalyst [14,15]. The photocatalyzed chemistry was extended by Wrighton and Schroeder [2] in 1973 to a variety of other dienes, and their results are summarized in Table 3-1. Their turnover numbers in the dark for hydrogenation after the catalyst is photogenerated clearly indicate that irradiation of $[Cr(CO)_6]$ generates a *thermally* active catalyst. Again it was observed that only 1,4-hydrogenation obtained, and *trans, trans*-2,4-hexadiene, for example, under D_2 gives 2,5-dideutero-*cis*-3-hexene as the exclusive product. Importantly, it was noted that the relative rate of hydrogenation was related to the ease with which the dienes adopt the S-cis conformation (Eq. 3-2). *cis*-1,3-Pentadiene, for example, cannot

$$\text{S-trans} \qquad\qquad \text{S-cis}$$

(3-2)

easily adopt the S-cis conformation; and it was not hydrogenated at a rate comparable to that of either *trans*-1,3-pentadiene or 2-methyl-1,3-butadiene, each of which can attain the S-cis conformation. It was stated [2] that $[Mo(CO)_6]$ and $[W(CO)_6]$ would effect comparable chemistry, but unlike $[Cr(CO)_6]$ these species also effect 1,3-diene isomerization (*vide infra*). In a related study, Wrighton and Schroeder [10] showed that $[Cr(CO)_6]$ also photocatalyzed the 1,4-hydrosilation of 1,3-dienes. The products observed from the various dienes and silanes employed are shown in Table 3-2, and it is likely that a mechanism similar to that for hydrogenation is operative.

It has since been realized that the initial role of light is to generate $[Cr(1,3\text{-}diene)(CO)_4]$ as the active catalyst precursor (Eq. 3-3) [3]. Koerner

$$[Cr(CO)_6] + 1,3\text{-diene} \xrightarrow{h\nu} [Cr(1,3\text{-diene})(CO)_4] + 2CO \qquad (3\text{-}3)$$

von Gustorf and co-workers [4], for example, isolated $[Cr(butadiene)(CO)_4]$ and $[Cr(trans,trans\text{-}2,4\text{-hexadiene})(CO)_4]$ from photolysis of $[Cr(CO)_6]$ in

TABLE 3-1

Selective M(CO)$_6$-Photoassisted Hydrogenation of 1,3-Diene Mixtures[a]

M(CO)$_6$ (M)	Starting mixture, %			Total diene, M	Reaction[b] time, h	Mixture after reaction, %			
Cr(CO)$_6$ (0.001)	49.27	1.05	49.67	0.1	1.0	43.46	0.93	49.57	6.04
Cr(CO)$_6$ (0.001)	49.16	50.83		0.2	0.5	46.88	51.63	1.49	
					1.0	45.22	50.68	4.10	
					1.5	43.37	50.43	6.20	
					2.0	40.70	50.58	8.72	
Cr(CO)$_6$ (0.002)	50.00	49.79	0.21	0.1	2.0	11.31	38.69	~0	49.34	0.65
Cr(CO)$_6$ (0.002)	0.42	49.58	49.39	0.61	0.1	2.0	0.20	8.36	0.12	49.68	2.14	39.50	~0
Cr(CO)$_6$ (0.01)	50.0	50.0		0.1	22.0[c]	0.4	50.0	49.6	~0

[a] Reprinted with permission from Wrighton and Schroeder [2], *J. Am. Chem. Soc.* **95**, 5764 (1973). Copyright by the American Chemical Society.

[b] Reaction carried out at 10°C with continuous UV irradiation in the presence of 1 atm of H$_2$, in benzene.

[c] Reaction for first 2 h at 10°C then allowed to warm to 25°C.

TABLE 3-2

Allylsilanes Synthesized by Cr(CO)$_6$-Photocatalyzed 1,4-Hydrosilation of 1,3-Dienes[a,b]

Starting diene	Starting silane	Product(s)
(1,3-butadiene)	HSiMe$_3$	(allyl)—SiMe$_3$
(1,3-butadiene)	H$_2$SiPh$_2$	(allyl)—SiHPh$_2$
(1,3-butadiene)	HSi(OEt)$_3$	(allyl)—Si(OEt)$_3$
(1,3-butadiene)	H$_2$SiMe$_2$	(allyl)—SiHMe$_2$
(2-methyl-1,3-butadiene)	HSiMe$_3$	—SiMe$_3$ (~40%) + Me$_3$Si— (~60%)
(2-methyl-1,3-butadiene)	DSiMe$_3$	D—SiMe$_3$ (~40%) + Me$_3$Si—D (~60%)
(1,3-pentadiene)	HSiMe$_3$	Me$_3$Si— (~90%) + —(~10%) SiMe$_3$
(2,3-dimethyl-1,3-butadiene)	HSiMe$_3$	—SiMe$_3$

[a] Reprinted with permission from Wrighton and Schroeder [10], *J. Am. Chem. Soc.* **96**, 6235 (1974). Copyright by the American Chemical Society.

[b] These reactions are carried out on the neat substrates in approximately 1:1 mole ratios on an approximately 0.05 mol scale. The deoxygenated solutions are saturated in Cr(CO)$_6$ and exposed for several days to the output of a black light equipped with two 15-W GE black light bulbs. The reaction temperature is approximately 30°C. The products were generally identified by GC ($\beta\beta'$-ODPN or OV-101 column), NMR, and IR spectra after distillation to remove catalyst and residual starting materials.

the presence of the respective dienes; they showed the latter to be an inter-mediate in the photocatalytic hydrogenation of *trans,trans*-2,4-hexadiene. It is also clear that the [Cr(diene)(CO)$_4$] intermediate must undergo absorption of another photon of light in order to induce dissociation of a ligand and generate the active thermal catalyst. The exact nature of this subsequent photoreaction has produced considerable discussion, however, and it does not yet appear to be resolved. Two possible mechanisms which can be written are shown in Fig. 3-1 [4].

Wilputte-Steinert and co-workers [3,5,6] have favored mechanism II in which photolysis of [Cr(diene)(CO)$_4$] induces dissociation of one end of the diene which then undergoes rotation about the C—C bond to open a coordination site. However, Koerner von Gustorf and co-workers [7] observed only CO dissociation when [Cr(*trans,trans*-2,4-hexadiene)(CO)$_4$]

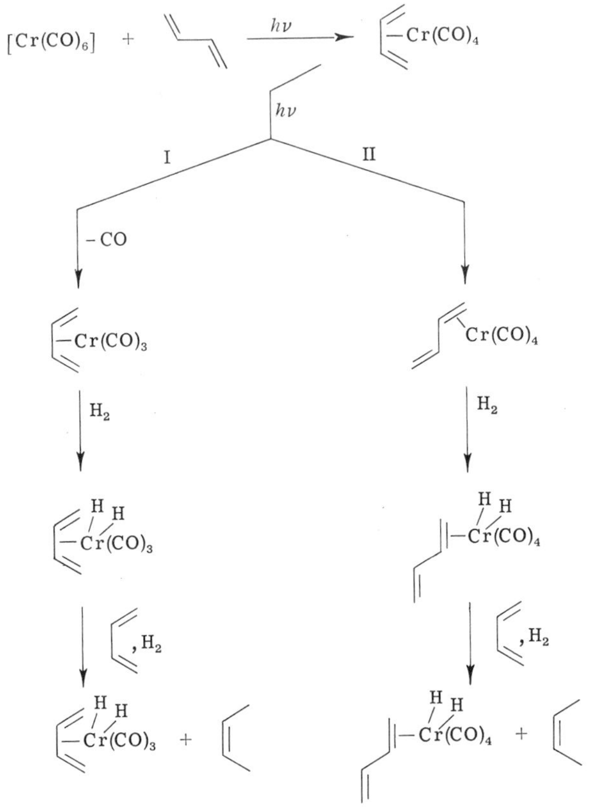

FIG. 3-1. Mechanistic schemes for $[Cr(CO)_6]$-photocatalyzed hydrogenation.

was irradiated in an argon matrix, and they obtained no evidence for dissociation of one end of the coordinated diene. Their results thus implicated $[Cr(diene)(CO)_3]$ as the key intermediate in the hydrogenation scheme. Further evidence for the intermediacy of a $[Cr(diene)(CO)_3]$ species comes from Schroeder and Wrighton's finding [9] that $[Cr(CO)_3(CH_3CN)_3]$ thermally catalyzes the 1,4-addition of hydrogen to 1,3-dienes in exactly the same manner as in the $[Cr(CO)_6]$ photocatalyzed hydrogenation. The thermal reaction presumably proceeds by dissociation of the labile aceto-nitrile ligands and addition of a diene to generate a $[Cr(diene)(CO)_3]$ intermediate. Likewise, it is noteworthy that $[Cr(\eta^6\text{-arene})(CO)_3]$ thermal catalysts give the same chemistry and probably involve $[Cr(CO)_3]$ as the repeating unit [14,15].

Another interesting $[Cr(CO)_6]$-photocatalyzed reaction is that shown in Eq. (3-4); the probable catalyst precursor $[Cr(norbornadiene)(CO)_4]$

catalyzes the hydrogenation of norbornadiene to nortricyclene and nor-bornene in a constant ratio of 2.80 [3].

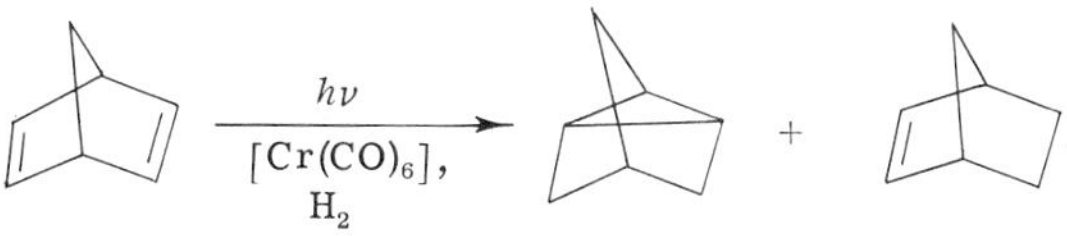

$$\text{(structure)} \xrightarrow[\substack{[Cr(CO)_6], \\ H_2}]{h\nu} \text{(structure)} + \text{(structure)} \tag{3-4}$$

The arguments of Wilputte-Steinert and co-workers concerning the likelihood of path II in Fig. 3-1 are based on their observation of the photo-stability of $[Cr(norbornadiene)(CO)_4]$ when it is irradiated in the absence of hydrogen and substrate and on their studies of photoinduced substitution of PPh_3 into $[Cr(norbornadiene)(CO)_4]$ [5]. The stability obtains even in solutions irradiated under a vigorous inert-gas purge [5], suggesting that CO elimination does not occur upon photolysis. If CO loss did occur, one might expect net CO loss and decomposition. The rate of recombination of CO with a photogenerated $[Cr(norbornadiene)(CO)_3]$ species, however, could be sufficiently fast to prevent it from being swept out of solution. Indeed, it was later reported [6] that degradation of the complex with CO loss did occur after several hours of irradiation. Photolysis of $[Cr(norbornadiene)\cdot(CO)_4]$ in the presence of excess PPh_3 has been shown to yield *mer*-$[Cr(norbornadiene)(CO)_3(PPh_3)]$ [5,6]. The quantum yields measured for disappearance of $[Cr(norbornadiene)(CO)_4]$ when irradiated in the presence of H_2, PPh_3, and of both reagents were 0.140, 0.105, and 0.104, respectively. In the presence of hydrogen alone, conversion to nortricyclene, norborna-diene, $[Cr(CO)_6]$, $[Cr(norbornadiene)(CO)_5]$, and chromium metal oc-curred [6]. The mechanism shown in Fig. 3-2 was proposed by Rietvelde and Wilputte-Steinert to account for their various observations [6]. They suggested that irradiation initially produces a $[Cr(norbornadiene)(CO)_4]$

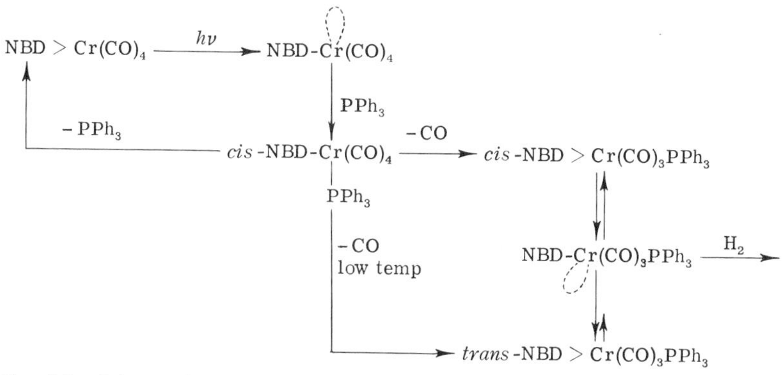

$$
\begin{array}{l}
NBD > Cr(CO)_4 \xrightarrow{\ h\nu\ } NBD\text{-}Cr(CO)_4 \\
\end{array}
$$

FIG. 3-2. Primary photoprocesses. Reprinted with permission from Rietvelde and Wilputte-Steinert [6].

species with only one of the norbornadiene double bonds coordinated. This intermediate can then add PPh_3 to give a coordinatively saturated complex which can, in turn, lose PPh_3 to yield the original complex or can eliminate CO to give the substituted $[Cr(norbornadiene)(CO)_3PPh_3]$ derivative.

More definitive work on the mechanism of the $[M(CO)_6]$-photoinduced hydrogenation of norbornadiene was recently described by Darensbourg and co-workers [8a]. They conducted detailed infrared and ^{13}C-NMR spectroscopic studies on the photoinduced incorporation of ^{13}CO into $[M(norbornadiene)(CO)_4]$ (M = Cr, Mo, W). For M = Mo, W it was found that ^{13}CO incorporation initially occurs cis to both coordinated norbornadiene double bonds but that the complexes subsequently thermally isomerize to give ^{13}CO in trans positions. Specifically, they proposed that the *thermal isomerization* proceeds via a $[M(norbornadiene)(CO)_4]$ (M = Mo, W) intermediate in which only one of the norbornadiene double bonds is coordinated to the metal. Spectral overlap prevented an unambiguous analysis for $[Cr(norbornadiene)(CO)_4]$, but it was proposed that similar photochemical and thermal processes obtain for this derivative. $[Mo(CO)_6]$ and $[W(CO)_6]$ were also observed to catalyze the hydrogenation of norbornadiene but in a manner different from that of $[Cr(CO)_6]$. Darensbourg and co-workers found that when $[Cr(CO)_6]$ was employed as the catalyst precursor, nortricyclene and norbornene were formed in a constant ratio of 1.9:1. This differs from the 2.8:1 ratio found by Platbrood and Wilputte-Steinert [3,3a], but different experimental conditions were employed. Norbornene, though, was by far the major product with $[Mo(CO)_6]$ and $[W(CO)_6]$. However, when $[M(norbornadiene)(CO)_4]$ (M = Mo, W) were the catalyst precursors, the product ratios of nortri-cyclene and norbornene were 0.25–0.50 and 2.0–4.0 for Mo and W, respectively. The shift is most dramatic for tungsten which gives principally norbornene if $[W(CO)_6]$ is used and principally nortricyclene if $[W(norbornadiene)(CO)_4]$ is employed as the catalyst precursor. It was further observed that thermal hydrogenation of norbornadiene was assisted by all the $[M(norbornadiene)(CO)_4]$ derivatives at 60°C, a temperature at which the isomerization studies showed that opening of the metal–olefin chelate occurs.

These workers interpreted all their observations in terms of the pathways outlined in Fig. 3-3 [8a]. They proposed that both pathways A and B, analogous to mechanisms I and II in Fig. 3-2, obtain in the hydrogenation process. It was specifically suggested that norbornene only arises through initial photoinduced cleavage of a metal–olefin bond whereas nortricyclene is formed via initial loss of CO. Thus, the two processes compete for the excitation energy. The relative importance of the two pathways is dependent

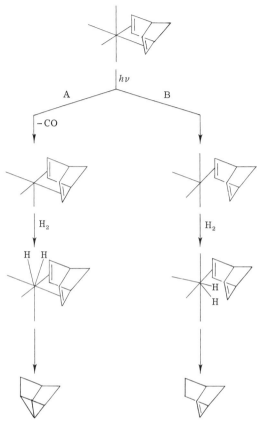

FIG. 3-3.   Proposed mechanism for [M(norbornadiene)(CO)₄]-photocatalyzed hydrogenation. Reprinted with permission from Darensbourg *et al.* [8a], *J. Am. Chem. Soc.* **99**, 896 (1977). Copyright by the American Chemical Society.

upon the metal, with pathway A dominating in the order of W > Cr > Mo. The anomaly between the [W(CO)₆] and [W(norbornadiene)(CO)₄] product ratios is nicely explained by logically assuming that irradiation of [W(CO)₆] in the presence of norbornadiene ultimately gives [W(norbornadiene)(CO)₄] species with only one norbornadiene double bond coordinated. This species is an intermediate in pathway B, and norbornene is produced almost exclusively, starting with [W(CO)₆]. Indeed, [W(norbornadiene)(CO)₅] was observed spectroscopically in large quantities upon irradiation of [W(CO)₆] in the presence of norbornadiene. If [W(norbornadiene)(CO)₄] is the catalyst precursor, photolysis apparently induces CO loss to generate [W(norbornadiene)(CO)₃], an intermediate in pathway A, and this leads to nortricyclene.

Jennings and Hill [16] have reported that $[Cr(CO)_6]$ catalyzed the dimerization of norbornadiene, and $[Cr(norbornadiene)(CO)_4]$ was implicated as a key intermediate. The three dimers (XVII), (XVIII), and (XIX) were produced in a ratio of 1.8:1.0:1.4.

(exo-trans-exo)	(endo-trans-exo)	(endo-trans-endo)
(XVII)	(XVIII)	(XIX)

B. $[M(CO)_6]$-PHOTOASSISTED OLEFIN ISOMERIZATION

Ultraviolet irradiation of solutions containing excess olefin and $[Mo(CO)_6]$ or $[W(CO)_6]$ has been shown to lead to double-bond migration and cis-trans isomerization of the olefin [17,18]. Some of the complex–olefin combinations that were studied are given in Table 3-3. Detailed studies were conducted mainly with 1-pentene and *cis*- and *trans*-2-pentene, each of which gives the other two isomers when irradiated in the presence of $[M(CO)_6]$ [17]. Irradiation of any of the three led to an equilibrium mixture which is approximately 3% 1-pentene, 17% *cis*-2-pentene, and 80% *trans*-2-pentene. *cis*-2-Pentene appeared to be substantially more reactive than *trans*-2-pentene. No thermal isomerization was detected at 30°C, and the absolute number of pentene molecules photoisomerized was greater than the number of $[M(CO)_6]$ molecules present. The system is thus photocatalytic with respect to complex, and further, the isomerization continued to a limited extent in the dark [17]. The linear hexenes were also investigated, and no evidence was obtained for direct interconversion between 1-hexene and the 3-hexenes, indicating that products arise only from 1,3-hydrogen shifts.

Metal carbonyl–olefin complexes are produced in the initial stages of irradiation, and infrared evidence indicated formation of $[W(CO)_5(1\text{-pentene})]$ [17]. Likewise, photolysis of $[W(CO)_6]$ in the presence of $C_2H_2D_2$ gave infrared evidence for $[W(CO)_5(C_2H_2D_2)]$ and $[W(CO)_4(C_2H_2D_2)_2]$ [17]. Prolonged photolysis in the presence of olefin led to loss of all metal carbonyl species, and in the case of ethylene an uncharacterized brown precipitate was obtained which had no activity toward $C_2H_2D_2$ photoisomerization. The $[W(CO)_5(1\text{-pentene})]$ complex was itself shown to be photosensitive in the presence of 1-pentene, presumably giving *trans*-$[W(CO)_4(1\text{-pentene})_2]$ with 313 and 366 nm quantum yields of 0.31 and 0.44, respectively [17].

TABLE 3-3

$[M(CO)_6]$-Photoassisted Reactions of Olefins

Metal complex used	Starting olefin	Product(s)	Reference
	Cis–trans isomerization		
$[W(CO)_6]$	cis-1,3-Pentadiene	trans-1,3-Pentadiene	
	trans-1,3-Pentadiene	cis-1,3-Pentadiene	
	trans,trans-2,4-Hexadiene	cis,trans-2,4-Hexadiene	
	cis,trans-2,4-Hexadiene	cis,cis-2,4-Hexadiene	[19]
		trans,trans-2,4-Hexadiene	
	cis,cis-2,4-Hexadiene	trans,trans-2,4-Hexadiene	
		cis-trans-2,4-Hexadiene	
	cis-Stilbene	trans-Stilbene	
	trans-Stilbene	cis-Stilbene	[18]
	trans-1,2-$C_2H_2D_2$	cis-1,2-$C_2H_2D_2$	
	trans-2-Pentene	cis-2-Pentene	
	cis-2-Pentene	trans-2-Pentene	
	trans-3-Hexene	cis-3-Hexene	[17]
	cis-3-Hexene	trans-3-Hexene	
$[Mo(CO)_6]$	cis-1,3-Pentadiene	trans-1,3-Pentadiene	
	trans-Stilbene	cis-Stilbene	[18]
	cis-Stilbene	trans-Stilbene	
	Hydrogen shift reactions		
$[W(CO)_6]$	1-Pentene	cis-2-Pentene	
		trans-2-Pentene	
	cis- or trans-2-pentene	1-Pentene	
	1-Hexene	trans-2-Hexene	
	cis-3-Hexene	cis-2-Hexene	
		trans-2-Hexene	
$[W(CO)_6]$	trans-3-Hexene	cis-2-Hexene	[17]
		trans-2-Hexene	
	cis-1,4-Hexadiene	cis,cis-2,4-Hexadiene	
	trans-1,4-Hexadiene	cis,trans-2,4-Hexadiene	
		trans,trans-2,4-Hexadiene	
	1,4-Pentadiene	trans-1,3-Pentadiene	
		cis-1,3-Pentadiene	
$[Mo(CO)_6]$	1-Pentene	cis-2-Pentene	

$[W(CO)_6]$ has also been shown to assist the isomerization of dienes. As summarized in Table 3-3, 1,4-pentadiene can be converted to a mixture of cis-and trans-1,3-pentadiene; and cis- and trans-1,4-hexadiene can be isomerized to a mixture of trans,trans-2,4-, cis,trans-2,4-, and cis,cis-2,4-hexadiene [17]. For the latter, no interconversion between the cis- and trans-1,4-hexadiene occurred, and 1,3-hexadienes were formed in amounts less than

1/10 that of the 2,4-hexadienes. These data indicate a much lower reactivity for the internal double bond compared to that of the terminal bond. Representative quantum yield data for the isomerization of the hexadienes are shown in Table 3-4 [19].

TABLE 3-4

Quantum Yields for the $[W(CO)_6]$-Photoassisted Isomerization of
0.088 M 2,4-Hexadienes in Pentane Solution[a]

Irradiation time, min	$\Phi_{tt \to ct}$[b]	$\Phi_{tt \to cc}$	$\Phi_{cc \to ct}$	$\Phi_{cc \to tt}$	$\Phi_{ct \to tt}$	$\Phi_{ct \to cc}$
175	0.038	c	0.053	0.019	0.046	0.005
460	0.026	c	0.085	0.019	0.065	0.010

[a] Reprinted with permission from Wrighton et al. [19], J. Am. Chem. Soc. **92**, 6068 (1970). Copyright by the American Chemical Society.
[b] tt = trans,trans; cc = cis,cis; ct = cis,trans.
[c] $\Phi_{tt \to cc}$ is less than 0.002.

Isomerization of *cis*- and *trans*-stilbene, (XX) and (XXi), is also photo-

cis
(XX)

trans
(XXI)

assisted by $[Mo(CO)_6]$ and $[W(CO)_6]$ [18]. An equilibrium mixture of *cis*- and *trans*-stilbene results starting from irradiation of solutions containing either isomer or a mixture of the two, as shown by the data summarized in Fig. 3-4. The production of *cis*-stilbene from *trans*-stilbene represents movement away from the thermodynamic ratio of the two isomers, thus revealing the direct intermediacy of an electronic excited state in the isomerization. The infrared and electronic absorption spectral changes (Fig. 3-5) which obtain during photolysis strongly indicate that $[W(CO)_5(stilbene)]$ is formed initially. Isomerization presumably arises through excitation of this species. As shown by the relative quantum yield data presented in Table 3-5, $[W(CO)_6]$ is most effective for assisting the isomerization, whereas no isomerization was observed with $[Cr(CO)_6]$ [18]. The overall quantum efficiencies for cis–trans isomerization are on the order of 0.01.

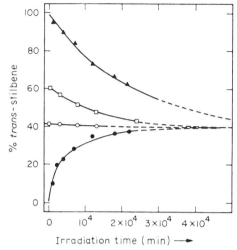

FIG. 3-4. [W(CO)$_6$]-photoassisted stilbene interconversion with different initial *trans/cis*-stilbene ratios. Total stilbene concentration is ~0.05 M, irradiation is at 366 nm at 25°C, and the [W(CO)$_6$] concentration is ~10^{-3} M. Reprinted with permission from Wrighton *et al.* [17].

FIG. 3-5. Spectral changes in the UV–visible region due to photochemical formation of [W(CO)$_5$(*cis*-stilbene)] (– – –) and [W(CO)$_5$(*trans*-stilbene)] (——) from 366 nm irradiation of [W(CO)$_6$] (···). Reprinted with permission from Wrighton *et al.* [18], *J. Am. Chem. Soc.* **93**, 3285 (1971). Copyright by the American Chemical Society.

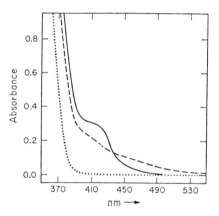

The overall mechanistic scheme which has been written [17] for the [M(CO)$_6$] photoassisted isomerizations is given in Eqs. (3-5)–(3-10).

$$[M(CO)_6] + \text{olefin} \xrightarrow{\ h\nu\ } [M(CO)_5(\text{olefin})] + CO \qquad (3\text{-}5)$$

$$[M(CO)_5(\text{olefin})] + \text{olefin} \xrightarrow{\ h\nu\ } [M(CO)_4(\text{olefin})_2] + CO \qquad (3\text{-}6)$$

$$[M(CO)_5(\text{olefin})] \xrightleftharpoons{\ h\nu \text{ or } \Delta\ } [M(CO)_5(\text{olefin}')] \qquad (3\text{-}7)$$

$$[M(CO)_4(\text{olefin})_2] \xrightleftharpoons{\ h\nu \text{ or } \Delta\ } [M(CO)_4(\text{olefin}')_2] \qquad (3\text{-}8)$$

TABLE 3-5

Observed Initial Relative Rates of
$[M(CO)_6]$-Photoassisted Isomerization of
cis- and trans-Stilbene[a,b]

$[M(CO)_6]$	cis → trans	trans → cis
$[W(CO)_6]$	1.000^c	0.18
$[Mo(CO)_6]$	0.04_9	0.07_3
$[Cr(CO)_6]$	—	0.00_0

[a] Reprinted with permission from Wrighton et al. [18], J. Am. Chem. Soc. **93**, 3285 (1971). Copyright by the American Chemical Society.

[b] 366 nm irradiation, 5.83×10^{-10} einstein/sec.

[c] Corresponds to observed quantum efficiency of 0.1_1.

$$[M(CO)_5(\text{olefin}')] + \text{olefin} \xrightarrow{hv \text{ or } \Delta} [M(CO)_5(\text{olefin})] + \text{olefin}' \tag{3-9}$$

$$[M(CO)_4(\text{olefin}')_2] + \text{olefin} \xrightarrow{hv \text{ or } \Delta} [M(CO)_4(\text{olefin})_2] + 2\ \text{olefin}' \tag{3-10}$$

Olefin′ represents the isomerized olefin, and only Eqs. (3-5) and (3-6) must occur by a photochemical process. Equations (3-8) and (3-9) are necessary to account for the presence of uncoordinated but isomerized olefin in the reaction solution. The key question, however, concerns the actual photoisomerization steps and the nature of any intermediates produced.

Irradiation could induce the formation of a π-allyl hydride, as shown in Eq. (3-11). This is the route by which transition-metal complexes are believed

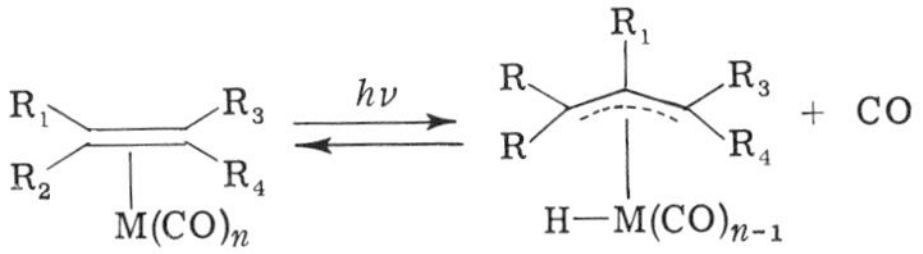

$$\tag{3-11}$$

to thermally catalyze olefin isomerization; the role of light could be simply to induce ligand dissociation and open a coordination site allowing the rearrangement to occur. Photoinduced loss of CO from $[M(CO)_5(\text{olefin})]$, or loss of olefin from $[M(CO)_4(\text{olefin})_2]$, would give coordinatively unsaturated $[M(CO)_4(\text{olefin})]$, and this intermediate could readily rearrange to the π-allyl hydride. A second possibility is that irradiation could produce an excited state in which the nature of the metal–olefin bonding is altered to an extent that free rotation about the olefinic bond could occur (Eq. 3-12).

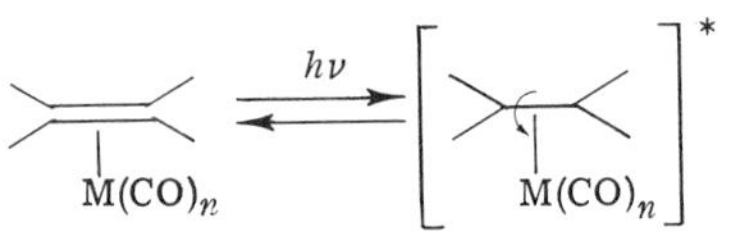

$$\tag{3-12}$$

In light of the thermal isomerization studies, the former process seems more reasonable, especially since in a few experiments isomerization continued for short periods after irradiation had ceased. The π-allyl route, however, cannot account for the cis–trans isomerization of $C_2H_2D_2$ and stilbene, since neither of these olefins possess allylic hydrogens and a π-allyl hydride cannot form. Thus, it was proposed [17,18] that at least for these latter two olefins, isomerization proceeds according to Eq. (3-12) in which the excited state was assumed to be a σ-bound diradical such as that shown in (XXII).

$$\cdot\ M(CO)_n$$

(XXII)

For stilbene, isomerization could also occur via a ligand localized excited state [17]. Consistent with this proposal is the observation that the photo-stationary state using $[W(CO)_6]$ is close to that using high-energy sensitizers in the absence of $[W(CO)_6]$ [20]. Such a mechanism would require internal conversion from the initially populated metal-centered excited states, presumably to the stilbene triplet state which lies below the energetic position of the first absorption in the complexes (Fig. 3-6). Such an energy migration mechanism cannot account for the alkene isomerizations since all ligand localized excited states lie at too high an energy to be populated.

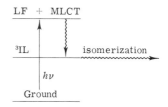

Fig. 3-6.   State diagram appropriate for $[W(CO)_5-$ (stilbene)].

C. $[W(CO)_5(trans\text{-}4\text{-Styrylpyridine})]$
AND $cis\text{-}[W(CO)_4(trans\text{-}4\text{-Styrylpyridine})_2]$

Although these two *trans*-4-styrylpyridine complexes, the former of which has structure (XXIII), do not strictly fall into the class of metal–olefin complexes, it is nevertheless appropriate to discuss them briefly here.

$$(CO)_5W\text{—}N$$

(XXIII)

It has been reported [21] that, whereas $[W(CO)_5(pyridine)]$ undergoes efficient photosubstitution of pyridine, the photosubstitution quantum yield of the analogous *trans*-4-styrylpyridine complex (XXIII) is greatly reduced due to competing cis–trans isomerization of the coordinated ligand. Although relatively few details were presented, it was noted that the 436 nm quantum

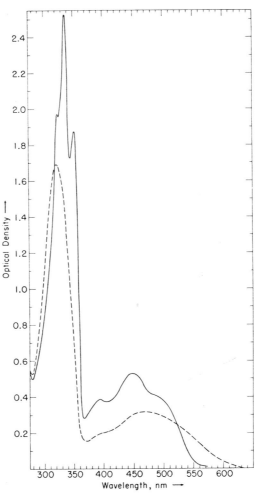

FIG. 3-7. Electronic absorption spectra of *cis*-$[W(CO)_4(trans$-4-styrylpyridine$)_2]$ in EPA at 298 (– – – –) and 77°K (———). The 77°K spectrum is that found upon cooling the 298°K solution and has not been corrected for solvent contraction. Reprinted with permission from Pdungsap and Wrighton [22].

yields for $[W(CO)_5(trans\text{-}4\text{-styrylpyridine})]$ were $\Phi_{\text{dissociation}} = 0.16$, $\Phi_{c \to t} = 0.31$, and $\Phi_{t \to c} = 0.08$. It was proposed that the isomerization occurs from an IL triplet state formed by internal conversion from the lowest lying metal-centered LF triplet.

The disubstituted $cis\text{-}[W(CO)_4(trans\text{-}4\text{-styrylpyridine})_2]$ derivative was investigated in somewhat more detail [22]. The electronic absorption spectrum of the complex is shown in Fig. 3-7, and band assignments were made on the basis of solvent shifts and comparison to free ligand and $cis\text{-}[W(CO)_4L_2]$ spectra. The two lowest bands at 500 and 450 nm were attributed to MLCT transitions, the band at about 390 nm was assigned as an LF transition, and the intense band at 310 nm with vibrational structure is probably an IL transition. Irradiation of the complex gives only trans → cis isomerization of the coordinated ligand, but the process is very wavelength-dependent. Efficient reaction occurs only under 313 nm irradiation, and the measured quantum yields are 0.001 (436 nm), 0.02 (366 nm), 0.1 (313 nm), and 0.05 (254 nm). The 436 nm quantum yield of photosubstitution is ≤0.001. The scheme illustrated in Fig. 3-8 was proposed to account for the observed excited-state processes [22]. It was proposed that isomerization occurs from a triplet IL state that is itself populated by internal conversion either from the singlet IL state or from the triplet LF state. The photosubstitution inertness implied that the lowest state is not the ^{3}LF state, known to be substitution-active, but rather the ^{3}IL or ^{3}MLCT state [22].

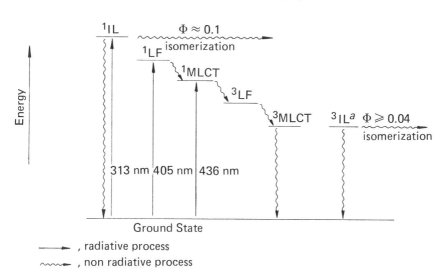

FIG. 3-8.   Excited-state processes in $cis\text{-}[W(CO)_4(trans\text{-}4\text{-styrylpyridine})_2]$. Reprinted with permission from Pdungsap and Wrighton [22].

IV. RHENIUM COMPLEXES

$[ReX(CO)_3L_2]$ (X = Cl, Br; L = *trans*-3-STYRYLPYRIDINE, *trans*-4-STYRYLPYRIDINE)

Wrighton, Morse, and Pdungsap [23] have studied the photochemical properties of these $[ReX(CO)_3L_2]$ complexes which have the structure (XXIV). Electronic absorption spectral measurements unambiguously dem-

$$
\begin{array}{c}
X \\
| \\
OC\text{-}\text{-}\text{-}\text{-}L \\
\diagup Re \diagup \\
Cl\text{-}\text{-}\text{-}\text{-}\text{-}L \\
| \\
CO
\end{array}
$$

(XXIV)

onstrate that the lowest lying, but intense, transition in each of these complexes is a spin-allowed singlet IL $\pi–\pi^*$ transition. Irradiation of the complexes leads to efficient trans → cis isomerization of the styrylpyridine ligands, and representative quantum yield data for reaction (3-13) are set out in Table 3-6. The similarity between the triplet-sensitized isomerization of

$$[ReX(CO)_3(trans\text{-}stpy)_2] \xrightarrow{\ hv\ } [ReX(CO)_3(trans\text{-}stpy)(cis\text{-}stpy)] \qquad (3\text{-}13)$$

stpy = styrylpyridine

the complexes and the triplet-sensitized isomerization of the free ligand led

TABLE 3-6

Direct Photolysis Isomerization of Coordinated Styrylpyridines[a,b]

Compound	$\Phi_{t\to c}$ 313 nm	$\Phi_{t\to c}$ 366 nm	Cis at PSS ± 2 (%) 313 nm	366 nm	436 nm
trans-3-Styrylpyridine	0.48	c	90	c	c
trans-4-Styrylpyridine	0.38	c	88	c	c
$[ReCl(CO)_3(trans\text{-}4\text{-}stpy)_2]$	0.49	0.54	84	90	99
$[ReBr(CO)_3(trans\text{-}4\text{-}stpy)_2]$	0.64	0.51	99	98	>99
$[ReCl(CO)_3(trans\text{-}3\text{-}stpy)_2]$	0.60	0.51	93	90	99

[a] Reprinted with permission from Wrighton *et al.* [23], *J. Am. Chem. Soc.* **97**, 2073 (1975). Copyright by the American Chemical Society.

[b] CH_2Cl_2 solutions at 25°C; light intensity $\sim 10^{-7}$ einstein/min; Φs are ±10%.

[c] No absorption at these wavelengths.

the authors to propose that the sensitized isomerization of the complexes occurs from a relatively nonperturbed triplet IL state. It is not clear whether the direct irradiation proceeds via an IL triplet or singlet state.

V.   IRON, RUTHENIUM, AND OSMIUM COMPLEXES

A.   Photocatalyzed Reactions of Alkenes Using the Binary Carbonyls as Catalyst Precursors

Irradiation of $[Fe(CO)_5]$ or $[M_3(CO)_{12}]$ (M = Fe, Ru, Os) in the presence of alkenes has been shown to lead to a variety of reactions presumably involving olefin complexes as the actual catalytically active species [2a,10a,b]. For example, irradiation of $[Fe(CO)_5]$ in the presence of 1-pentene results in the formation of *cis*- and *trans*-2-pentene with initial quantum yields which exceed 400! Ultimately, equilibration of the linear pentenes can be achieved by irradiation of $[Fe(CO)_5]$ in the presence of any one of the three isomers. Visible light excitation of $[Fe_3(CO)_{12}]$ produces photodeclusterification and, when carried out in the presence of 1-pentene, isomerization does result. $[Ru_3(CO)_{12}]$ is somewhat less effective, but $[Os_3(CO)_{12}]$ seems to be qualitatively poor as an alkene isomerization photocatalyst.

Irradiation of $[Fe(CO)_5]$ in the presence of an alkene and H_2 yields the corresponding alkane for a fairly large number of alkenes [2a]. Also, alkene reactions with $HSiR_3$ (R = Me, Et) can be photocatalyzed to yield a mixture of the corresponding alkane, (*n*-alkyl)SiR_3, and several isomers of (alkenyl)SiR_3 [10c]. All Si-containing products apparently result from the addition of the $-SiR_3$ moiety to a terminal carbon of the alkene. Hydrogenation and reaction of alkenes with silicon hydrides can be thermally catalyzed using $[Fe(CO)_5]$ but only at elevated temperatures. Since light induces the dissociation of CO from $[Fe(CO)_5]$, the role of light is effectively to lower the activation energy for the generation of the actual catalytically repeating unit which is believed to be $[Fe(CO)_3]$.

The $[M_3(CO)_{12}]$ species also yield reactions of alkenes with silicon hydrides [10b]. For M = Fe or Ru, there is evidence to suggest that the actual catalyst is a mononuclear species from photodeclusterification, but for M = Os it appears that the Os_3 unit remains intact during the photocatalysis. The retention of the Os cluster during the catalysis may account for the qualitative differences in the distribution of Si-containing products [(alkyl)SiR_3 vs. (alkenyl)SiR_3] for M = Os compared to M = Fe or Ru. In all cases there is a strong preference for the hydrosilation of terminal vs. internal double bonds.

B. [Fe(CO)₄(Olefin)]

Koerner von Gustorf and co-workers [24–32] have studied the photo-chemical properties of a series of $[Fe(CO)_4(olefin)]$ complexes in which the olefin is *cis*- or *trans*-1,2-dibromoethylene, *cis*-1-bromo-2-fluoroethylene, *cis*- or *trans*-1,2-dichloroethylene, dimethylmaleate, dimethylfumarate, or methylacrylate. These olefins all have electron-withdrawing groups which yield thermally inert $[Fe(CO)_4(olefin)]$ complexes. By comparison to the simple alkene complexes, such species can be better viewed as having a metallocyclopropane structure, (XXVb). The electronic absorption spectral

(XXVa)   (XXVb)

data for these complexes are set out in Table 3-7, and a qualitative molecular orbital diagram which has been used to interpret the spectra is shown in

TABLE 3-7

Electronic Absorption Spectral Data for Some $[Fe(CO)_4L]$
Complexes in Hexane Solution

L	$\nu \times 10^{-3}$ cm^{-1} (ε, M^{-1} cm^{-1})			
cis-1,2-Dibromoethylene	~28.5 sS (470)	36 S (5400)	~43 sS (14,600)	46 (17,000)
trans-1,2-Dibromoethylene	~28 sS (490)	34.7 M (5680)	~40.5 sS (10,200)	~46 (17,000)
cis-1-Bromo-2-fluoroethylene	~29.5 sS (700)	~37 sS (5100)		
trans-1-Bromo-2-fluoroethylene	~29 sS (420)	~36.5 sS (4500)	~40.5 sS (8500)	46 (17,500)
cis-1,2-Dichloroethylene	~29 sS (550)	37 S (4900)		~46 (18,300)
trans-1,2-Dichloroethylene	~29 sS (290)	36.5 S (4550)	~43 sS (15,000)	46 (17,500)
Maleic acid dimethylester	~29 sS (630)	38.5 S (7700)		46 (20,400)
Fumaric acid dimethylester	~28.5 sS (710)	36.0 M (9100)		46 (21,600)
Acrylic acid methylester	~29 sS (700)	38.5 S (7600)		46 (20,000)
CO		35.5 (3800)	41.5 (10,200)	50 (37,000)

[a] Reprinted with permission from Grevels and Koerner von Gustorf [31].
[b] sS = Weak shoulder; S = shoulder; M = maximum.

Fig. 3-9. Each of the olefin complexes shows a band between 34,700 and 38,500 cm^{-1} with $\varepsilon = 4500$–9430, and this band has been attributed to an M $\rightarrow$ olefin CT transition [31]. The shoulder at low energy is presumably an LF transition.

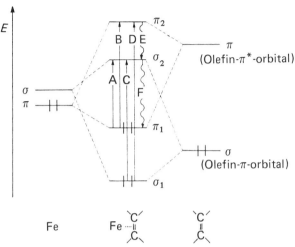

FIG. 3-9. Qualitative molecular orbital diagram for an [Fe(CO)$_4$(olefin)] complex. Reprinted with permission from Grevels and Koerner von Gustorf [31].

Unfortunately, few details of the photochemical studies have been published in the journal literature, and we can only summarize the available information. Irradiation of [Fe(CO)$_4$(dimethylmaleate)] has been reported [27,28] to lead to photoexchange with labeled dimethylmaleate and to cis $\rightarrow$ trans isomerization of the coordinated olefin. The latter reaction is of low efficiency and gives [Fe(CO)$_4$(dimethylfumarate)] as the product, whereas the former occurs with quantum yields between 0.27 and 0.47 [28]. It was suggested that photoisomerization occurs from the Fe $\rightarrow$ olefin CT state and olefin exchange from the lower lying LF state. The low efficiency of the photoisomerization presumably arises because of competing deactivation via efficient internal conversion to the LF state. It was subsequently argued [29] that an MLCT transition which populates an olefin π^* orbital can lead to isomerization because of the reduced olefin bond order.

Irradiation of [Fe(CO)$_4$(methylacrylate)] in the presence of excess methylacrylate at $-30°$C has been reported [30] to give [Fe(CO)$_3$(methylacrylate)$_2$]. The latter was stable at low temperature but established the equilibrium shown in Eq. (3-14) when warmed to $-5°$C. Irradiation of [Fe(CO)$_4$(methylacrylate)] and excess methylacrylate at $+20°$C instead of $-30°$C led to two

$$[\text{Fe(CO)}_3(\text{CH}_3\!=\!\text{CH}\!-\!\text{COOCH}_3)_2] \rightleftharpoons$$
$$[\text{Fe(CO)}_3(\text{CH}\!=\!\text{CH}\!-\!\text{COOCH}_3)] + \text{CH}_2\!=\!\text{CH}\!-\!\text{COOCH}_3 \quad (3\text{-}14)$$

isomeric complexes with the composition $[Fe(CO)_4(C_4H_6O_2)_2]$. These were shown by spectral studies to be the ferracyclopentane derivatives (XXVIa) and (XXVIb).

$$
\begin{array}{c}
\text{R}_1 \quad \text{R}_2 \\
\text{H} \\
\text{H} \\
\text{H} \quad \text{H} \quad \text{Fe(CO)}_4 \\
\text{H} \\
\overset{|}{\text{COOCH}_3}
\end{array}
$$

(XXVIa) $R_1 = COOCH_3$, $R_2 = H$
(XXVIb) $R_1 = H$, $R_2 = COOCH_3$

The dihaloolefin complexes are also photosensitive and the net photo-reaction is that illustrated in Eq. (3-15) [31]. This transformation was

$$
\text{(CO)}_4\text{Fe}\!-\!\overset{\overset{\text{H}\diagdown\;\diagup\text{Br}}{\text{C}}}{\underset{\overset{\text{C}}{\text{H}\diagup\;\diagdown\text{Br}}}{\|}} \quad \xrightarrow{h\nu} \quad \text{(CO)}_3\text{Fe}\!-\!\!-\!\text{Fe(CO)}_3
$$
(3-15)

proposed to arise via the reactions represented in Eqs. (3-16)–(3-19) [29,31].

$$
\text{(CO)}_4\text{Fe} \xrightarrow{h\nu} \text{(CO)}_3\text{Fe} + CO
$$
(3-16)

$$
\longrightarrow \text{Fe(CO)}_4 + \text{olefin}
$$
(3-17)

$$
\text{(CO)}_3\text{Fe} \longrightarrow \underset{\text{(CO)}_3\text{Fe}}{\text{C}=\text{C}} \xrightarrow{CO} \underset{\text{(CO)}_4\text{Fe}}{\text{C}=\text{C}}
$$
(3-18)

$$
\underset{\text{(CO)}_4\text{Fe}}{\text{C}=\text{C}} + \text{Fe(CO)}_4 \longrightarrow \text{(CO)}_3\text{Fe}\!-\!\text{Fe(CO)}_3 + 2\,CO
$$
(3-19)

Irradiation presumably can lead to labilization of both carbon monoxide and olefin. If the former occurs, the coordinatively unsaturated $[Fe(CO)_3(olefin)]$ complex can undergo oxidative addition across an olefin–halogen bond to give an $[Fe(CO)_3X(vinyl)]$ complex. The latter can then react with both CO and the $[Fe(CO)_4]$ fragment photochemically produced by labilization of olefin, and the dimerization product results.

The same product obtains whether *cis*- or *trans*-dihaloolefin complexes are irradiated, implying that cis–trans isomerization about the olefinic bond also occurs in the process. Isomerization presumably arises from the MLCT state. It has also been argued that the oxidative-addition reaction could occur directly from the initially populated MLCT state [31]. Such a state would give an oxidized metal center and a nucleophilic olefin, and an intermediate such as shown in reaction (3-20) has been drawn [31].

$$(CO)_4Fe \overset{H}{\underset{CHX}{\overset{\diagdown C \diagup X}{\|}}} \xrightarrow[\text{MLCT}]{h\nu} \left[(CO)_4\overset{\oplus}{Fe} \overset{H}{\underset{CHX}{\overset{\diagdown \overset{\ominus}{C} \diagup X}{\|}}} \right] \longrightarrow (CO)_4Fe \overset{X}{\underset{\underset{H}{\diagdown}C=CHX}{\diagup}} \qquad (3\text{-}20)$$

It could also be surmised that the photochemical reaction could occur from two different excited states in the complex. The MLCT state believed to be responsible for the intense band near 37,000 cm^{-1} could lead to the $[Fe(CO)_4(X)(—CHCHX)]$ intermediate through reaction (3-20), whereas an LF state could give loss of olefin and generation of $[Fe(CO)_4]$. The two intermediates could then combine to give the observed product. The low-energy shoulders near 29,000 cm^{-1} in the spectra are suggestive of an LF transition, and its presence is consistent with MLCT and LF states competing for the excitation energy. Unfortunately, the necessary wavelength-dependence studies have not yet been reported, and it is not now possible unambiguously to distinguish between the two routes.

C. $[Fe(CO)_3COT]$

Schwartz [33] has shown that irradiation of $[Fe(CO)_3COT]$ (COT = cyclooctatetraene) in degassed hexane solution gives a clean and high-yield production of $[Fe_2(CO)_5COT]$ (Eq. 3-21). The same product was shown to

$$[Fe(CO)_3COT] \xrightarrow{h\nu} \underset{\substack{O \\ \| \\ C}}{\overset{\diagdown}{Fe}} \!\!-\!\! \overset{\diagdown}{Fe} \quad + \text{ COT } + \text{ CO} \qquad (3\text{-}21)$$

result from irradiation of the $[Fe_2(CO)_6COT]$ dimer (XXVII), but only

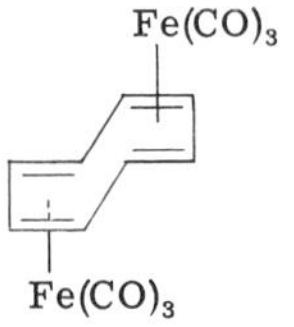

(XXVII)

traces of (XXVII) were found even in cases of low photochemical conversion of $[Fe(CO)_3COT]$. This finding suggested that the reaction does not proceed through (XXVII) but that the initially produced $[Fe(CO)_2COT]$ intermediate reacts with another molecule of $[Fe(CO)_3COT]$ to give the product directly.

Schrauzer and co-workers [34,35] have reported that prolonged irradiation of $[Fe(CO)_3COT]$ in the presence of excess cyclooctatetraene gives two isomeric products of the formula $[Fe(CO)_3C_{10}H_{16}]$. These two compounds are believed to have structures (XXVIII) and (XXIX), and it was suggested that they result from reaction of uncoordinated C_8H_8 with an intermediate such as (XXX) which could be formed photochemically from $[Fe(CO)_3COT]$ (Eq. 3-22).

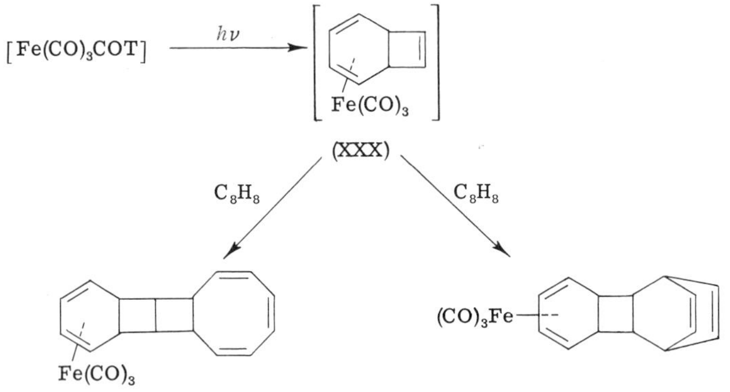

(3-22)

(XXVIII)   (XXIX)

D. $[Fe(CO)_3(\eta^4\text{-}C_4H_4)]$

Several reports concerning the photochemical properties of cyclobuta-dienetricarbonyliron have appeared [36–42]. The first was that by Gunning and co-workers [36] who obtained mass spectral evidence for the production of uncoordinated C_4H_4 when $[Fe(CO)_3(\eta^4\text{-}C_4H_4)]$ was flash-photolyzed in the vapor phase. They thus concluded that the primary photochemical step was that shown in Eq. (3-23). It was further noted that photolysis of the

$$[Fe(CO)_3(\eta^4\text{-}C_4H_4)] \xrightarrow{h\nu} [Fe(CO)_3] + C_4H_4 \qquad (3\text{-}23)$$

complex in ether solution in the presence of dimethylacetylenedicarboxylate gave a 17% yield of dimethylphthalate. It was implied that this reaction proceeded by the thermal interaction of photochemically generated C_4H_4 with the acetylene. Irradiation in the presence of O_2 generated furan.

Chapman and co-workers [37] subsequently irradiated $[Fe(CO)_3(\eta^4\text{-}C_4H_4)]$ isolated in a krypton matrix with $\lambda > 2800$ Å and observed no evidence for production of C_4H_4 but rather only the reaction shown in Eq. (3-24). It had been previously noted [38] that the photochemical reac-

$$[Fe(CO)_3(\eta^4\text{-}C_4H_4)] \xrightarrow{h\nu} [Fe(CO)_2(\eta^4\text{-}C_4H_4)] + CO \qquad (3\text{-}24)$$

tion between $[Fe(CO)_3(\eta^4\text{-}C_4H_4)]$ and dimethylacetylenedicarboxylate observed by Gunning et al. [36] could proceed through initial coordination of the acetylene to photochemically generated $[Fe(CO)_2(\eta^4\text{-}C_4H_4)]$; thus Chapman's study casts doubt on the validity of reaction (3-23). It has also been observed that photolysis in a N_2 matrix leads to formation of $[Fe(CO)_2(N_2)(\eta^4\text{-}C_4H_4)]$ [39]. Braterman et al. [41], however, correctly pointed out that the rigid matrix cage could retard the displacement of relatively bulky C_4H_4, and hence the matrix isolation experiments must be treated with caution.

More recent results do tend to suggest that loss of CO according to Eq. (3-24) is the primary photochemical process for $[Fe(CO)_4(\eta^4\text{-}C_4H_4)]$. Koerner von Gustorf and co-workers [40] found that irradiation of $[Fe(CO)_4(\eta^4\text{-}C_4H_4)]$ with $\lambda \geq 280$ nm in THF solution at $-40°C$ led to a product formulated as $[Fe_2(CO)_3(\eta^4\text{-}C_4H_4)_2]$ (Eq. 3-25). This reaction is only consistent with labilization of CO since it is unlikely that free C_4H_4

$$[Fe(CO)_4(\eta^4\text{-}C_4H_4)] \underset{+CO}{\overset{h\nu,\ -CO}{\rightleftarrows}} \qquad (3\text{-}25)$$

would live long enough to recombine with the iron fragments. It was further shown that irradiation of $[Fe(CO)_4(\eta^4\text{-}C_4H_4)]$ in the presence of $P(OMe)_3$ led only to $[Fe(CO)_2(\eta^4\text{-}C_4H_4)P(OMe)_3]$ and that the quantum yield of CO loss must be 10 times that of loss of C_4H_4 [40]. Irradiation of $[Fe(CO)_3(\eta^4\text{-}C_4Me_4)]$ in hexane in the presence of C_2HF_3 has been reported to lead to insertion of the olefin into the complex to give the isomeric products (XXXI) and (XXXII) [42]. These products presumably arise through

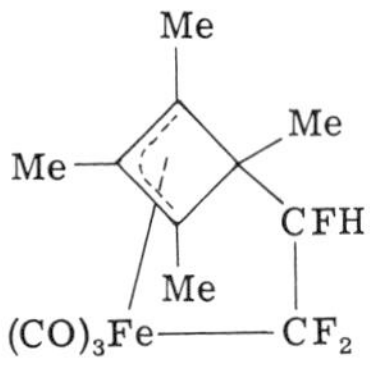

(XXXI) 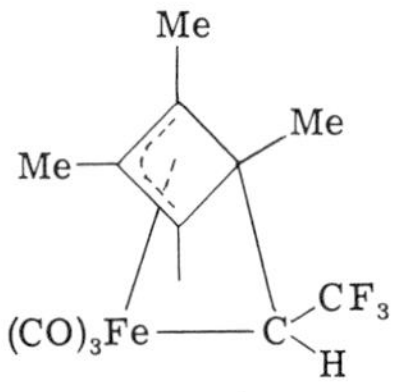   (XXXII)

initial coordination of C_2HF_3 to photogenerated $[Fe(CO)_2(\eta^4\text{-}C_4Me_4)]$.

E. $[Fe(CO)_3C(CH_2)_3]$

Irradiation of trismethylenemethyltetracarbonyliron (XXXIII) in pentane,

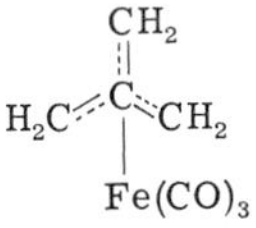

(XXXIII)

cyclopentane, and cyclopentadiene solutions has been shown to lead to 20, 16, and 16 products, respectively [43]. Space does not permit presentation of these products and a discussion of how they might arise, but it was proposed that many derived from thermal reactions of photochemically liberated $C(CH_2)_3$ [43].

F. $[Fe(CO)_3(1,4\text{-}Diphenylbutadiene)]$

Iron pentacarbonyl has been reported to photoassist the dimerization of norbornadiene [44]. Although no mechanistic details were given, it is likely that $[Fe(CO)_3NBD]$ complexes are formed first, either thermally or photochemically, and irradiation than gives loss of CO to open a coordination site for subsequent norbornadiene addition. Under the assumption that an $[Fe(CO)_3diene]$ intermediate might play a key role in the dimerization, the photochemical properties of $[Fe(CO)_3(1,4\text{-diphenylbutadiene})]$ were examined by Jennings and co-workers [45]. Irradiation of the complex in the presence of excess norbornadiene led to stereospecific production of only three dimers, in contrast to the large number of products obtained with $[Fe(CO)_5]$ and other metal carbonyls.

G. $[Fe(C_2H_4)(Ph_2PCHCHPPh_2)_2]$

Ultraviolet irradiation of the title complex has been reported [46] to induce loss of C_2H_4 and formation of the ortho-metallated derivative shown in Eq. (3-26). The product presumably arises through initial generation of highly reactive, 16-valence-electron $[Fe(Ph_2PCHCHPPh_2)_2]$ which satisfies

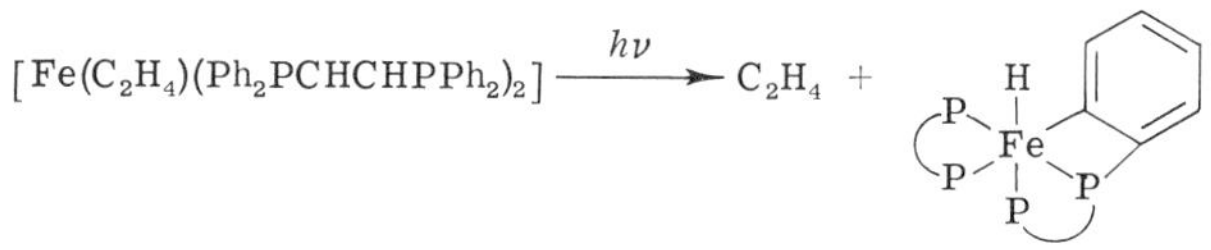

$$[Fe(C_2H_4)(Ph_2PCHCHPPh_2)_2] \xrightarrow{h\nu} C_2H_4 \; + \qquad\qquad (3\text{-}26)$$

its requirement to obtain 18-valence electrons by insertion into one of its ligands. The insertion reaction is reversible since treatment of the ortho-metallated product with C_2H_4 regenerated the original complex. Under HCl and H_2, the complexes $[FeClH(Ph_2PCHCHPPh_2)_2]$ and $[FeH_2(Ph_2PCHCHPPh_2)_2]$ were formed [46].

H. $[Ru(bipy)_2(4\text{-}\text{STYRYLPYRIDINE})_2]^{2+}$
AND $[RuCl(bipy)_2(4\text{-}\text{STYRYLPYRIDINE})]^+$

Whitten and co-workers [47,48] have studied the photochemistry of the four complexes $[Ru(bipy)_2(cis\text{-}4\text{-styrylpyridine})_2]^{2+}$, $[Ru(bipy)_2(trans\text{-}4\text{-}$styrylpyridine$)_2]^{2+}$, $[RuCl(bipy)_2(trans\text{-}4\text{-styrylpyridine})]^+$, and $[RuCl\text{-}$(bipy$)_2(cis\text{-}4\text{-styrylpyridine})]^+$. The complexes all show similar, but not identical, absorption spectra; and for illustration the spectrum of $[Ru(bipy)_2(trans\text{-}4\text{-styrylpyridine})_2]^{2+}$ is shown in Fig. 3-10. The bands in the 350–500 nm region were assigned as MLCT transitions, some of which are Ru $\rightarrow$ styrylpyridine CT in nature. The intense band near 300 nm was ascribed to an IL 4-styrylpyridine $\pi \rightarrow \pi^*$ transition because of its similarity to the absorption band shown by the free ligand.

Irradiation of each of the complexes at 570, 436, 366, or 313 nm leads only to cis–trans isomerization of the styrylpyridine ligands. The cis:trans ratio and the isomerization quantum yields, however, are highly dependent

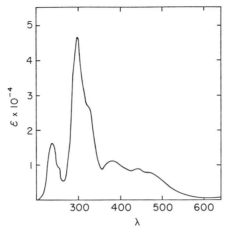

FIG. 3-10. Electronic absorption spectrum of [Ru(bipy)₂(*trans*-4-styrylpyridine)₂]BF₄ in ethanol solution. Reprinted with permission from Zarnegar *et al.* [48], *J. Am. Chem. Soc.* **95**, 4367 (1973). Copyright by the American Chemical Society.

upon the wavelength of irradiation (Tables 3-8 and 3-9). Long-wavelength irradiation gives stationary states rich in complexes of the trans ligand, whereas 313 nm photolysis gives complexes rich in the cis ligand. This observation was attributed to the presence of two different but weakly interacting excited states, each of which leads to the same photoreaction. Irradiation at 313 nm presumably populates the IL $\pi \rightarrow \pi^*$ state. The 0.96

TABLE 3-8

Photostationary States for Direct Irradiation of
Ruthenium(II)–4-Styrylpyridine Complexes[a,b]

Wavelength, nm	Trans isomer in photostationary state (%)	
	Complexes (1) and (2)	Complexes (3) and (4)
313	35 ± 2.3	23.6 ± 1.1
366	88 ± 0.3	
436	85 ± 1.5	98.6 ± 0.9
570	88 ± 2.3	

[a] Reprinted with permission from Zarnegar *et al.* [48], *J. Am. Chem. Soc.* **95**, 4367 (1973). Copyright by the American Chemical Society.

[b] Butyronitrile solution, $T = 25°C$. (1) = [Ru(bipy)₂(*trans*-4-styrylpyridine)₂]²⁺; (2) = [Ru(bipy)₂(*cis*-4-styrylpyridine)₂]²⁺; (3) = [RuCl(bipy)₂(*trans*-4-styrylpyridine)]⁺; (4) = [RuCl(bipy)₂(*cis*-4-styrylpyridine)]⁺.

TABLE 3-9

Measured and Calculated Quantum Yields for Direct Isomerization of
Ruthenium(II)–4-Styrylpyridine Complexes[a,b]

Wavelength, nm	Complex (1) $\Phi_{t \to c}$	Complex (2) $\Phi_{c \to t}$	Complex (3) $\Phi_{t \to c}$	Complex (4) $\Phi_{c \to t}$
313	0.15	$0.15_6{}^c$	0.09	$0.12_3{}^c$
366	0.05	0.15		
436	0.05^c	0.51	$0.03_5{}^c$	0.66_5

[a] Reprinted with permission from Zarneger *et al.* [48], *J. Am. Chem. Soc.* **95**, 4367 (1973). Copyright by the American Chemical Society.
[b] Butyronitrile solutions, $T = 25°C$; (1) = [Ru(bipy)$_2$(*trans*-4-styrylpyridine)$_2$]$^{2+}$; (2) = [Ru(bipy)$_2$(*cis*-4-styrylpyridine)$_2$]$^{2+}$; (3) = [RuCl(bipy)$_2$(*trans*-4-styrylpyridine)]$^+$; (4) = [RuCl(bipy)$_2$(*cis*-4-styrylpyridine)]$^+$.
[c] Calculated from the stationary state.

cis:trans ratio at the stationary state resulting from 313 nm irradiation is close to the 1.01–1.36 values obtained for direct excitation of free 4-styrylpyridine. Longer wavelength irradiation populates MLCT states in which the metal is oxidized and the ligand system reduced. Importantly, the cis:trans ratio of 0.04 obtained for 546 nm irradiation is remarkably similar to the thermodynamic ratio of 0.006 obtained for the free 4-styrylpyridine. The evidence is thus strong that cis–trans isomerization occurs via two different excited states, IL $\pi \to \pi^*$ and MLCT, which apparently do not undergo efficient internal conversion. The latter was attributed not to a slow internal conversion process but rather an extremely short lifetime ($\leq 2 \times 10^{-13}$ sec) of the IL state.

VI. RHODIUM COMPLEXES

[RhCl(COD)]$_2$

Srinivasan [49] originally noted that [RhCl(COD)]$_2$ (COD = 1,5-cyclooctadiene) (XXXIV) photoassisted the isomerization of 1,5-cycloocta-diene to 1,3-cyclooctadiene, 1,4-cyclooctadiene, and bicyclo[4.2.0]octene-7 (XXXIV) when irradiated with 254 nm light. In the absence of excess cyclo-octadiene the same products were observed, but a brown precipitate—of inorganic origin—was deposited. In the presence of excess cyclooctadiene the precipitate was not obtained. It was suggested that (XXXV) was formed

(XXXIV)   (XXXV)

by direct photolysis of 1,3-cyclooctadiene and that the 1,3-cyclooctadiene was itself formed by a series of C—C bond shifts [49].

Salomon and El Sanadi [50] extended these studies and demonstrated the formation of cyclooctene and bicyclo[3.3.0]octene-2 (XXXVI) in addition

(XXXVI)

to the products observed by Srinivasan [49]. It was again proposed that (XXXV) formed by direct excitation of 1,3-cyclooctadiene and further that (XXXII) results from direct photolysis of 1,4-cyclooctadiene through the route shown in Eq. (3-27). It was further demonstrated that 1,4-cyclooctadiene

$$\overset{h\nu}{\longrightarrow} \qquad \overset{[Rh]}{\longrightarrow} \qquad \qquad (3\text{-}27)$$

was the exclusive primary photoproduct from 1,5-cyclooctadiene and that all the other products arise from its rearrangement. The overall mechanism shown in Fig. 3-11 was suggested. The initial isomerization step was proposed to proceed by initial photodissociation of one of the two coordinated olefin bonds to give a coordinatively unsaturated Rh(I) intermediate which

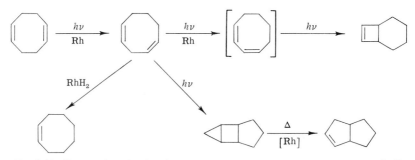

FIG. 3-11. Suggested mechanism for Rh-photoassisted reactions of cyclooctadiene [50].

could then form a π-allyl hydride according to Eq. (3-28). Collapse of the π-allyl hydride could give either the starting 1,5-cyclooctadiene or the

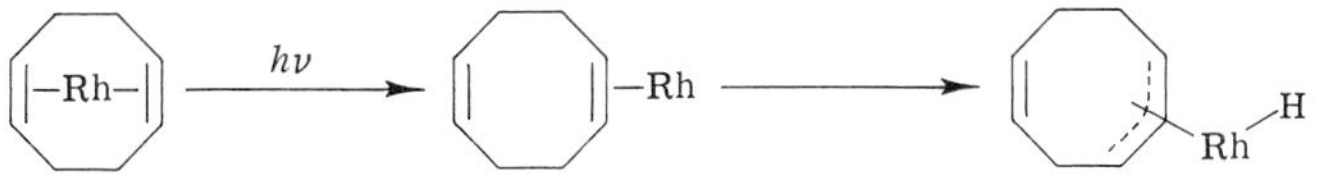

$$\text{(3-28)}$$

coordinated 1,4-cyclooctadiene as a final product.

VII. PLATINUM COMPLEXES

A. $K[PtCl_3(C_2H_4)]$

Zeise's salt $K[PtCl_3(C_2H_4)]$ is historically a very important compound for it was the first characterized olefin complex. The complex is reasonably stable in acidic solution although it decomposes in alkaline or neutral solution. The ethylene ligand exhibits a strong trans effect, however, and in aqueous solution the equilibrium shown in Eq. (3-29) rapidly obtains [51]. Thermal aquation of a cis chloride has also been reported (Eq. 3-30), but

$$[PtCl_3(C_2H_4)]^- \xrightleftharpoons{K\,=\,3\,\times\,10^{-3}} \textit{trans-}[PtCl_2(OH_2)(C_2H_4)] + Cl^- \qquad \text{(3-29)}$$

to a much lesser extent than trans aquation. [52].

$$[PtCl_3(C_2H_4)]^- \rightleftharpoons \textit{cis-}[PtCl_2(OH_2)(C_2H_4)] + Cl^- \qquad \text{(3-30)}$$

Natarajan and Adamson [52] have studied the photochemical properties of Zeise's salt. The interesting fact is that ethylene aquation can be observed photochemically but not thermally. Chloride aquation was also found (Eq. 3-31).

$$[PtCl_3(C_2H_4)]^- \xrightarrow{h\nu} \begin{cases} [PtCl_3(OH_2)]^- + C_2H_4 \\ \\ \textit{cis-}[PtCl_2(OH_2)(C_2H_4)] + Cl^- \end{cases} \qquad \text{(3-31)}$$

However, the back anation reaction shown in Eq. (3-29) is so rapid that they were not able to observe trans aquation and only cis chloride aquation was studied. The electronic absorption spectrum of Zeise's salt is shown in Fig. 3-12. The quantum yields of ethylene aquation are given in Table 3-10 and are represented by the relative height of the bars in Fig. 3-12. The quantum yield for ethylene aquation is essentially zero at longer wavelengths but

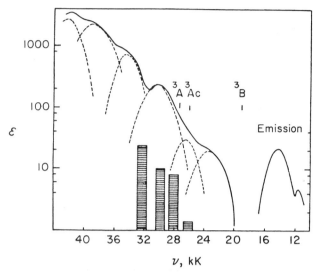

$\mathcal{E}$

Fig. 3-12. Absorption and emission spectra of Zeise's salt. Emission intensity, in arbitrary units, is plotted on a logarithmic scale. The superimposed bar diagram shows the quantum yields for ethylene aquation over the indicated width of wavelengths; the bars are linearly proportional to the quantum yields. Reprinted with permission from Natarajan and Adamson [52], *J. Am. Chem. Soc.* **93**, 5599 (1971). Copyright by the American Chemical Society.

TABLE 3-10

Ethylene Photoaquation of Zeise's Salt[a]

Irradiation wavelength, nm	Filter combination[b]	Concentration, M		T, °C	Reaction,[c] %	$\Phi_3{}^d$
		Complex	HCl			
420	CS-3-73 + BL-410	0.01	0.75	25	0	$<10^{-4}$
		2.5×10^{-3}	0.065	5	0	$<10^{-4}$
410	CS-4-67 + BL-410	0.01	0.75	25	0	$<10^{-4}$
385	CS-4-96 + BL-380	2.5×10^{-3}	0.75	25	0.93	0.008
345	CS-4-96 + BL-520	2.5×10^{-3}	0.75	25	3.5	0.066
340	BL-520 + BL-670	2.5×10^{-3}	0.75	25	3.72	0.078
		2.5×10^{-3}	0.50	20	8.6	0.057
		2.5×10^{-3}	0.50	28	9.0	0.060
		2.5×10^{-3}	0.50	36	9.0	0.090
305	OT-305	2.5×10^{-3}	0.75	25	5.1	0.100

[a] Reprinted with permission from Natarajan and Adamson [52], *J. Am. Chem. Soc.* **93**, 5599 (1971). Copyright by the American Chemical Society.
[b] CS, Corning glass filter; BL, Bausch and Lomb interference filter; OT, Optics Technology interference filter.
[c] Percent of ethylene aquation.
[d] Quantum yield for ethylene aquation.

becomes noticeable at 385 nm and reaches a maximum at 305 nm. The ratio of chloride aquation to ethylene aquation was 1.8 at 340 nm irradiation and 1.4 at 305 nm. The sums of the individual quantum yields were 0.22 and 0.24 at the two wavelengths. Although the electronic absorption spectrum could not be definitively assigned, it was argued on the basis of tentative assignments and sensitization data that both ethylene and chloride aquation probably occurred from a singlet LF state that possesses near-tetrahedral geometry. This was rationalized on the basis of previous studies which had argued that certain excited states in planar d^8 complexes of this type should adopt a tetrahedral geometry and further that the olefin–metal π bonding should be greatly weakened in a tetrahedral environment. More recent SCF-Xα calculations by Johnson and co-workers [53], however, suggest that it is a Pt $\rightarrow$ C$_2$H$_4$ CT transition, populating a strongly antibonding metal–olefin orbital, that actually gives rise to ethylene aquation. Such a CT transition makes the metal more nucleophilic and the reaction may proceed via an associative process.

B.   [Pt(PPh$_3$)$_2$(DICYANOACETYLENE)] AND
     [Pt(PPh$_3$)$_2$(TETRACYANOETHYLENE)]

Baddley and co-workers [54] have reported that photolysis of [Pt(PPh$_3$)$_2$(NCC$_2$CN)] for 3 h with a sunlamp gives rise to the rearrangement shown in Eq. (3-32). The product was characterized by x-ray crystallog-

$$
\begin{array}{cc}
\underset{\text{Ph}_3\text{P}}{\overset{\text{Ph}_3\text{P}}{\diagdown}}\text{Pt}\!\!-\!\!\underset{\text{C}}{\overset{\text{C}}{\parallel}}\underset{\diagdown\text{CN}}{\overset{\diagup\text{CN}}{}}
\quad\xrightarrow{\;h\nu\;}\quad
\underset{\text{Ph}_3\text{P}}{\overset{\text{Ph}_3\text{P}}{\diagdown}}\text{Pt}\underset{\diagdown\text{C}\equiv\text{C}}{\overset{\diagup\text{CN}}{}}\underset{\diagdown\text{CN}}{}
& (3\text{-}32)
\end{array}
$$

raphy and represented the first characterization of the —C≡C—CN group as a ligand. No further details concerning the photolysis were presented.

It was later observed that irradiation of the related tetracyanoethylene complex led to the reaction shown in Eq. (3-33) [55]. The 313 nm quantum yield for disappearance of the starting complex was 0.01, and spectral

$$
\begin{array}{cc}
\underset{\text{Ph}_3\text{P}}{\overset{\text{Ph}_3\text{P}}{\diagdown}}\text{Pt}\!\!-\!\!\underset{\underset{\text{NC}}{\overset{\text{NC}}{|}}}{\overset{\text{C}}{\underset{\text{C}}{\parallel}}}\underset{\diagdown\text{CN}}{\overset{\diagup\text{CN}}{}}
\quad\xrightarrow{\;h\nu\;}\quad
\underset{\text{Ph}_3\text{P}}{\overset{\text{Ph}_3\text{P}}{\diagdown}}\text{Pt}\underset{\text{C}}{\overset{\diagup\text{CN}}{}}\underset{\text{NC}\;\;\;\text{CN}}{\overset{\diagdown\text{C}\diagup\text{CN}}{}}
& (3\text{-}33)
\end{array}
$$

evidence showed that the reaction proceeds through the initial generation of TCNE^{-}.

C. $[Pt(C_2H_4)(PPh_3)_2]$

Traverso and co-workers [56] have briefly described the photochemical properties of $[Pt(C_2H_4)(PPh_3)_2]$. Irradiation with 280 nm was shown to induce ethylene dissociation with a quantum yield of 0.85, measured in ethanol solution (Eq. 3-34). The 14-valence-electron $[Pt(PPh_3)_2]$ intermediate is quite reactive. In $CHCl_3$ solution the final product obtained was

$$[Pt(C_2H_4)(PPh_3)_2] \xrightarrow{280\ nm} [Pt(PPh_3)_2] + C_2H_4 \qquad (3\text{-}34)$$

$[PtClH(PPh_3)_2]$, apparently resulting from abstraction of HCl from $CHCl_3$. In ethanol solutions a red compound was isolated, but it was incompletely characterized. It appeared to be a dimeric species resulting from ortho-metallation of $[Pt(PPh_3)_2]$ followed by loss of benzene. Irradiation of 254 nm in CH_2Cl_2 solution produced a different reaction (Eq. 3-35). It was suggested that 254 nm did not induce ethylene loss but rather ortho-metallation of one

$$(3\text{-}35)$$

of the PPh_3 ligands. The resultant ethylene–hydride complex would be expected to collapse to the isolated ethyl complex. However, $[Pt(PPh_3)_2]$ which would result from ethylene elimination would itself be expected to undergo ortho-metallation. The resultant ortho-metallated complex could then add the photoreleased ethylene to give the intermediate shown in Eq. (3-35).

D. $trans\text{-}[PtCl_2(\text{Olefin})(\text{Amine})]$

Irradiation of $trans\text{-}[PtCl_2(\text{olefin})(\text{amine})]$ (olefin = ethylene, styrene, t-butylethylene, $trans$-phenyl-1-propene, stilbene; amine = pyridine, substituted pyridines, t-butylamine, piperidine, dimethylamine, and p-anisidine) complexes have been reported to lead to the dimerization reaction shown in Eq. (3-36) [57, 57a]. The photolyses were conducted at 336, 313, and 254 nm;

$$trans\text{-}[PtCl_2(\text{olefin})(\text{amine})] \xrightarrow{hv} [PtCl_2(\text{amine})]_2 + 2\ \text{olefin} \qquad (3\text{-}36)$$

and it was noted that the more highly substituted olefins appeared to be expelled with a higher quantum yield. The reaction was suggested to arise from a Pt → olefin CT state [57a]. The corresponding cis complexes undergo cis–trans photoisomerization apparently prior to olefin loss [57a].

VIII.   COPPER COMPLEXES

A.   COPPER-1,5-CYCLOOCTADIENE COMPLEXES

Srinivasan [49] originally reported that 254 nm irradiation of diethyl ether solutions containing CuCl and 1,5-cyclooctadiene leads to the formation of tricyclooctane (Eq. 3-37). The electronic absorption spectrum of a solution of

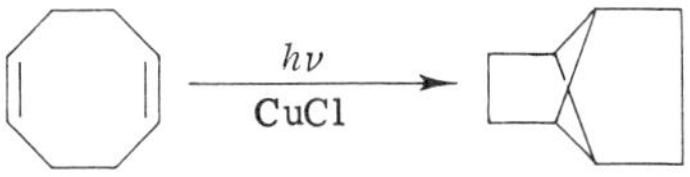

$$\text{(structure)} \xrightarrow[\text{CuCl}]{h\nu} \text{(structure)} \qquad (3\text{-}37)$$

CuCl and 1,5-cyclooctadiene was simply a summation of the spectra of the two individual components, and it was clear from the concentrations employed that the diene was the actual light absorber. It was proposed that the function of the CuCl was merely to stabilize the excited state of the diene once formed. The overall mechanism in Eq. (3-38) was suggested. Haller and Srinivasan [58] later demonstrated that the isomerization did not proceed through the intermediacy of a free-radical species as had been suggested by others [59].

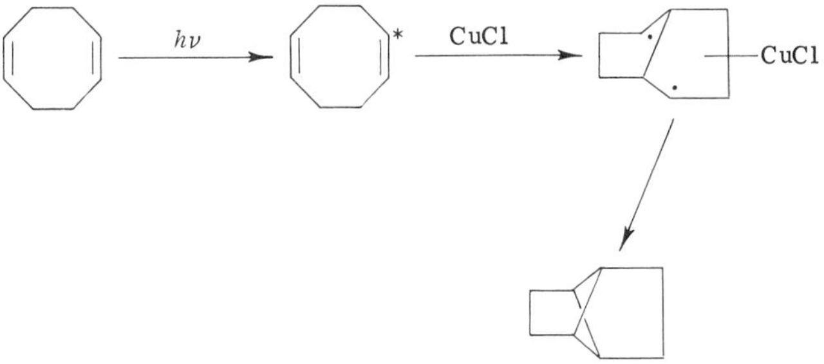

$$(3\text{-}38)$$

The most definitive study of the mechanism of this process was performed by Whitesides and co-workers [60] who conducted their investigations using $[\text{CuCl(COD)}]_2$ (XXXVII) in pentane solution. It was first shown that com-

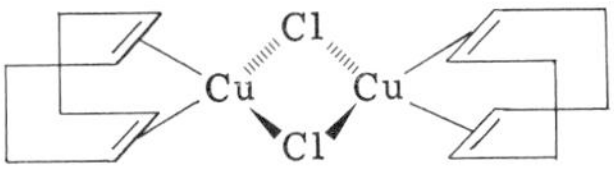

(XXXVII)

plex (XXXVII) undergoes appreciable dissociation to free *cis,cis*-1,5-cyclo-octadiene in solution and that under the experimental conditions used the free diene was the principal light absorber. In addition to tricyclooctane and *cis,-cis*-1,5-cyclooctadiene, the *cis,trans*-1,5-cyclooctadiene and *trans,trans*-cyclooctadiene isomers were isolated from 254 nm irradiated solutions. Rate data showed that tricyclooctane was not formed directly from *cis-cis*-1,5-cyclooctadiene but that *cis,trans*-1,5-cyclooctadiene was an intermediate in the conversion. These workers preferred the overall mechanism shown in Fig. 3-13 in which the role of the copper is to shift the position of the photoequilibria toward that of the more reactive *cis,trans*- and *trans,trans*-1,5-cyclooctadiene isomers [60]. It was noted that the major part of the tricyclooctane produced could arise from direct photolysis of an intermediate CuCl–*cis,trans*-1,5-cyclooctadiene complex, or it could form through photochemical conversion of the complex to *trans,trans*-1,5-cyclooctadiene which itself could photochemically convert to the product [60].

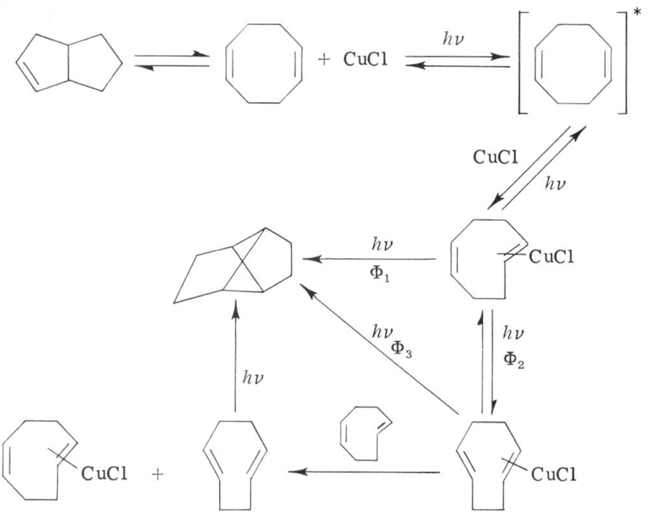

FIG. 3-13. Proposed mechanism for CuCl-photoassisted reactions of cyclooctadiene. Reprinted with permission from Whitesides *et al.* [60], *J. Am. Chem. Soc.* **91**, 2608 (1969). Copyright by the American Chemical Society.

B. COPPER(I)-PHOTOASSISTED DIMERIZATION OF CYCLIC OLEFINS

Trecker and co-workers [61,62] first reported that 254 nm irradiation of norbornene in solutions containing copper(I) halides led to dimerization of the olefin. With CuBr, for example, norbornene was converted stereospecifically to the exo, trans, exo diene in an overall yield of 38.4%. The electronic absorption spectra of free norbornene, CuBr, and CuBr with norbornene in ether solution (Fig. 3-14) clearly indicate the formation of a Cu(I)–olefin complex and that this complex is the principal 254 nm light-absorbing species. On the basis of the invariance of the band shape or intensity of the 239 nm absorption band as the concentration of olefin was varied, these workers concluded that only a 1:1 Cu–olefin complex was formed. This, coupled with the quantum yield dependence on norbornene concentration, led to the proposed reaction mechanism shown in Eqs. (3-39)–(3-41).

$$\text{Cu–olefin} + h\nu \longrightarrow \text{Cu–olefin*} \qquad (3\text{-}39)$$

$$\text{Cu–olefin*} \xrightarrow{k} \text{Cu–olefin} + \Delta \text{ (or } h\nu') \qquad (3\text{-}40)$$

$$\text{Cu–olefin*} + 2 \text{ olefin} \longrightarrow \text{dimer} + \text{Cu–olefin (or Cu + olefin)} \qquad (3\text{-}41)$$

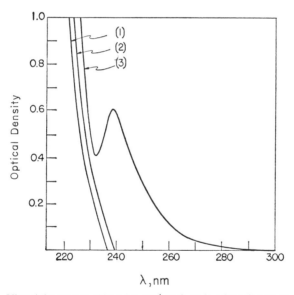

FIG. 3-14.   Ultraviolet spectra of (1) 7.6×10^{-4} M CuBr in ether; (2) 1.06 M norbornene in ether, and (3) 5.85×10^{-5} M CuBr and 5.3×10^{-2} M norbornene in ether. Reprinted with permission from Trecker *et al.* [61], *J. Am. Chem. Soc.* **87**, 3261 (1965). Copyright by the American Chemical Society.

Kochi and co-workers [63,64] later reinvestigated this reaction using Cu(I) trifluoromethanesulfonate (CuOTf) as the photocatalyst. The same reaction was observed except that the exo, trans, exo dimer was obtained in 88% isolated yield in contrast to the 38% previous yield [62] with CuBr as the catalyst. These workers had previously prepared and characterized several

Cu(I)–olefin complexes containing one, two, three, and four coordinated double bonds and noted that the electronic absorption spectra were rather invariant with respect to the number of bound olefins [65]. All the spectra exhibited two strong UV absorption bands at 233–241 nm ($\varepsilon = 2500$–3600) and 272–282 nm ($\varepsilon = 1500$–2100). These observations cast serious doubt on the earlier conclusion [62] that only a 1:1 Cu–olefin complex was formed in the norbornene solutions; furthermore, NMR evidence indicated the presence of several different complexes, most likely including both the 1:1 and 1:2 Cu–olefin species [63].

The CuOTf complex is much more soluble in ether than is CuBr, and, consequently, Kochi and co-workers were able to study the reaction over a much greater range of complex and olefin concentrations. The quantum yields which were obtained under various conditions are given in Table 3-11. Analysis of this and other data led to the strong conclusion that a 1:2 Cu–olefin complex is the direct precursor of the dimer, and the overall scheme in Fig. 3-15 was suggested. The mechanistic details of the collapse of the excited bis(olefin) complex to the product could not be unambiguously determined, but the intermediate bis-σ-bonded species shown in Fig. 3-15 could give the dimer by reductive elimination and coupling of the two alkyl ligands. It was

TABLE 3-11

Quantum Yields for the Photodimerization of Norbornene
Catalyzed by Copper(I) Triflate[a,b]

Norbornene (M)	$\Phi_{xx}{}^c$	$1/\Phi_{xx}$	$\Phi_{nx}{}^d$	$1/\Phi_{nx}$	$1/[nb]^e$	$1/[nb]^2$
0.10	0.060	16.7	0.012	83.3	10.00	100.00
0.15	0.091	11.00	0.017	58.8	6.67	44.49
0.20	0.103	9.71	0.019	52.6	5.00	25.00
0.30	0.138	7.25	0.026	38.5	3.33	11.09
0.40	0.162	6.17	0.031	32.3	2.50	6.25
0.60	0.178	5.62	0.036	27.8	1.67	2.79
0.80	0.255	3.92	0.051	19.6	1.25	1.56
1.20	0.323	3.10	0.065	15.4	0.83	0.69
1.60	0.370	2.70	0.068	14.7	0.63	0.40
5.00	0.610	1.64	0.102	9.80	0.20	0.04

[a] Reprinted with permission from Salomon and Kochi [63], *J. Am. Chem. Soc.* **96**, 1137 (1974). Copyright by the American Chemical Society.
[b] In ethereal solutions containing 3.0×10^{-2} CuOTf at 25°C.
[c] Overall quantum yield for the formation of *exo-trans-exo*-norbornene dimer.
[d] *endo-trans-exo*-Norbornene dimer.
[e] nb = norbornene.

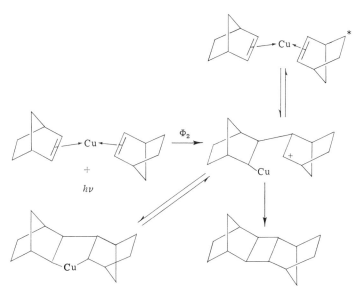

FIG. 3-15.   Proposed mechanism for the [CuOTf]-photoassisted dimerization of nor-bornene. Reprinted with permission from Salomon and Kochi [63], *J. Am. Chem. Soc.* **96**, 1137 (1974). Copyright by the American Chemical Society.

noted that the copper ion could also simply facilitate the absorption of light by the otherwise weakly absorbing olefins and act as a template for the allowed photochemical (2 + 2)-cycloaddition.

It was further pointed out that the CuOTf salt is a much superior catalyst in reactions of this type [63]. The Cu halide catalysts are unstable and form an opaque insoluble deposit on the walls of the photoreactor which must be removed and Cu halide replenished during the course of the reaction. No decomposition was noted with CuOTf, and the reaction mixture remained light and clear during the course of the reactions. Furthermore, the high solubility of the CuOTf salt in ether and the weak coordinating ability of the OTf^- ion lead to high concentrations of the Cu–olefin complexes.

Kochi and co-workers [63,64] also demonstrated that CuOTf catalyzed the photodimerization of *endo*-dicyclopentadiene, cyclopentene, cyclohexene, and cycloheptene and the cross-dimerization of norbornene and cyclooctene

$$\text{CuOTf} \atop h\nu$$

(3-42)

(48%)

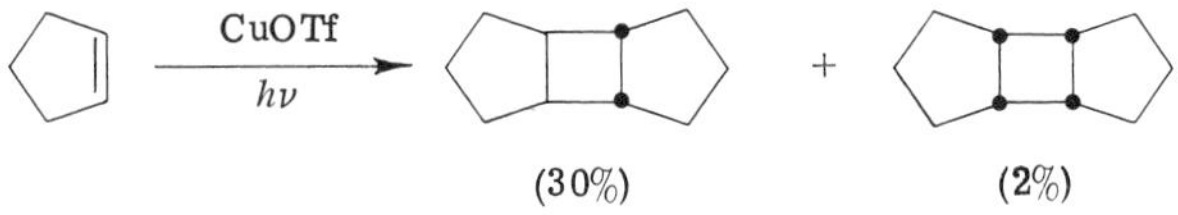

(3-43)

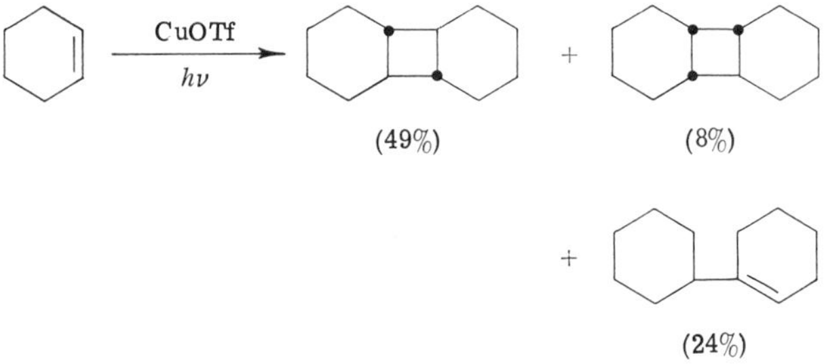

(30%)   (2%)

(3-44)

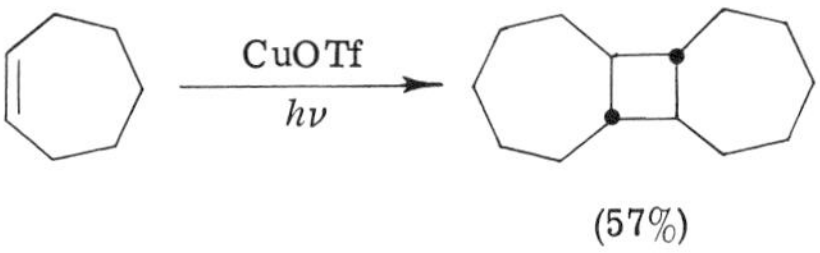

(49%)   (8%)

(24%)

(3-45)

(57%)

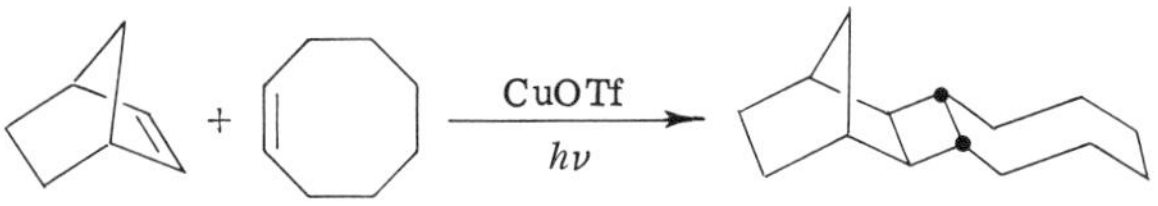

(3-46)

[Eqs. (3-42)–(3-46)]. Copper halide salts do not photoassist the dimerization of the simple unstrained cyclic olefins, but CuOTf does. This was attributed to the stronger coordinating ability of olefins with the aqueous ion in order to compensate for the very weak coordinating ability of OTf^- [64].

C. COPPER(I)-PHOTOASSISTED VALENCE ISOMERIZATION OF NORBORNADIENE

Schwendiman and Kutal [11,11a] have reported the first detailed study of the copper(I) halide photoassisted conversion of norbornadiene to quadri-cyclane (Eq. 4-47). The conversion proceeds with high quantum efficiency

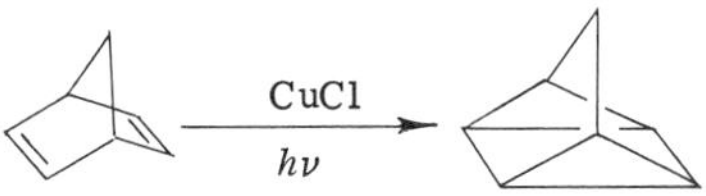

(3-47)

but shows a marked solvent dependence. No reaction was observed in acetonitrile, presumably due to competition of CH_3CN for the bonding site on copper, while the 313 nm quantum yields in $CHCl_3$ and ethanol solutions were 0.3–0.4 and 0.2–0.3, respectively. Since > 200 moles of quadricyclane was produced per mole of CuCl, the overall process is catalytic in CuCl. The reaction proceeds cleanly to conversions > 90%. The electronic absorption spectrum of solutions of CuCl and norbornadiene (Fig. 3-16) clearly indicate complex formation, and analysis of the data suggests a 1:1 complex. The intense electronic absorption band at 248 nm was described as a metal–olefin CT transition, but the direction of charge movement was not defined. Although a metal–olefin CT seems more likely to us, the authors noted that a transfer in either direction would weaken the two olefin bonds and promote bonding across the molecule. That is, population of Ψ_3 (Fig. 3-17), the lowest unoccupied molecular orbital of norbornadiene, or depopulation of Ψ_2, the highest occupied orbital, should have qualitatively the same overall effect.

It was suggested that the norbornadiene–quadricyclane conversion could be developed into a solar energy storage system [12]. One could, for example, irradiate solutions of norbornadiene and CuCl and drive the conversion to quadricyclane. In the presence of a suitable metal catalyst, quadricyclane spontaneously reverts to norbornadiene, releasing a considerable amount of energy in the process. Thus, light energy is stored and can be released later in the form of heat. It was noted that the advantages of this system include the low cost of the norbornadiene starting material, the high storage capacity of 230–260 cal/g of quadricyclane, and storage as chemical energy rather than

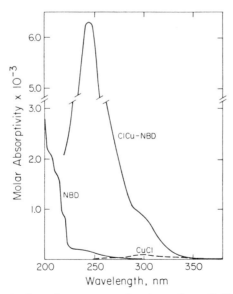

FIG. 3-16. Electronic absorption spectrum of norbornadiene, CuCl, and norbornadiene and CuCl in ethanol solution. Reprinted with permission from Schwendiman and Kutal [11], *Inorg. Chem.* **16**, 719 (1977). Copyright by the American Chemical Society.

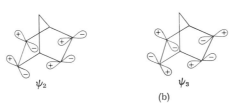

FIG. 3-17. (a) Proposed mechanism of the CuCl-photoassisted isomerization of norbornadiene to quadricyclane. (b) Representation of the highest filled (Ψ_2) and lowest unoccupied (Ψ_3) π-molecular orbitals of norbornadiene. Reprinted with permission from Schwendiman and Kutal [11], *Inorg. Chem.* **16**, 719 (1977). Copyright by the American Chemical Society.

thermal energy, thereby reducing insulation needs. The major disadvantage is the very low percentage of visible light that the systems absorb, but this work does suggest that search should be made for related systems which absorb a greater fraction of the solar spectrum.

In a subsequent report Grutsch and Kutal [65] described similar experiments using $[Cu(BH_4)(PPh_3)_2]$, $[Cu(BH_4)(PPh_2Me)_3]$, and $[Cu(BH_4) \cdot (Ph_2PCH_2CH_2PPh_2)]$ as the photoassistance agents. These complexes produce quadricyclane with 313 nm quantum yields of 0.18, 0.27, and 0.003, respectively. The mechanism of production of quadricyclane using these Cu–phosphine complexes appeared to be qualitatively different from that of CuCl. No evidence for formation of a Cu–NBD complex was obtained, for example. It was suggested that either path A or path B outlined in Fig. 3-18

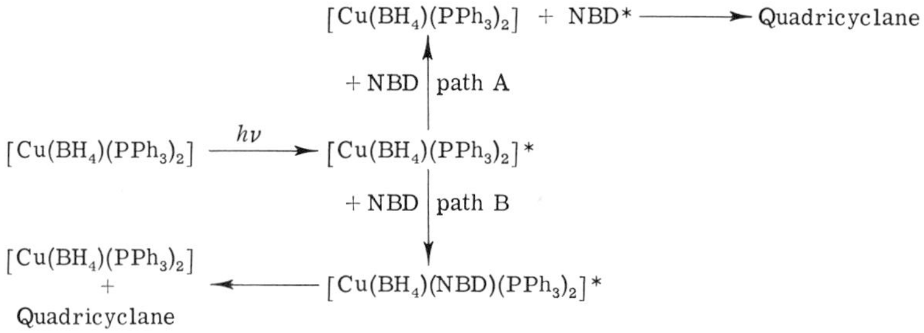

$$[Cu(BH_4)(PPh_3)_2] \ + \ NBD^* \longrightarrow Quadricyclane$$

$$\uparrow \ {\small + NBD} \ \big| \ path \ A$$

$$[Cu(BH_4)(PPh_3)_2] \ \xrightarrow{\ h\nu\ } \ [Cu(BH_4)(PPh_3)_2]^*$$

$$\downarrow \ {\small + NBD} \ \big| \ path \ B$$

$$\begin{array}{c} [Cu(BH_4)(PPh_3)_2] \\ + \\ Quadricyclane \end{array} \longleftarrow [Cu(BH_4)(NBD)(PPh_3)_2]^*$$

FIG. 3-18. Proposed mechanism for the $[Cu(BH_4)(PPh_3)_2]$-photoassisted isomerization of norbornadiene [65].

could account for the valence isomerization. Quadricyclane could be produced by direct intermolecular energy transfer from the excited complex to a
$\pi \to \pi^*$ state of norbornadiene as in path A. The latter has been shown [66]
to efficiently relax to quadricyclane. Alternatively, complexation of norbornadiene could occur in the excited state of the Cu(I) complex as in path B,
and this species could then relax to the product. Although not explicitly
stated, path A seems more reasonable in view of the decreased quantum yield
shown by $[Cu(BH_4)(Ph_2PCH_2CH_2PPh_2)]$ relative to $[Cu(BH_4)(PPh_3)_2]$
and $[Cu(BH_4)(PPh_2Me)_3]$. The electronic absorption spectrum of the
diphosphine complex shows a much lower energy absorption feature than do
the spectra of the other two complexes, and it is attractive to assume that the
low quantum yield results because of less efficient energy transfer from the
lower lying excited state of the diphosphine complex.

D.   COPPER(I)-PHOTOASSISTED REARRANGEMENT
     AND FRAGMENTATION OF 7-METHYLENENORCARANE

Irradiation of 7-methylenenorcarane in the presence of CuOTf has been
shown to lead to a variety of products arising from rearrangement and
fragmentation of the olefin (Eq. 3-48) [67]. The overall mechanistic scheme

(3-48)

which was drawn for the rearrangement reaction is shown in Fig. 3-19 [67].
The fragmentations to cyclohexene and acetylene were thought to arise via
the paths shown in Fig. 3-20. The key intermediates in both of these schemes
are copper(I) carbenium ions formed through a *photocupration* reaction.
A recent study [68] of the CuOTf-catalyzed photorearrangements and
photofragmentation reactions of the methylenecyclopropanes 7-methylene-
bicyclo[4.1.0]heptane and 8-methylenebicyclo[5.1.0]octane also suggested
similar Cu(I) carbenium intermediates.

Although in no case has a definite assignment of a Cu(I)–olefin electronic
absorption spectrum been made, the spectra which have been presented
(Fig. 3-14 and 3-16) are suggestive of a metal–olefin CT for the predominant

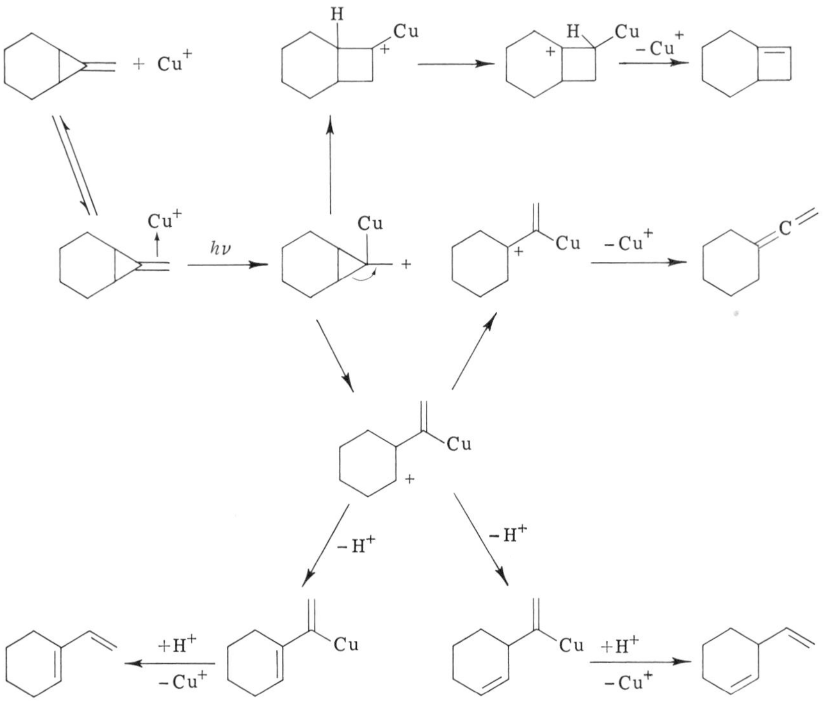

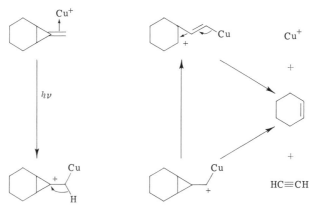

FIG. 3-19.   Proposed mechanism for [CuOTf]-photoassisted reactions of 7-methylene-norcarane. Reprinted with permission from Salomon and Salomon [67], *J. Am. Chem. Soc.* **98**, 7454 (1976). Copyright by the American Chemical Society.

FIG. 3-20.   Proposed mechanism for the [CuOTf]-photoassisted fragmentation of 7-methylenenorcarane. Reprinted with permission from Salomon and Salomon [67], *J. Am. Chem. Soc.* **98**, 7454 (1976). Copyright by the American Chemical Society.

band. Such a transition could lead to the simple conversion shown in Eq. (3-49). The radical intermediate is consistent with the mechanisms in Fig. 3-13

$$
\begin{array}{ccc}
\underset{H}{\overset{R_1}{\diagdown}}\underset{C}{\overset{R_2}{\diagup}} & & \\
\underset{H}{\overset{\|}{\diagup}}\overset{}{C}\overset{}{\diagdown}_H\!-\!Cu & \xrightarrow[\text{MLCT}]{h\nu} &
\end{array}
$$

(3-49)

and 3-15 and is also consistent with the $[W(CO)_6]$-photoassisted isomerization of C_2H_4 and stilbene discussed earlier in this chapter.

E.   OTHER COPPER(I)–OLEFIN PHOTOREACTIONS

Sato *et al.* [69] have described the Cu(I)-photoassisted oxidation of dypnone and substituted dypnones (Eq. 3-50). No reaction occurs in the

$$
\begin{array}{ccc}
\underset{H_3C}{\overset{Ar}{\diagdown}}\overset{H}{\underset{C}{\diagup}}\overset{}{=}\overset{H}{\underset{C-Ar}{\diagup}} & \xrightarrow[\text{O}_2/\text{metal}]{h\nu} &
\end{array}
$$

(3-50)

absence of Cu(I), but the mechanism is unclear since it is evident from the experimental details that dypnone is the principal light absorber. Nozaki *et al.* [70] have reported the copper halide-photoassisted interconversion of several 12-membered cyclic olefins.

IX.   SUMMARY

The several studies which have been conducted indicate that olefin complexes can undergo a wide variety of reactions ranging from substitution of the olefin to substitution of another ligand, isomerization of a coordinated olefin, and skeletal rearrangement. Unfortunately, no definitive excited-state assignments have yet been made, and it is not possible to correlate the nature of the excited states with the observed reactivity. Darensbourg's recent studies [8,8a] have led to a reasonably clear understanding of the mechanisms of the photoreactions of the $[M(\text{diene})(CO)_4]$ complexes, but it has not yet been determined exactly how excited metal–olefin complexes lead to isomerization of the olefin. The π-allyl hydride mechanism shown in Eq. (3-11) seems attractive for most olefins, but it cannot account for isomerization of ethylene and stilbene. The latter were proposed [17,18] to proceed through an intermediate such as that shown in (XXII). Finally, the overall details of many of the Cu(I)-assisted olefin photoreactions have yet to be worked out although some synthetically useful transformations have been observed.

REFERENCES

1. J. Nasielski, P. Kirsch, and L. Wilputte-Steinert, *J. Organomet. Chem.* **27**, C13 (1971).
2. M. Wrighton and M. A. Schroeder, *J. Am. Chem. Soc.* **95**, 5764 (1973).
2a. M. A. Schroeder and M. S. Wrighton, *J. Am. Chem. Soc.* **98**, 551 (1976).
3. G. Platbrood and L. Wilputte-Steinert, *J. Organomet. Chem.* **70**, 393 (1974); see also M. J. Mirbach, D. Steinmetz, and A. Faus, *ibid.* **168**, C13 (1979).
3a. G. Platbrood and L. Wilputte-Steinert, *J. Organomet. Chem.* **70**, 407 (1974).
4. I. Fischler, M. Budzwait, and E. A. Koerner von Gustorf, *J. Organomet. Chem.* **105**, 325 (1976).
5. G. Platbrood and L. Wilputte-Steinert, *J. Organomet. Chem.* **85**, 199 (1975).
6. D. Rietvelde and L. Wilputte-Steinert, *J. Organomet. Chem.* **118**, 191 (1976).
7. I. Fischler, R. Wagner, M. Budzwart, R. N. Pertuz, and E. A. Koerner von Gustorf, *Proc. Int. Conf. Organomet. Chem., 7th, 1975* p. 255 (1975).
8. D. J. Darensbourg and H. H. Nelson, *J. Am. Chem. Soc.* **96**, 6511 (1974).
8a. D. J. Darensbourg, H. H. Nelson, and M. A. Murphy, *J. Am. Chem. Soc.* **99**, 896 (1977).
9. M. A. Schroeder and M. S. Wrighton, *J. Organomet. Chem.* **74**, C29 (1974).
10. M. S. Wrighton and M. A. Schroeder, *J. Am. Chem. Soc.* **96**, 6235 (1974).
10a. M. A. Schroeder and M. S. Wrighton, *J. Organomet. Chem.* **128**, 345 (1977).
10b. R. G. Austin, R. S. Paonessa, P. J. Giordano, and M. S. Wrighton, *Adv. Chem. Ser.* **168**, 189 (1978); see also J. L. Graff, R. D. Sanner, and M. S. Wrighton, *J. Am. Chem. Soc.* **101**, 273 (1979).
11. D. P. Schwendiman and C. Kutal, *Inorg. Chem.* **16**, 719 (1977).
11a. D. P. Schwendiman and C. Kutal, *J. Am. Chem. Soc.* **99**, 5677 (1977).
12. R. C. Kerber and E. A. Koerner von Gustorf, *J. Organomet. Chem.* **110**, 345 (1976), and references therein.
13. N. I. Kirillova, A. I. Gusev, A. A. Pasynskii, and Y. T. Struchkov, *J. Organomet. Chem.* **63**, 311 (1973).
14. M. Cais, E. N. Frankel, and A. Rejoan, *Tetrahedron Lett.* p. 1919 (1968).
15. E. N. Frankel, E. Selke, and C. A. Glass, *J. Am. Chem. Soc.* **90**, 2446 (1968).
16. W. Jennings and B. Hill, *J. Am. Chem. Soc.* **92**, 3199 (1970).
17. M. Wrighton, G. S. Hammond, and H. B. Gray, *J. Organomet. Chem.* **70**, 283 (1974).
18. M. Wrighton, G. S. Hammond, and H. B. Gray, *J. Am. Chem. Soc.* **93**, 3285 (1971).
19. M. Wrighton, G. S. Hammond, and H. B. Gray, *J. Am. Chem. Soc.* **92**, 6068 (1970).
20. G. S. Hammond, J. Saltiel, A. A. Lamola, N. J. Turro, J. S. Bradshaw, D. O. Cowan, R. C. Counsell, V. Vogt, and C. Dalton, *J. Am. Chem. Soc.* **86**, 3197 (1964).
21. M. Wrighton, G. S. Hammond, and H. B. Gray, *Mol. Photochem.* **5**, 179 (1973).
22. L. Pdungsap and M. S. Wrighton, *J. Organomet. Chem.* **127**, 337 (1977).
23. M. S. Wrighton, D. L. Morse, and L. Pdungsap, *J. Am. Chem. Soc.* **97**, 2073 (1975).
24. F. W. Grevels, D. Schulz, and E. A. Koerner von Gustorf, *Int. Conf. Photochem., 7th, 1973*.

25. J. Buchkremer, F. W. Grevels, O. Jaenicke, P. Kirsch, R. Knoesel, E. A. Koerner von Gustorf, and J. Shields, *Int. Conf. Photochem., 7th, 1973.*

26. E. A. Koerner von Gustorf, J. Buchkremer, F. W. Grevels, J. M. Kelly, and R. Knoesel, *164th Nat. Meet., Am. Chem. Soc., 1972,* Abstr. Inorg. 44.

27. F. W. Grevels, Dissertation, University of Bochum (1970).

28. F. W. Grevels, *in* "Concepts of Inorganic Photochemistry" (A. W. Adamson and P. O. Fleischauer, eds.). Wiley, New York, 1975.

29. C. R. Bock and E. A. Koerner von Gustorf, *Adv. Photochem.* **10**, 221 (1977).

30. F. W. Grevels, D. Schulz, and E. A. Koerner von Gustorf, *Angew. Chem., Int. Ed. Engl.* **13**, 534 (1974).

31. F. W. Grevels and E. Koerner von Gustorf, *Justus Leibigs. Ann. Chem. 1975*, 547 (1975).

32. C. Kruger, Y. H. Tsay, F. W. Grevels, and E. Koerner von Gustorf, *Isr. J. Chem.* **10**, 201 (1972).

33. J. Schwartz, *Chem. Commun.* p. 814 (1972).

34. G. N. Schrauzer and P. W. Glockner, *J. Am. Chem. Soc.* **90**, 2800 (1968).

35. G. N. Schrauzer and S. Eichler, *Angew Chem.* **74**, 585 (1962).

36. W. J. R. Tyerman, M. Kato, P. Kebarle, S. Masamune, O. P. Strausz and H. E. Gunning, *Chem. Commun.* p. 497 (1967).

37. O. L. Chapman, J. Pacansky, and P. W. Wojtkowski, *Chem. Commun.* p. 681 (1973).

38. J. S. Ward and R. Pettit, *J. Am. Chem. Soc.* **93**, 262 (1971).

39. A. Rest, personal communication to E. Koerner von Gustorr, quoted in Fischler *et al.* [40].

40. I. Fischler, K. Hildenbrand, and E. Koerner von Gustorf, *Angew. Chem., Int. Ed. Engl.* **14**, 54 (1975).

41. J. D. Black, M. J. Boylan, P. S. Braterman, and W. J. Wallace, *J. Organomet. Chem.* **63**, C21 (1973).

42. A. Bond, M. Green, and S. H. Taylor, *Chem. Commun.* p. 112 (1973).

43. A. C. Day and J. T. Powell, *Chem. Commun.* p. 1241 (1968).

44. C. W. Bird, D. L. Colinese, R. C. Cookson, J. Hudec, and R. O. Williams, *Tetrahedron Lett.* p. 373 (1961).

45. B. Hill, K. Math, D. Pillsbury, G. Voecks, and W. Jennings, *Mol. Photochem.* **5**, 195 (1973).

46. G. Hata, H. Kondo, and A. Miyake, *J. Am. Chem. Soc.* **90**, 2278 (1968).

47. P. P. Zarnegar and D. G. Whitten, *J. Am. Chem. Soc.* **93**, 3776 (1971).

48. P. P. Zarnegar, C. R. Bock, and D. G. Whitten, *J. Am. Chem. Soc.* **95**, 4367 (1973).

49. R. Srinivasan, *J. Am. Chem. Soc.* **86**, 3318 (1964).

50. R. G. Salomon and N. El-Sanadi, *J. Am. Chem. Soc.* **97**, 6214 (1975).

51. I. Leden and J. Chatt, *J. Chem. Soc.* p. 2936 (1955).

52. P. Natarajan and A. W. Adamson, *J. Am. Chem. Soc.* **93**, 5599 (1971).

53. N. Rosch, R. P. Messmer, and K. H. Johnson, *J. Am. Chem. Soc.* **96**, 3855 (1974).

54. W. H. Baddley, C. Panattoni, G. Bandoli, D. A. Clemente, and V. Belluco, *J. Am. Chem. Soc.* **93**, 5590 (1971).

55. O. Traverso, V. Carassiti, M. Graziani, and V. Belluco, *J. Organomet. Chem.* **57**, C22 (1973).

56. S. Sostero, O. Traverso, M. Lenarda, and M. Graziani, *J. Organomet. Chem.* **134**, 259 (1977).

57. P. Courtot, A. Peron, R. Rumin, J. C. Chottard, and D. Mansuy, *J. Organomet. Chem.* **99**, C59 (1975).

57a. P. Courtot, R. Rumin, and A. Peron, *J. Organomet. Chem.* **144**, 357 (1978).

58. I. Haller and R. Srinivasan, *J. Am. Chem. Soc.* **88**, 5084 (1966).

59. J. E. Baldwin and R. H. Greeley, *J. Am. Chem. Soc.* **87**, 4514 (1965).

60. G. M. Whitesides, G. L. Goe, and A. C. Cope, *J. Am. Chem. Soc.* **91**, 2608 (1969).

61. D. J. Trecker, J. P. Henry, and J. E. McKeon, *J. Am. Chem. Soc.* **87**, 3261 (1965).

62. D. J. Trecker, R. S. Foote, J. P. Henry, and J. E. McKeon, *J. Am. Chem. Soc.* **88**, 3021 (1966).

63. R. G. Salomon and J. K. Kochi, *J. Am. Chem. Soc.* **96**, 1137 (1974).

64. R. G. Salomon, K. Folting, W. E. Streib, and J. K. Kochi, *J. Am. Chem. Soc.* **96**, 1145 (1974).

65. P. A. Grutsch and C. Kutal, *J. Am. Chem. Soc.* **99**, 6460 (1977).

66. S. Murov and G. S. Hammond, *J. Phys. Chem.* **72**, 3797 (1968).

67. R. G. Salomon and M. F. Salomon, *J. Am. Chem. Soc.* **98**, 7454 (1976).

68. R. G. Salomon, A. Sinha, and M. F. Salomon, *J. Am. Chem. Soc.* **100**, 520 (1978).

69. T. Sato, K. Tamura, K. Maruyama, and O. Ogawa, *Tetrahedron Lett.* p. 4221 (1973).

70. H. Nozaki, Y. Nisikawa, M. Kawanisi, and R. Noyori, *Tetrahedron* **23**, 2173 (1967).

Arene Complexes

4

I. INTRODUCTION

A very large number of metal–arene complexes have been prepared and characterized, and an excellent review by Silverthorn [1] discusses the chemical and physical properties of members of this class of compounds. Yearly reviews on the subject appear in the Specialist Periodical Reports on Organometallic Chemistry [2]. Although the thermal chemistry of metal–arene complexes is now well documented, relatively few investigations into the photochemical properties of these complexes have been conducted. Before those few studies are discussed in detail, it is appropriate to review briefly the bonding and electronic structure of the most representative compound, bis(benzene)chromium.

In the solid state $[Cr(\eta^6\text{-}C_6H_6)_2]$ has been shown by x-ray [3] and neutron [4] diffraction to possess the D_{6h} structure (XXXVIII) with eclipsed rings. Electron diffraction studies [5] have shown that the D_{6h} symmetry is preserved in the gas phase, and this is further supported by infrared

(XXXVIII)

analysis [6,7]. Bonding of the arenes to the chromium atom occurs principally through interaction of d_{xz} and d_{yz} with the filled e_1 arene π orbitals and through a weak interaction of $d_{x^2-y^2}$ and d_{xy} with the empty e_2 arene π orbitals. This bonding is reflected in the molecular orbital diagram shown in Fig. 4-1, and this diagram is consistent with the most recent photoelectron spectroscopy results [8]. The highest occupied molecular orbital is of a_{1g} symmetry, principally d_{z^2}, and is essentially nonbonding. Photoelectron spectroscopy [8] has shown that ionization from a_{1g} occurs at about 5.4 eV and from e_{2g} at 6.4 eV.

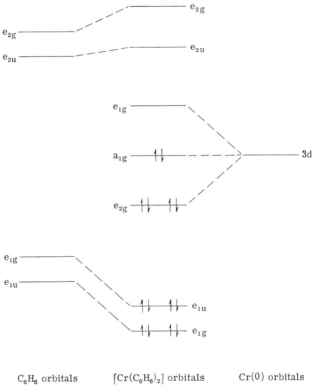

C_6H_6 orbitals $[Cr(C_6H_6)_2]$ orbitals $Cr(0)$ orbitals

FIG. 4-1. Molecular orbital energy level diagram for $[Cr(\eta^6\text{-}C_6H_6)_2]$. Reprinted with permission from Silverthorn [1].

The molecular orbital diagram in Fig. 4-1 predicts that LF transitions corresponding to the one-electron $a_{1g} \rightarrow e_{1g}$ and $e_{2g} \rightarrow e_{1g}$ excitations should be observed at relatively low energy and that both MLCT and LMCT transitions should be seen at higher energy. The electronic absorption spectrum of the complex has been measured and is shown in Fig. 4-2 [9].

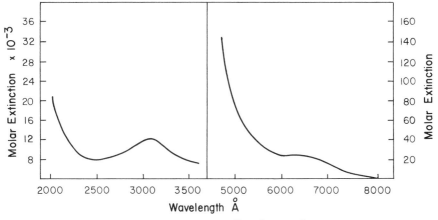

FIG. 4-2. Electronic absorption spectrum of $[Cr(\eta^6\text{-}C_6H_6)_2]$ measured in cyclohexane solution. Reprinted with permission from Feltham [9].

Although the resolution in this spectrum is poor and no detailed assignment has been presented, it is attractive to assume that the weak band at 650 nm arises from an LF transition. The intensity of the broad band at 308 nm suggests charge transfer, but it is not possible to distinguish between MLCT and LMCT. The 308 nm band is noticeably asymmetric, and a shoulder has been resolved at 392 nm ($\varepsilon \sim 1000$) [10]. Definitive assignments of the electronic transitions of $[Cr(\eta^6\text{-}C_6H_6)_2]$ must await further experimentation.

II.   CHROMIUM, MOLYBDENUM, AND TUNGSTEN COMPLEXES

A. $[Cr(\eta^6\text{-}C_6H_6)_2]$

In parallel to its established [1, 11–13] high thermal stability, $[Cr(\eta^6\text{-}C_6H_6)_2]$ has been shown to be rather inert photochemically. The most definitive study of this complex was conducted by Koerner von Gustorf and co-workers [10]. They observed no ligand exchange when the complex was irradiated in deaerated benzene-d_6 or benzene-d_6/cyclohexane solutions, although they did detect a small amount of uncoordinated benzene in solution after prolonged photolysis, amounting to no more than 10% decomposition of the complex. This lack of photosensitivity of $[Cr(\eta^6\text{-}C_6H_6)_2]$ was also verified by Borrell and Henderson [14] who placed an upper limit of 10^{-4} on the 254 nm quantum yield of decomposition of the complex in cyclohexane solution. No evidence has been obtained for any intermediate which could account for the apparently very efficient deactivation of excited $[Cr(\eta^6\text{-}C_6H_6)_2]$, although complexes arising from partial "loosening" of the metal–arene bonds (XXXIX) or from benzene valence bond isomerization (XL) were considered. However, Koerner von Gustorf and co-workers [10] were not able to trap any intermediates such

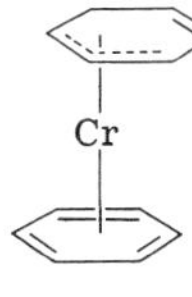

(XXXIX) 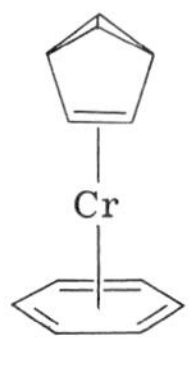   (XL)

as (XXXIX) and (XL) by photolysis in the presence of strong ligands such as CO or dimethylfumarate, nor could they be detected by flash photolysis.

The only report of any photoactivity of $[Cr(\eta^6\text{-}C_6H_6)_2]$ was a brief statement that irradiation of bis(arene)chromium complexes in the presence of alkyl halides gave formation of bis(arene)chromium(I) halides [15]. No details concerning the photolysis were given, nor has this reaction been examined by other workers.

Unlike ferrocene, bis(benzene)chromium does not sensitize the cis–trans isomerization of stilbene or piperylene, nor does it sensitize dimerization of isoprene [10]. It does, however, quench the 18,640 cm^{-1} [16] triplet state of fluorenone at a diffusion-controlled rate [10,17]. This quenching was shown to arise through energy transfer rather than from electron transfer [17], and it indicates that the energy of the lowest lying excited state of $[Cr(\eta^6\text{-}C_6H_6)_2]$ is less than 18,500 cm^{-1}. The apparent efficient deactivation of the complex following excitation may arise through rapid internal conversion from a very low-lying excited state to the ground state. Unfortunately, the lack of a detailed analysis of the electronic absorption spectrum, and hence of the electronic structure of $[Cr(\eta^6\text{-}C_6H_6)_2]$, prevents a thorough discussion of these decay paths.

B. $[Cr(\eta^6\text{-}C_6H_6)_2]^+$

Examination of the molecular orbital diagram shown in Fig. 4-1 suggests that oxidation of $[Cr(\eta^6\text{-}C_6H_6)_2]$ to its cation $[Cr(\eta^6\text{-}C_6H_6)_2]^+$ should occur by removal of one electron from the a_{1g} orbital, leading to a

$^2A_{1g}$ ($e_{2g}{}^4a_{1g}{}^1$) ground state for $[Cr(\eta^6\text{-}C_6H_6)_2]^+$. LF transitions of $[Cr(\eta^6\text{-}C_6H_6)_2]^+$ which derive from $e_{2g} \rightarrow a_{1g}$ excitations should be observed at lower energy. While the MLCT state should be higher in energy owing to the oxidized central metal, the LMCT state should be at lower energy compared to the neutral complex. ESR studies [18] have confirmed the $^2A_{1g}$ ground state, and the weak band at 1160 nm in the electronic absorption spectrum of $[Cr(\eta^6\text{-}C_6H_6)_2]^+$ [9,19] (Fig. 4-3) has been assigned [19] as the $^2A_{1g} \rightarrow {}^2E_g$ $[e_{2g}{}^4a_{1g}{}^1] \rightarrow [e_{2g}{}^3a_{1g}{}^2]$ LF transition. Although the intense bands in the visible and ultraviolet region are almost certainly CT, no detailed assignment has yet been presented.

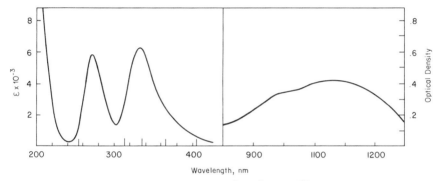

FIG. 4-3. Electronic absorption spectrum of $[Cr(\eta^6\text{-}C_6H_6)_2]^+$ in aqueous solution. Reprinted with permission from Feltham [9] and Gilbert et al. [10].

The photochemistry of $[Cr(\eta^6\text{-}C_6H_6)_2]^+$ was examined [19] under 254, 334, 365, and 404 nm irradiation and was found to be qualitatively the same at each wavelength. In aqueous solution, irradiation led to a decrease in the $[Cr(\eta^6\text{-}C_6H_6)_2]^+$ absorbance with a corresponding appearance and increase in intensity of the characteristic benzene vibrational structure in the 250–260 nm region. Some of the quantum yields measured for disappearance of $[Cr(\eta^6\text{-}C_6H_6)_2]^+$ ($\Phi_{Cr(bz)_2{}^+}$) and for appearance of benzene (Φ_{bz}) are summarized in Table 4-1. At each wavelength $\Phi_{Cr(bz)_2{}^+}$ is essentially half Φ_{bz}. Prolonged photolysis led to spectral evidence for formation of $[Cr(H_2O)_6]^{3+}$.

Photolysis of N_2-purged solutions also led to disappearance of $[Cr(\eta^6\text{-}C_6H_6)_2]^+$ and to formation of benzene, but a brown precipitate was observed. The precipitate proved to be $[Cr(\eta^6\text{-}C_6H_6)_2]$, and its quantum yield of formation ($\Phi_{Cr(bz)_2}$) was measured by extraction of the complex into cyclohexane. The quantum yields measured under these "deaerated" conditions are also given in Table 4-1, and in each case $\Phi_{Cr(bz)_2{}^+} = \frac{1}{2}\Phi_{bz} + \Phi_{Cr(bz)_2}$. Irradiation of both aerated and deaerated solutions of

TABLE 4-1

Quantum Yield Values for the Photochemical Reactions of
$[Cr(\eta^6\text{-}C_6H_6)_2]^+$ in Aqueous Solutions[a]

	Wavelengths of irradiation, nm			
	254	313	365	404
Aerated solutions				
$\Phi_{Cr(bz)_2^+}$ (disappearance)	0.10	0.064	0.055	0.075
Φ_{bz} (formation)	0.21	0.13	0.12	0.15
Deaerated solutions				
$\Phi_{Cr(bz)_2^+}$ (disappearance)	0.18	0.10	0.094	0.14
Φ_{bz} (formation)	0.21	0.13	0.11	0.15
$\Phi_{Cr(bz)_2^+}$ (formation)	0.08	0.04	0.04	0.06

[a] Data from Traverso *et al.* [19].

$[Cr(\eta^6\text{-}C_6H_6)_2]^+$ with $\lambda = 1010\text{--}1370$ nm gave essentially no reaction, and an upper limit of 10^{-4} was placed on the disappearance quantum yield of $[Cr(\eta^6\text{-}C_6H_6)_2]^+$ under these conditions.

A strong complexing ability for chromium in a variety of oxidation states is exhibited by 2,2'-bipyridine. Irradiation of $[Cr(\eta^6\text{-}C_6H_6)_2]^+$ in deaerated solutions containing 2,2'-bipyridine led to formation of dark blue $[Cr(bipy)_3]^+$ [19]. If the 2,2'-bipyridine was added to the irradiated solutions after extraction of $[Cr(\eta^6\text{-}C_6H_6)_2]$, only red $[Cr(bipy)_3]^{2+}$ was obtained.

The complete reaction mechanism which was suggested to be consistent with the observed photochemistry and the quantum yield relationships is given in Eqs. (4-1)–(4-3). The primary photoreaction was suggested to

Primary photochemical process:

$$[Cr(\eta^6\text{-}C_6H_6)_2]^+ \xrightarrow{h\nu} Cr^+ + 2C_6H_6 \qquad (4\text{-}1)$$

Secondary thermal reactions:

$$[Cr(\eta^6\text{-}C_6H_6)_2]^+ + Cr^+ \longrightarrow [Cr(\eta^6\text{-}C_6H_6)_2] + Cr^{2+} \qquad (4\text{-}2)$$

$$Cr^+ + \tfrac{1}{2}O_2 + H_2O \longrightarrow Cr^{3+} + 2OH^- \qquad (4\text{-}3)$$

be decomposition into benzene and Cr^+ ions. When the photolysis was conducted in aerated solutions, the Cr^+ was rapidly scavenged by O_2 and oxidized to Cr^{3+}, giving $\Phi_{bz} = 2\Phi_{Cr(bz)_2^+}$. However, when $[Cr(\eta^6\text{-}C_6H_6)_2]^+$ was irradiated in "deaerated" solutions, a secondary redox reaction between Cr^+ and another molecule of $[Cr(\eta^6\text{-}C_6H_6)_2]^+$ produced $[Cr(\eta^6\text{-}C_6H_6)_2]$ and Cr^{2+}. Irradiation in the presence of 2,2'-bipyridine allowed the Cr^+ ion to be trapped as $[Cr(bipy)_3]^+$, whereas addition of 2,2'-bipyridine *after* the apparently rapid redox between Cr^+ and $[Cr(\eta^6\text{-}C_6H_6)_2]^+$ occurred gives only $[Cr(bipy)_3]^{2+}$.

A similar photoreactivity has been reported [20] for the $[Cr(\eta^6\text{-}C_6H_5\text{—}C_6H_5)_2]^+$ cation. Irradiation of this complex in the presence of 2,2'-bipyridine gave the reaction shown in Eq. (4-4). In the absence of 2,2'-bipyridine, irradiation gave $[Cr(\eta^6\text{-}C_6H_5\text{—}C_6H_5)_2]$, $CrCl_2$, and unco-

$$2[Cr(\eta^6\text{-}C_6H_5\text{—}C_6H_5)_2]Cl \xrightarrow[\text{bipy}]{h\nu} [Cr(\eta^6\text{-}C_6H_5\text{—}C_6H_5)_2] + [Cr(bipy)_3]Cl_2 + 2C_6H_5\text{—}C_6H_5$$

(4-4)

ordinated biphenyl. Although a photolysis mechanism was not proposed, it must be similar to that of $[Cr(\eta^6\text{-}C_6H_6)_2]^+$.

No detailed excited-state arguments have yet been presented to rationalize the observed photochemistry of $[Cr(\eta^6\text{-}C_6H_6)_2]^+$. It is interesting to note that irradiation into the low-lying LF band at 1160 nm leads to no detectable reaction, implying that this LF state is either unreactive toward Cr–arene bond cleavage or that unusually rapid deactivation to the ground state occurs. The latter possibility certainly seems reasonable in view of the relatively low energy of this state. Although the higher energy CT transitions have not been assigned, it is attractive to propose an MLCT state as the active excited state. This assignment is principally due to the nature of the metal–arene bonding. The strong bonding interaction is between filled d_{xz}, d_{yz}, and the empty arene π orbitals. Increasing charge on the metal contracts the d orbitals and lessens their availability for this type of π delocalization. An MLCT transition would generate a Cr(II) center with greatly weakened Cr–arene bonding, whereas an LMCT state would produce Cr(0) with strong π bonding similar to that of photoinert $[Cr(\eta^6\text{-}C_6H_6)_2]$.

C. $[Cr(\eta^6\text{-}C_6H_6)(CO)_3]$

The extensive carbonyl photosubstitution chemistry of $[Cr(\eta^6\text{-}C_6H_6)\cdot(CO)_3]$ was discussed in detail in Chapter 2. Photolysis of the complex in the presence of various donor ligands L has proved to be an extremely versatile technique for the preparation of specific $[Cr(\eta^6\text{-}C_6H_6)(CO)_2L]$ derivatives. In addition to CO photosubstitution, $[Cr(\eta^6\text{-}C_6H_6)(CO)_3]$ has also been claimed [10,21] to undergo exchange of the arene ring. Strohmeier and von Hobe [21] in early work used ^{14}C labeling to demonstrate that photolysis gave exchange of the arene in $[Cr(\eta^6\text{-}C_6H_6)(CO)_3]$, $[Cr(\eta^6\text{-}ClC_6H_5)(CO)_3]$, and $[Cr(\eta^6\text{-}CH_3C_6H_5)(CO)_3]$. Their observations were recently confirmed and extended by the definitive studies [10] of Koerner von Gustorf and his co-workers.

It was shown [10] that when $[\mathrm{Cr}(\eta^6\text{-}C_6H_6)(CO)_3]$ was irradiated for 2 h in C_6D_6 solution, 70% of the complex underwent exchange of C_6H_6 for C_6D_6. It was further observed that addition of carbon monoxide greatly suppressed the exchange process. These observations led to the suggestion that a complex such as $[\mathrm{Cr}(\eta^6\text{-}C_6H_6)(CO)_2]$, resulting from photoinduced loss of CO, is a common intermediate both for substitution of CO and for arene exchange (Eq. 4-5). In particular these experiments tend to rule out

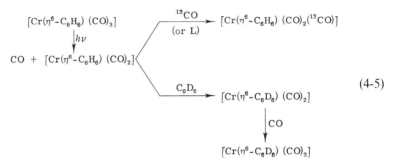

$$(4\text{-}5)$$

arene exchange proceeding through competition for a di- or tetrahapto coordinated arene intermediate or through generation of $(\mathrm{Cr}(CO)_3)$.

D. $[\mathrm{Mo}(\eta^6\text{-}CH_3C_6H_5)(CO)_3]$ AND $[\mathrm{W}(\eta^6\text{-}CH_3C_6H_5)(CO)_3]$

These complexes have also been reported by Strohmeier and von Hobe [21] to undergo arene exchange with ^{14}C-labeled toluene. Under comparable 366 nm irradiation conditions $[\mathrm{Mo}(\eta^6\text{-}CH_3C_6H_5)(CO)_3]$ is about twice as reactive as is $[\mathrm{Cr}(\eta^6\text{-}CH_3C_6H_5)(CO)_3]$. It is necessary, however, to use 254 nm irradiation to induce a significant reaction from $[\mathrm{W}(\eta^6\text{-}CH_3C_6H_5)\cdot(CO)_3]$. No mechanistic discussion has been presented, but presumably arene exchange occurs through photoinduced loss of CO in analogy to $[\mathrm{Cr}(\eta^6\text{-arene})(CO)_3]$ [10].

III. IRON AND RUTHENIUM COMPLEXES

A. $[\mathrm{Fe}(\eta^5\text{-}C_5H_5)(\eta^6\text{-}C_6H_6)]^+$

Nesmeyanov and co-workers [22] have reported that irradiation of $[\mathrm{Fe}(\eta^5\text{-}C_5H_5)(\eta^6\text{-}C_6H_6)]^+$ with ultraviolet light leads to formation of ferrocene according to the stoichiometry given in Eq. (4-6). The reaction is greatly dependent upon the nature of the solvent, as illustrated by the

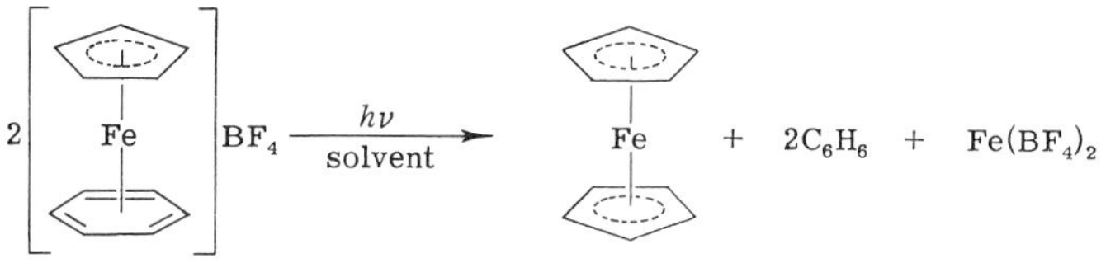

$$2\begin{bmatrix} \text{Fe} \end{bmatrix} BF_4 \xrightarrow[\text{solvent}]{h\nu} \text{Fe} + 2C_6H_6 + Fe(BF_4)_2 \qquad (4\text{-}6)$$

ferrocene yields given in Table 4-2. The highest yields were obtained in solvents which can function as σ-donor ligands, and it was suggested that the photodisproportionation occurred through photoinduced electron transfer from solvent to complex. Although no experimental evidence was presented to support this mechanism, it does correlate well with Nesmeyanov's previous studies [23] which showed that the same disproportionation can be induced thermally by treatment of the complex with a number of reducing agents. Similar photodisproportionation occurred with complexes containing substituted arenes, but these reactions gave less than the expected 1:2 ratio of ferrocene/benzene.

TABLE 4-2

Photodisproportionation Yields of $[Fe(\eta^5\text{-}C_5H_5)(\eta^6\text{-}C_6H_6)]BF_4{}^a$

Solvent	Ferrocene yield (%)	Solvent	Ferrocene yield (%)
THF	76	Acetic anhydride	8
Dioxane	42	Ethyl acetate	Trace
Acetonitrile	20	Water	Trace
Dimethoxyethane	15	Benzene	Trace
Acetone	15	Acetic acid	0
Diethyl ether	10	Methanol	0

^{a} Reprinted with permission from Nesmeyanov et al. [22].

B. $[Fe(2,3\text{-Dimethylbutadiene})(\eta^6\text{-phenyl-ethylphenylphosphine})]$

Fischler and Koerner von Gustorf [24] reported that the title compound, when irradiated in the presence of excess 2,3-dimethylbutadiene, underwent the photoreaction shown in Eq. (4-7). No mechanistic or quantum yield details were presented.

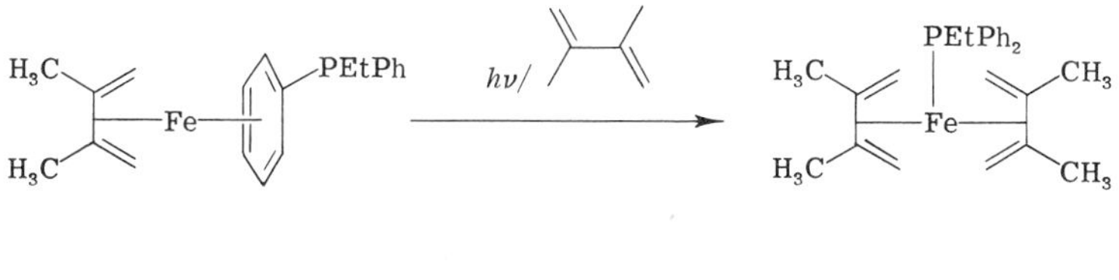

$$\text{(4-7)}$$

C. $[RuCl_2(\eta^6\text{-ARENE})PR_3]$

Bennett and Smith [25] reported that irradiation of solutions of $[RuCl_2(\eta^6\text{-}$ arene)$PR_3]$ led to partial or complete exchange of the coordinated arene with aromatic solvent molecules, in addition to general decomposition. A summary of their various experiments is presented in Table 4-3. Although no mechanism was proposed for the reaction, it is likely that the exchange occurs through photoinduced loss of PR_3 followed by coordination of an aromatic solvent molecule which subsequently replaces the original arene, much as has been proposed [10] for arene exchange in $[Cr(\eta^6\text{-arene})(CO)_3]$.

TABLE 4-3

Results of Arene Exchange Induced by Ultraviolet Irradiation of $[RuCl_2(\eta^6\text{-Arene})(PR_3)]$[a]

Complex	Solvent	Exchange, %[b]	Recovery (product + starting material), %
$[RuCl_2(\eta^6\text{-}p\text{-}MeC_6H_4CHMe_2)(P\text{-}n\text{-}Bu_3)]$	Benzene	57[c]	25
$[RuCl_2(\eta^6\text{-}p\text{-}MeC_6H_4CHMe_2)(P\text{-}n\text{-}Bu_3)]$	Toluene	65[d]	31
$[RuCl_2(\eta^6\text{-}p\text{-}MeC_6H_4CHMe_2)(P\text{-}n\text{-}Bu_3)]$	Ethylbenzene	~ 50[e]	18
$[RuCl_2(\eta^6\text{-}p\text{-}MeC_6H_4CHMe_2)(P\text{-}n\text{-}Bu_3)]$	o-Xylene	48	44
$[RuCl_2(\eta^6\text{-}p\text{-}MeC_6H_4CHMe_2)(P\text{-}n\text{-}Bu_3)]$	m-Xylene	52	36
$[RuCl_2(\eta^6\text{-}p\text{-}MeC_6H_4CHMe_2)(P\text{-}n\text{-}Bu_3)]$	p-Xylene	63	51
$[RuCl_2(\eta^6\text{-}p\text{-}MeC_6H_4CHMe_2)(P\text{-}n\text{-}Bu_3)]$	Mesitylene	25[f]	51
$[RuCl_2(\eta^6\text{-}p\text{-}MeC_6H_4CHMe_2)(P\text{-}n\text{-}Bu_3)]$	Cumene	100	13
$[RuCl_2(\eta^6\text{-}p\text{-}MeC_6H_4CHMe_2)(P\text{-}n\text{-}Bu_3)]$	Anisole	23	29
$[RuCl_2(\eta^6\text{-}p\text{-}MeC_6H_4CHMe_2)(P\text{-}n\text{-}Bu_3)]$	Ethyl benzoate	0	20
$[RuCl_2(\eta^6\text{-}p\text{-}MeC_6H_4CHMe_2)(P\text{-}n\text{-}Bu_3)]$	Chlorobenzene	0	35
$[RuCl_2(\eta^6\text{-}p\text{-}MeC_6H_4CHMe_2)(P\text{-}n\text{-}Bu_3)]$	Trifluorotoluene	0	37
$[RuCl_2(\eta^6\text{-}p\text{-}MeC_6H_4CHMe_2)(PPh_3)]$	Benzene	70	46
$[RuCl_2(\eta^6\text{-}C_6H_6)(P\text{-}n\text{-}Bu_3)]$	p-Cymene	0	35
$[RuCl_2(\eta^6\text{-}C_6H_6)(P\text{-}n\text{-}Bu_3)]$	Toluene	11	40
$[RuCl_2(\eta^6\text{-}C_6H_6)(P\text{-}n\text{-}Bu_3)]$	p-Xylene	11	42
$[RuCl_2(\eta^6\text{-}C_6H_6)(P\text{-}n\text{-}Bu_3)]$	Cumene	0	50
$[RuCl_2(\eta^6\text{-}C_6H_6)(P\text{-}n\text{-}Bu_3)]$	Anisole	23	34
$[RuCl_2(\eta^6\text{-}C_6H_6)(P\text{-}n\text{-}Bu_3)]$	Trifluorotoluene	0	38

[a] Reprinted with permission from Bennett and Smith [25].
[b] 4 h irradiation at 14°C, unless otherwise stated.
[c] 62% after 6 h, 10% recovery.
[d] 100% if solution not cooled during irradiation.
[e] 5 h irradiation.
[f] 85% after 8 h, 50% recovery.

IV. SUMMARY

Although relatively few studies of the photochemistry of metal–arene complexes have been conducted, the work reviewed in this chapter does lead to several general conclusions. Population of LF excited states does not appear to lead to direct loss of an arene ligand. However, loss of another ligand such as CO or PR_3 from LF excited states does occur, and substitution of arene can then proceed via a thermal process with the photogenerated coordinatively unsaturated intermediate. Arene elimination can apparently occur following CT excitation, and we specifically propose that MLCT states should be most active since they give a more positive metal center which has decreased capacity for π bonding with the ligands.

REFERENCES

1. W. E. Silverthorn, *Adv. Organomet. Chem.* **13**, 47 (1975).
2. E. W. Abel and F. G. A. Stone, eds., "Organometallic Chemistry, A Specialist Periodical Report," Vols. 1–4, Chem. Soc., London.
3. K. Keulen and F. Jellinek, *J. Organomet. Chem.* **5**, 490 (1966).
4. G. Albrecht, E. Forster, D. Sippel, F. Eichkorn, and E. Kurras, *Z. Chem.* **8**, 311 (1968).
5. A. Haaland, *Acta Chem. Scand.* **19**, 41 (1965).
6. L. H. Ngai, F. E. Stafford, and L. Schafer, *J. Am. Chem. Soc.* **91**, 48 (1969).
7. L. Schafer, J. F. Southern, S. J. Cyvin, and J. Brunvoll, *J. Organomet. Chem.* **24**, C13 (1970).
8. S. Evans, J. C. Green, and S. E. Jackson, *J. Chem. Soc., Faraday Trans. 2* **68**, 249 (1972).
9. R. D. Feltham, *J. Inorg. Nucl. Chem.* **16**, 197 (1961).
10. A. Gilbert, J. M. Kelly, M. Budzwait, and E. Koerner von Gustorf, *Z. Naturforsch., Teil B* **31**, 1091 (1976).
11. F. Hein and K. Kartte, *Z. Anorg. Allg. Chem.* **307**, 22 (1960).
12. E. O. Fischer and K. Ofele, *Chem. Ber.* **90**, 2532 (1957).
13. T. Kruck, *Chem. Ber.* **97**, 2018 (1964).
14. P. Borrell and E. Henderson, *Inorg. Chim. Acta* **12**, 215 (1975).
15. G. A. Razuvaev and G. A. Domrachev, *Tetrahedron* **19**, 341 (1963).
16. P. S. Engel and B. M. Monroe, *Adv. Photochem.* **8**, 245 (1971).
17. A. Gilbert, J. M. Kelly, and E. Koerner von Gustorf, *Mol. Photochem.* **6**, 225 (1974).
18. S. E. Anderson and R. S. Drago, *Inorg. Chem.* **11**, 1564 (1972).
19. O. Traverso, F. Scandola, V. Balzani, and S. Valcher, *Mol. Photochem.* **1**, 289 (1969).
20. F. Hein and H. Scheel, *Z. Anorg. Allg. Chem.* **312**, 264 (1961).
21. W. Strohmeier and D. von Hobe, *Z. Naturforsch., Teil B* **18**, 981 (1963).
22. A. N. Nesmeyanov, N. A. Vol'kenau, and L. S. Shilovtesa, *Dokl. Akad. Nauk SSR* **190**, 857 (1970).
23. A. N. Nesmeyanov, N. A. Vol'kenau, and L. S. Shilovtesa, *Dokl. Akad. Nauk SSSR* **190**, 354 (1970); *Izv. Akad. Nauk SSSR, Ser. Khim.* p. 726 (1969).
24. I. Fischler and E. A. Koerner von Gustorf, *Z. Naturforsch., Teil B* **30**, 291 (1975).
25. M. A. Bennett and A. K. Smith, *J. Chem. Soc., Dalton Trans.* p. 233 (1974).

Cyclopentadienyl Complexes

5

I. INTRODUCTION

Cyclopentadienyl complexes constitute a historically important class of organometallics, for it was the discovery and subsequent characterization of ferrocene $[Fe(\eta^5\text{-}C_5H_5)_2]$ which led to the development of modern organometallic chemistry. A large number of cyclopentadienyl complexes have been prepared and characterized, and—unlike some of the other classes of organometallics—many have had their photochemical properties examined. In the majority of these investigations, however, cyclopentadienyl has served as an innocent ligand that is not at all involved in the primary photochemical process. In this chapter we describe only those studies in which the cyclopentadienyl ligand plays a key role in the photochemistry and those studies which do not conveniently fit into other chapters of this book. Many cyclopentadienylcarbonyl complexes have been shown to lose CO upon irradiation, and they were discussed in detail in Chapter 2. Likewise, the alkyl complexes $[W(\eta^5\text{-}C_5H_5)_2(CH_3)_2]$, $[Ti(\eta^5\text{-}C_5H_5)_2(CH_3)_2]$, $[Zr(\eta^5\text{-}C_5H_5)_2\cdot(CH_3)_2]$, and $[Hf(\eta^5\text{-}C_5H_5)_2(CH_3)_2]$ are discussed in Chapter 8, and the hydrides $[Mo(\eta^5\text{-}C_5H_5)_2H_2]$ and $[W(\eta^5\text{-}C_5H_5)_2H_2]$ are presented in Chapter 9. Several previous reviews of the photochemistry of cyclopentadienyl complexes are available [1–3], the most recent being the excellent survey by Bock and the late Koerner von Gustorf [3].

II. TITANIUM AND ZIRCONIUM COMPLEXES

A. $[Ti(\eta^5\text{-}C_5H_5)_2X_2]$ AND ITS DERIVATIVES

The titanocene dihalides $[Ti(\eta^5\text{-}C_5H_5)_2X_2]$ are easily prepared, thermally stable compounds and are the starting materials employed in the synthesis of a large variety of organometallics. An x-ray diffraction study [4] has shown that $[Ti(\eta^5\text{-}Me_5C_5)_2Cl_2]$ possesses the distorted tetrahedral geometry shown in (XLI), and all evidence points to similar structures for the $[Ti(\eta^5\text{-}C_5H_5)_2X_2]$

(XLI)

derivatives. The electronic structures of tetrahedral bis(cyclopentadienyl) complexes of this type have been widely discussed, and several different molecular orbital schemes have been proposed [5–8]. The electronic absorption spectra of a series of $[Ti(\eta^5\text{-}C_5H_5)_2X_2]$ and $[Ti(\eta^5\text{-}Me_5C_5)_2X_2]$ derivatives are shown in Fig. 5-1 [9]. Each spectrum shows a low-energy band with $\varepsilon = 200$–600 followed by a more intense band with $\varepsilon = 600$–4000. Since these are formally complexes of the d^0 Ti(IV), the observed bands must represent charge transfer transitions.

The photoelectron spectrum of $[Ti(\eta^5\text{-}C_5H_5)_2Cl_2]$ has been measured independently by two groups of workers and interpreted with the aid of calculated molecular orbital diagrams [7,10]. Unfortunately, different calculational procedures were used for deriving the molecular orbital diagrams, and different level orderings resulted. Petersen *et al.* [7] obtained the level scheme shown in Fig. 5-2, giving the highest filled orbitals mainly Cl^- character. On the other hand, Condorelli *et al.* [10] derived an energy level ordering with the highest filled orbitals of mainly C_5H_5 character. The former diagram would lead to the assignment of the bands in the electronic absorption spectrum as Cl^--to-Ti charge transfer, whereas the latter would support cyclopentadienyl-to-titanium charge transfer. The photochemistry of the complexes strongly suggests the latter interpretation as the lowest excited state.

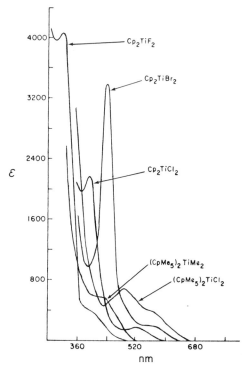

FIG. 5-1.   Electronic absorption spectra of various $[Ti(\eta^5\text{-}C_5H_5)_2X_2]$ derivatives measured in $CHCl_3$ solution. Reprinted with permission from Harrigan *et al.* [9].

The photochemistry of $[Ti(\eta^5\text{-}C_5H_5)_2X_2]$ (X = Cl^-, Br^-, I^-) derivatives was first described by Harrigan, Hammond, and Gray [9]. These workers observed that irradiation of the dihalide complexes in halogen-containing solvents gave apparent cleavage of the cyclopentadienyl–titanium bond and formation of $[Ti(\eta^5\text{-}C_5H_5)X_3]$ derivatives, through the sequence of reactions shown in Eqs. (5-1)–(5-2). The electronic absorption spectral

$$[Ti(\eta^5\text{-}C_5H_5)_2X_2] \xrightarrow{h\nu} [Ti(\eta^5\text{-}C_5H_5)X_2] + C_5H_5 \qquad (5\text{-}1)$$

$$[Ti(\eta^5\text{-}C_5H_5)X_2] + RX \longrightarrow [Ti(\eta^5\text{-}C_5H_5)X_3] + R \qquad (5\text{-}2)$$

changes obtained for photolysis of $[Ti(\eta^5\text{-}Me_5C_5)_2Cl_2]$ in degassed $CHCl_3$ solution are representative of all the complexes examined and are shown in Fig. 5-3. The isosbestic points in all the spectra suggest clean conversion to products. Rapid photoreaction occurred only when the medium contained a source of halide, and similar spectral changes were reported in CCl_4, $CHCl_3$, and HCl-saturated benzene solutions. It was reported [9] that very prolonged photolysis (3 weeks) of neat degassed benzene solutions gave slow formation

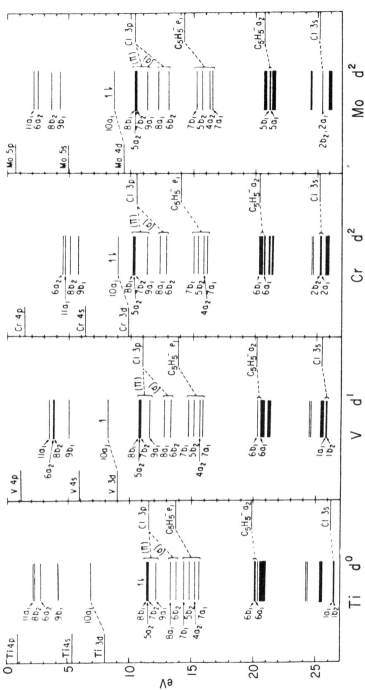

FIG. 5-2. Molecular orbital energy level scheme for [Ti(η^5-C$_5$H$_5$)$_2$Cl$_2$] and related complexes. Reprinted with permission from Petersen *et al.* [7]. *J. Am. Chem. Soc.* **97**, 6433 (1975). Copyright by the American Chemical Society.

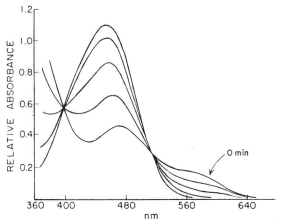

FIG. 5-3.   Electronic spectral changes during $\lambda > 545$ nm photolysis of $[Ti(\eta^5\text{-}C_5Me_5)_2Cl_2]$ in degassed $CHCl_3$ solution. Reprinted with permission from Harrigan *et al.* [9].

of an air-sensitive solid identified as $[Ti(\eta^5\text{-}C_5H_5)Cl_2]$, but this latter observation was not reproducible by other workers [11,12]. No evidence for cleavage of the titanium–halide bond was obtained. Irradiation of $[Ti(\eta^5\text{-}C_5H_5)_2Br_2]$ in $CHCl_3$, for example, gave no evidence for production of $[Ti(\eta^5\text{-}C_5H_5)_2ClBr]$ or $[Ti(\eta^5\text{-}C_5H_5)_2Cl_2]$, but rather produced a complex believed to be $[Ti(\eta^5\text{-}C_5H_5)ClBr_2]$.

The observations of Harrigan *et al.* [9] are consistent with the notion that the lowest excited states in these complexes are of cyclopentadienyl-to-titanium charge transfer character. The progressive red shift of the bands as the halide was changed from fluoride to chloride to bromide was proposed to reflect the reduced π donation of the halide p electrons to the metal d orbitals [9]. Such an LMCT state would lead to formal reduction of the metal and oxidation of the ring system. It was proposed that this state dissociates, giving free C_5H_5 radicals and $[Ti(\eta^5\text{-}C_5H_5)X_2]$. The latter could then abstract halide from a solvent molecule, as illustrated in Eqs. (5-3) and (5-4) for $[Ti(\eta^5\text{-}C_5H_5)_2Cl_2]$.

$$[Ti(\eta^5\text{-}C_5H_5)_2Cl_2] \xrightarrow{h\nu} [Ti(\eta^5\text{-}C_5H_5)Cl_2] + C_5H_5 \qquad (5\text{-}3)$$

$$[Ti(\eta^5\text{-}C_5H_5)Cl_2] + CHCl_3 \longrightarrow [Ti(\eta^5\text{-}C_5H_5)Cl_3] + \cdot CHCl_2 \qquad (5\text{-}4)$$

Formation of $[Ti(\eta^5\text{-}C_5H_5)Cl_2]$ as an intermediate was supported by their reported [9] isolation of this complex following prolonged photolysis in degassed benzene solution. The quantum yields of decomposition of the $[Ti(\eta^5\text{-}C_5H_5)_2X_2]$ derivatives could not be accurately measured but were of the order of 0.02 or lower. The apparent quantum yield of $[Ti(\eta^5\text{-}Me_5C_5)_2Cl_2]$ was at least 15 times greater than that of $[Ti(\eta^5\text{-}C_5H_5)_2Cl_2]$. It was proposed by Harrigan *et al.* [9] that the primary quantum yield of metal–ring cleavage is quite high but that the overall low yield reflects the efficiency of the recombination reaction. This notion is supported by the very slow decomposition of $[Ti(\eta^5\text{-}C_5H_5)_2Cl_2]$ in nonhalogenated solvents and by the increase in quantum yield with the Me_5C_5 derivative, owing to less efficient recombination of the Me_5C_5 radical with the Ti(III) center. The latter was attributed to increased steric hindrance in the recombination process. No experiments were reported describing attempts to trap the C_5H_5 radicals or to observe them using flash spectroscopy.

The results of Harrigan *et al.* [9] are further corroborated by the studies [11–13] of Brubaker and co-workers who extended the photochemistry into a more synthetically useful area. These workers showed that irradiation could lead to exchange of η^5-cyclopentadienyl ligands between different metallocenes. For example, photolysis of an equimolar mixture of perdeuterotitanocene dichloride and titanocene dichloride gave equilibration to titanocene-d_5 dichloride according to Eq. (5-5) [12,13]. The experimentally

$$[Ti(\eta^5\text{-}C_5H_5)_2Cl_2] + [Ti(\eta^5\text{-}C_5D_5)_2Cl_2] \underset{h\nu}{\overset{h\nu}{\rightleftharpoons}} 2[Ti(\eta^5\text{-}C_5H_5)(\eta^5\text{-}C_5D_5)Cl_2] \quad (5\text{-}5)$$

determined photostationary state for reaction (5-5) was close to the theoretically predicted value of 4, assuming no isotope effect. The quantum yield measured at 313 nm is 0.02 [12] although originally reported [13] as 0.22. Photoinduced ligand exchange was also observed between $[Ti(\eta^5\text{-}C_5D_5)_2Cl_2]/$ $[Ti(\eta^5\text{-}C_5H_5)_2Cl]_2$ [12,13], between $[Ti(\eta^5\text{-}C_5D_5)_2Cl_2]/[Ti(\eta^5\text{-}C_5H_5)_2Cl]_2$ [11], and between titanocene dichloride and trimethylenetitanocene dichloride [11]. The latter reaction led to the bridged derivative shown in Eq. (5-6) and suggests that

polymeric systems could be produced by photolysis of appropriately designed monomers [11].

Brubaker and co-workers [11,12] were unable to observe any production of $[Ti(\eta^5\text{-}C_5H_5)Cl_2]$ when they irradiated highly purified, dry and degassed benzene solutions of $[Ti(\eta^5\text{-}C_5H_5)_2Cl_2]$ for prolonged periods. They were thus initially hesitant to accept the proposed mechanism of Harrigan *et al.* [9] which invokes photoinduced cleavage of a titanium–ring bond and generation of free cyclopentadienyl radicals. Instead, they suggested that irradiation leads to a "reduction of hapticity," perhaps as indicated in Eq. (5-7). However, in a more recent study they were able to show that

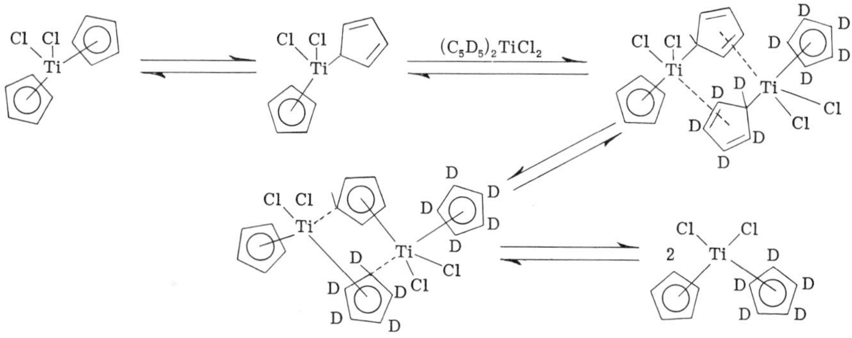

$$(5\text{-}7)$$

irradiation does indeed lead to homolysis of the $Ti-C_5H_5$ bond since they trapped the photoliberated C_5H_5 radical with the spin-trap nitrosodurene and observed both the C_5H_5-nitrosodurene adduct and the $[Ti(\eta^5\text{-}C_5H_5)Cl_2]$ product by ESR spectroscopy [13a].

Brubaker and co-workers also reported [12,13] that methanolysis of $[Ti(\eta^5\text{-}C_5H_5)_2Cl_2]$ can be photoinduced. Irradiation of the complex in methanol solution gave the reaction shown in Eq. 5-8. The 313 nm quantum

$$[Ti(\eta^5\text{-}C_5H_5)_2Cl_2] + MeOH \xrightarrow{h\nu} [Ti(\eta^5\text{-}C_5H_5)Cl_2(OMe)] + C_5H_6 \qquad (5\text{-}8)$$

yield varied with the concentration of reactants and was about 0.08 for 8.1×10^{-3} M $[Ti(\eta^5\text{-}C_5H_5)_2Cl_2]$ and 1.0 M MeOH and 0.05 for 4.05×10^{-3} M $[Ti(\eta^5\text{-}C_5H_5)_2Cl_2]$ and 0.25 M MeOH. No secondary thermal reactions were observed, and it was noted that previous reports of the thermal methanolysis reactions were incorrect as they must have proceeded through room-light photolysis. Irradiation is the most convenient method for preparation of the methanolysis products since they can only be obtained thermally through more difficult procedures. Prolonged photolysis or irradiation in solutions of high methanol concentrations gave other products which were not characterized.

B. $[Zr(\eta^5\text{-}C_5H_5)_2Cl_2]$ AND $[Hf(\eta^5\text{-}C_5H_5)_2Cl_2]$

Peng and Brubaker [14] have recently demonstrated that photolysis of $[Zr(\eta^5\text{-}C_5H_5)_2Cl_2]$ also leads to ring exchange as illustrated by the reaction shown in Eq. (5-9). The 313 nm quantum yield for the exchange reaction was

$$[Zr(\eta^5\text{-}C_5H_5)_2Cl_2] + [Zr(\eta^5\text{-}C_5D_5)_2Cl_2] \underset{h\nu}{\overset{h\nu}{\rightleftarrows}} 2[Zr(\eta^5\text{-}C_5H_5)(\eta^5\text{-}C_5D_5)Cl_2] \qquad (5\text{-}9)$$

0.021, and the measured photostationary state for the reaction was 2.8. The difference between this and the theoretical value of 4 was attributed to some decomposition which occurred as a side reaction upon photolysis. The $[Hf(\eta^5\text{-}C_5H_5)_2Cl_2]$ was also shown to undergo ring exchange as in Eq. (5-9), and the 313 nm quantum yield is 0.02 [14a].

III.   VANADIUM COMPLEXES

$[V(\eta^5\text{-}C_5H_5)_2Cl_2]$

Vitz and Brubaker have reported [11] that 313 nm photolysis of a mixture of vanadocene dichloride and bis(methylcyclopentadienyl)vanadium dichloride gave exchange of the cyclopentadienyl ligands according to Eq. (5-10). This reaction is exactly analogous to that described above for

$$[V(\eta^5\text{-}MeC_5H_4)_2Cl_2] + [V(\eta^5\text{-}C_5H_5)_2Cl_2] \underset{h\nu}{\overset{h\nu}{\rightleftharpoons}} 2[V(\eta^5\text{-}C_5H_5)(\eta^5\text{-}MeC_5H_4)Cl_2]$$

$$(5\text{-}10)$$

$[Ti(\eta^5\text{-}C_5H_5)_2Cl_2]$. The authors also observed significant photodecomposition accompanying the ligand exchange, amounting to 34% during 24 h photolysis. Similar photoinduced ligand exchange was absent between $[V(\eta^5\text{-}C_5H_5)_2Cl_2]/[V(\eta^5\text{-}C_5D_5)_2Cl_2]$, $[V(\eta^5\text{-}C_5H_5)_2Cl]/[V(\eta^5\text{-}C_5D_5)_2Cl]$, and $[V(\eta^5\text{-}C_5H_5)_2]/[V(\eta^5\text{-}C_5D_5)_2]$ [14a]. The 313 nm quantum yield for the vanadocene dichloride system was 0.028. Other cyclopentadienyl complexes, including $[Ta(\eta^5\text{-}C_5H_5)_2Cl_2]$, were observed to photodecompose readily upon photolysis [11].

IV.   IRON AND RUTHENIUM COMPLEXES

A.   FERROCENE $[Fe(\eta^5\text{-}C_5H_5)_2]$

1. Electronic Spectra and Structure

Before discussing the excited-state properties of ferrocene, it is appropriate to examine briefly its electronic structure. The electronic absorption spectrum of ferrocene is shown in Fig. 5-4 along with spectra of other d^6 metallocenes

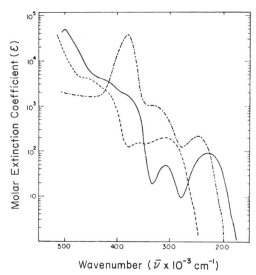

Wavenumber ($\bar{\nu}$ x 10^{-3} cm^{-1})

FIG. 5-4.   Electronic absorption spectra of $[Fe(\eta^5\text{-}C_5H_5)_2]$ (——) and $[Ru(\eta^5\text{-}C_5H_5)_2]$ (— — —) in isopentane solution and $[Co(\eta^5\text{-}C_5H_5)_2]ClO_4$ (—·—··—) in aqueous solution. Reprinted with permission from Sohn *et al.* [15], *J. Am. Chem. Soc.* **93**, 3603 (1971). Copyright by the American Chemical Society.

[15]. The two low-energy bands at 22,700 cm^{-1} ($\varepsilon = 91$) and 30,800 cm^{-1} ($\varepsilon = 49$) can be logically attributed to LF transitions, but their specific assignments have generated much discussion in the literature [16]. The molecular orbital diagram which is appropriate for ferrocene is given in Fig. 5-5, and it has been the relative ordering of the $1e_{2g}$ and $2a_{1g}$ levels which has led to controversy [16]. Although the ordering $2a_{1g} < 1e_{2g}$ for the highest occupied orbitals was inferred from intensity arguments applied to the photoelectron spectrum of ferrocene [18,19], a detailed examination by Sohn *et al.* [15] of the UV–visible spectrum of ferrocene crystals at $4.2°$K has led to the ordering $1e_{2g} < 2a_{1g}$ as shown in Fig. 5-5. Importantly, these workers were able to resolve the lowest band in the spectrum into a two-band pattern as illustrated in Fig. 5-6. These three bands were assigned as the three spin-allowed LF transitions; specifically, band IIa (Fig. 5-6) is $^1A_{1g} \rightarrow a^1E_{1g}$, band IIb is $^1A_{1g} \rightarrow {}^1E_{2g}$ and band III is $^1A_{1g} \rightarrow b^1E_{1g}$. This molecular orbital diagram and corresponding spectral interpretation now appears to be generally recognized as correct. A recent SCF-Xα scattered wave calculation is totally consistent with the level ordering shown in Fig. 5-5 [17].

Sohn *et al.* [15] were unable to assign completely the high-energy charge transfer bands although the band at 50,000 cm^{-1} ($\varepsilon = 51,000$) was attributed to the LMCT transition $^1A_{1g} \rightarrow {}^1A_{2u}(1e_{1u} \rightarrow 2e_{1g})$. Rösch and Johnson's

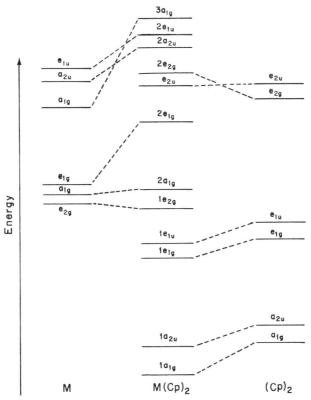

FIG. 5-5. Molecular orbital energy level diagram for $[Fe(\eta^5-C_5H_5)_2]$. Reprinted with permission from Sohn et al. [15], J. Am. Chem. Soc. **93**, 3603 (1971). Copyright by the American Chemical Society.

SCF-Xα calculation has led them to propose that the absorption edge-centered at 37,700 cm^{-1} is the onset of an LMCT transition but that the shoulders at 41,000 and 42,800 cm^{-1} are probably associated with MLCT [17].

The location of the LF singlet-to-triplet transitions and the corresponding energies of the triplet states are still unresolved. Scott and Becker [20] originally reported three spin-forbidden transitions, but the existence of the first two of these could only be produced by gaussian analysis. The third band at 18,900 cm^{-1} has been characterized by iodine perturbation experiments [20,21] and was observed in the crystal study of Sohn et al. [15]. Furthermore, detailed LF analysis of the three spin-allowed transitions allowed Sohn et al. to predict the positions of the three spin-forbidden bands as 18,600 cm^{-1}, 20,900 cm^{-1}, and 22,400 cm^{-1}, respectively. The

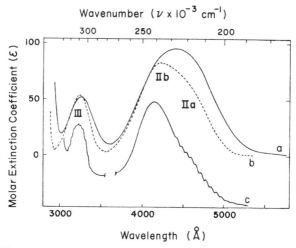

FIG. 5-6.   Electronic absorption spectra of $[\text{Fe}(\eta^5\text{-}C_5H_5)_2]$ at various temperatures: (a) EPA solution at $300°$K, (b) EPA glass at $77°$K, (c) single crystal at $4.2°$K. The intensity of curve (c) is not to scale. (EPA $= 5:5:2$ Et_2O:isopentane:EtOH.) Reprinted with permission from Sohn *et al.* [15], *J. Am. Chem. Soc.* **93**, 3603 (1971). Copyright by the American Chemical Society.

TABLE 5-1

Rate Constants Measured for Energy Transfer of Ferrocene in Benzene Solution at $22°C^{a,b}$

Sensitizer	E_T, kcal/mol	Sensitizer concentration, M	Ferrocene k_q
Triphenylene	(66.6)	4.0×10^{-5}	5.9×10^9
2-Acetonaphthone	(59.3)	4.0×10^{-5}	5.4×10^9
9-Fluorenone	(53)	4.0×10^{-5}	5.1×10^9
Benzanthrone	(47.0)	4.0×10^{-5}	5.7×10^9
Anthracene	(42.6)	4.0×10^{-5}	4.4×10^9
3,4:9,10-Dibenzopyrene	(40.2)	4.0×10^{-5}	3.0×10^9
5-Methyl-3,4:9,10-dibenzopyrene	(38.5)	1.2×10^{-5}	1.8×10^9
5,8-Dimethyl-3,4:9,10-dibenzopyrene	(37.3)	1.2×10^{-5}	1.1×10^9
3,4:8,9-Dibenzopyrene	(34.4)	2.5×10^{-5}	6.8×10^8
Anthanthrene	(33.8)	2.0×10^{-5}	4.8×10^8
Tetracene	(29.3)	2.5×10^{-5}	1.0×10^8
Pyranthrene	(26.9)	5.0×10^{-7}	1.7×10^8
Zinc phthalocyanine	(26.1)	4.0×10^{-6}	2.6×10^7
Violanthrene	(25)	5.0×10^{-7}	1.4×10^7
Isoviolanthrene	(24)	5.0×10^{-7}	1.3×10^7
Pentacene	(23)	5.0×10^{-6}	4.0×10^7
Singlet oxygen	(22.5)	$\sim 1 \times 10^{-3}$	9.0×10^6

^{a} Reprinted with permission from Herkstroeter [25], *J. Am. Chem. Soc.* **97**, 4161 (1975). Copyright by the American Chemical Society.
^{b} Units of the rate constants are M^{-1} sec^{-1}.

SCF-Xα calculations of Rösch and Johnson [17] are also consistent with the position of the lowest spin-forbidden band at about 18,900 cm^{-1}.

The major difficulty arises from quenching experiments. Ferrocene is an efficient quencher of triplet excited states, and several careful studies [22–25] have led to the conclusions that quenching occurs by energy transfer and that the lowest triplet level of ferrocene lies around 15,000 cm^{-1}. For illustration, data obtained by Herkstroeter [25] using flash kinetic spectroscopy is tabulated in Table 5-1 and plotted in Fig. 5-7. It can be seen that donors with $E_T > 43$ kcal/mol (15,000 cm^{-1}) are quenched by ferrocene at a diffusion-controlled rate and that the break in the curve occurs at about 40–42 kcal/mol. A similar set of data was independently obtained by Kikuchi et al. [22]. In a subsequent article [26] these latter workers proposed that the quenching by ferrocene of molecules with $E_T < 15,000$ cm^{-1} occurs by electron transfer interaction, although Herkstroeter [25] provides evidence that the quenching occurs through energy transfer via a very distorted ferrocene triplet. Finally, ferrocene has often been reported to show weak luminescence, but these reports have been largely discounted and ascribed to impurities.

In 1957, Brand and Snedden reported [27] that a new absorption feature around 307 nm appears in the ferrocene spectrum when measured in various halocarbon solvents. This is illustrated in Fig. 5-8 which shows the effect of increasing the CCl$_4$ concentration in CCl$_4$/EtOH solvent mixtures [28].

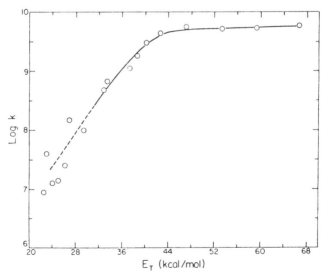

FIG. 5-7. Plot of the logarithms of the rate constants for energy transfer to ferrocene vs. the triplet energies of selected sensitizers. Reprinted with permission from Herkstroeter [25], J. Am. Chem. Soc. **97**, 4161 (1975). Copyright by the American Chemical Society.

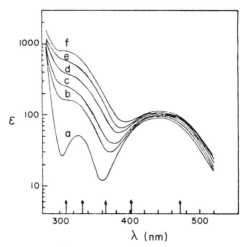

FIG. 5-8. Electronic absorption spectra of $[Fe(\eta^5\text{-}C_5H_5)_2]$ in CCl_4–ethanol solvent mixtures: (a) pure ethanol; (b) 15% CCl_4; (c) 25% CCl_4; (d) 50% CCl_4; (e) 75% CCl_4; (f) pure CCl_4. Reprinted with permission from Traverso and Scandola [28].

The new band has been attributed to a ferrocene-to-solvent charge transfer (CTTS) transition, and analysis of the spectral data suggests a strong association between ferrocene and solvent molecules. Although no real discussion of the nature of this associated complex has been presented, Borrell and Henderson noted that the association constant for the analogous ruthenocene–CCl_4 complex is only slightly greater than the association constants for aromatic hydrocarbons with CCl_4 [29]. It is possible that association is through the aromatic cyclopentadienyl ring and that a structure such as (XLII) results.

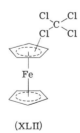

(XLII)

2. Photooxidation

Ferrocene is apparently completely photoinert in solvents such as cyclohexane, decalin, acetone, methanol, 1-propanol, and 2-propanol [30,31]. Brand and Snedden, however, observed that exposure of CCl_4 or CCl_3Br

solutions of ferrocene to light gave deposition of a precipitate which they showed contained $FeCl_3$ [27]. These workers proposed that the photo-oxidation occurred via the CTTS transition which these solutions exhibit and that the primary step is electron transfer from ferrocene to solvent. This observation was later confirmed by Koerner von Gustorf and co-workers [32] who were able to show that the precipitate was $[Fe(\eta^5\text{-}C_5H_5)_2][FeCl_4]$ which they could obtain in high yield.

Perhaps the most significant advance in the understanding of the photo-chemistry of ferrocene in halocarbon solvents came with Traverso and Scandola's 1970 report [28]. These workers observed that irradiation of ferrocene for very short periods in CCl_4 or $CHCl_3$ solutions gave a rapid increase in absorbance at 617 nm but also a rapid clouding of the solution due to the insolubility of the photoproduct. They noted, however, that the product was soluble in solutions containing 40% ethanol. Under these conditions they were able to identify the product as $[Fe(\eta^5\text{-}C_5H_5)_2]^+$ by its absorption spectrum, and they obtained good zero-order plots of its concentration vs. irradiation times. Under these zero-order conditions, Cl^- ions were detected but Fe^{3+} was not found. Longer irradiation periods led to decomposition of the ferricenium ion. Only the CTTS band of the $[Fe(\eta^5\text{-}C_5H_5)_2]/RX$ system near 307 nm was photoactive, and no decomposition of ferrocene was observed upon irradiation into the 325 and 445 nm LF bands. Quantum yields measured at several different wavelengths and corrected for the fraction of light absorbed by the CTTS band are given in Table 5-2. Experiments were conducted in the presence of acrylamide which would detect and trap any radicals produced. Formation of polyacrylamide was observed and the ferrocene decomposition yields decreased markedly. The yields in the presence of acrylamide are given in Table 5-2 and labeled Φ_p.

Traverso and Scandola's results led them to argue [28] for the process detailed in Eqs. (5-11)–(5-12). They proposed that the primary reaction was formation of $[Fe(\eta^5\text{-}C_5H_5)_2]Cl$ and generation of the organic radical

$$[Fe(\eta^5\text{-}C_5H_5)_2] + RCl \xrightarrow{h\nu} [Fe(\eta^5\text{-}C_5H_5)_2]^+ + Cl^- + R \qquad (5\text{-}11)$$

$$[Fe(\eta^5\text{-}C_5H_5)_2] + R \longrightarrow [Fe(\eta^5\text{-}C_5H_5)_2]^+ + R^- \qquad (5\text{-}12)$$

through CTTS excitation. The radicals which are produced can be efficiently trapped by acrylamide, preventing the occurrence of reaction (5-12), and the quantum yields obtained under these conditions can be taken as the primary quantum yields of Eq. (5-11). As illustrated in Table 5-2, they are nearly 1.0 at all wavelengths examined. In the absence of acrylamide, however, the quantum yields are considerably higher, indicating that ferrocene can be further oxidized in a subsequent thermal step by the radical produced, and reaction (5-12) is thus quite important in the overall process. It should

TABLE 5-2

Quantum Yields of Formation of the Ferricenium Cation Following
Photolysis of $[Fe(\eta^5\text{-}C_5H_5)_2]$ in CCl_4–Ethanol Solutions[a,b]

	Wavelength of irradiation, nm									
	313		334		365		404		472	
Solvent	Φ	$\Phi_p{}^c$	Φ	$\Phi_p{}^c$	Φ	$\Phi_p{}^c$	Φ	$\Phi_p{}^c$	Φ	$\Phi_p{}^c$
15% CCl_4	1.5	1.1	1.4	1.1	1.4	1.1	d	d	d	d
25% CCl_4	1.8	1.1	1.5	1.1	1.6	1.1	1.1^e	0.78^e	d	d
50% CCl_4	1.9	1.0	1.4	1.1	1.5	1.1	1.2^e	0.80^e	d	d
15% $CHCl_3$	1.5	1.2	1.3	0.94	d	d	d	d	d	d
25% $CHCl_3$	1.7	1.2	1.5	1.1	1.2^e	0.78^e	d	d	d	d
50% $CHCl_3$	1.6	1.3	1.6	1.1	1.1^e	0.85^e	d	d	d	d

[a] Reprinted with permission from Traverso and Scandola [28].

[b] Ferrocene concentration: 7.5×10^{-3} M; deoxygenated solutions. Values given are corrected for the fraction of light absorbed for the CTTS band.

[c] Quantum yield values in the presence of 7.5×10^{-2} acrylamide monomer.

[d] In these experimental conditions, the fraction of light absorbed by the intermolecular CT band is almost negligible.

[e] Owing to the small amount of CT absorption, these values are affected by a considerable error.

be noted that Traverso and Scandola showed the formation of $[Fe(\eta^5\text{-}C_5H_5)_2]Cl$ by its electronic absorption spectrum and not by actual isolation of the products.

In 1973, Sugimori and co-workers [33–34a] confirmed Traverso and Scandola's qualitative observations but also were able to isolate and characterize the products of the photolysis. They were able to show that photolysis of ferrocene in halocarbon/ethanol solvent mixtures gave not only formation of $[Fe(\eta^5\text{-}C_5H_5)_2]Cl$ but also products derived by substitution of one of the ferrocene rings (Eq. 5-13). Similar substitution products were obtained from

$$\text{Fe} + \text{R—Cl} \xrightarrow[\text{EtOH}]{h\nu} \text{Fe} \qquad (5\text{-}13)$$

R = —CCl₃   R′ = —CO₂Et
R = —CHCl₂   R′ = —CHO
R = —CH₂Cl   R′ = —CH₂OEt

other halocarbons, and yields of all the reactions are given in Table 5-3. The

TABLE 5-3

Photochemical Reactions of Ferrocene in Halogenated Hydrocarbon–Ethanol 1:1 Solutions[a]

Run	FeH[b] (g)	Solvent (ml)	Light (W)	FeH reacted (%)	Product	Yield[c] (%)
1	3.7	CCl_4–EtOH (280)	LP (16)[d] 48 h	57	FeCOOEt	20
2	3.0	CCl_4–EtOH (140)	HP (150)[e] 8 h	35	FeCOOEt	38
3	1.9	CCl_4–EtOH (400)	HPpy (100)[f] 7 h	34	FeCOOEt	32
4	3.0	$CHCl_3$–EtOH (220)	LP (120) 7 h	84	FeCHO	42
5	2.0	$CHCl_3$–EtOH (140)	HP (150) 5 h	35	FeCHO	trace[g]
6	4.0	CH_2Cl_2–EtOH (220)	LP (120) 7 h	72	$FeCH_2OEt$ $(C_5H_4CH_2OEt)_2Fe$[h]	43 36
7	1.0	CH_2Cl_2–EtOH	HP (150) 3 h	9	$FeCH_2OEt$ $(C_5H_4CH_2OEt)_2Fe$	53 12

[a] Reprinted with permission from Akiyama et al. [33].
[b] Fe: $C_5H_5FeC_5H_4$—.
[c] Yield based on unrecovered ferrocene.
[d] LP: Low-pressure mercury lamp.
[e] HP: High-pressure mercury lamp.
[f] HPpy: High-pressure mercury lamp with a Pyrex filter.
[g] Formylferrocene is sensitive to 313 nm light.
[h] 1,1′-Diethoxymethylferrocene.

various substitution products apparently arise through ethanolysis of a halocarbon-substituted ferrocene, as illustrated in Eq. (5-14) for CCl_4.

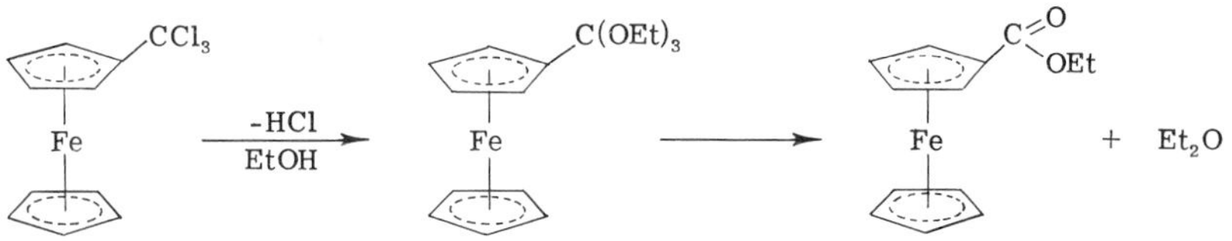

(XLIII)

$$(5\text{-}14)$$

In the second paper in the series, Sugimori and co-workers studied the mechanism of the substitution reaction [34]. Quantum yields measured at various wavelengths are summarized in Table 5-4 and are consistent with photochemistry occurring from the CTTS state characterized by Brand and Snedden [27]. From flash-photolysis studies they were able to conclude that in neat CCl_4 solution $FeCl_3$ is generated very rapidly, within 10 μsec of the flash, and that apparently it forms by decomposition of intermediate $[Fe(\eta^5\text{-}C_5H_5)(\eta^5\text{-}C_5H_4CCl_3)]$ (XLIII). The formation of $FeCl_3$ was completely quenched by addition of ethanol, diethylamine, or DMSO to solutions. Sugimori and co-workers [34] thus proposed that substitution occurs in the sequence of reactions (5-15)–(5-18). Irradiation initially leads to

$$[Fe(\eta^5\text{-}C_5H_5)_2] \xrightarrow[CCl_4]{h\nu} [Fe(\eta^5\text{-}C_5H_5)_2]Cl \ + \ \cdot CCl_3 \qquad (5\text{-}15)$$

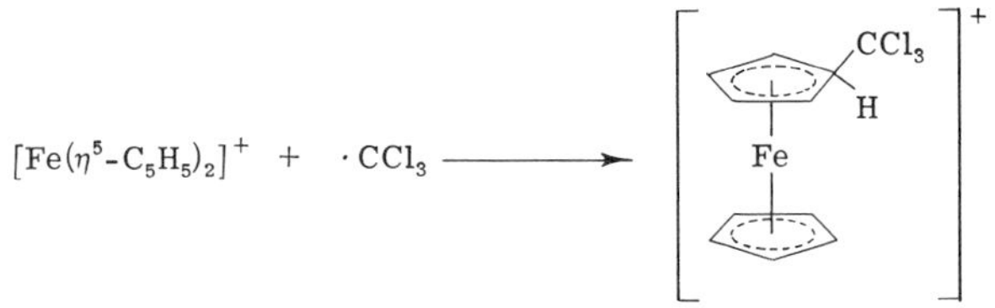

$$(5\text{-}16)$$

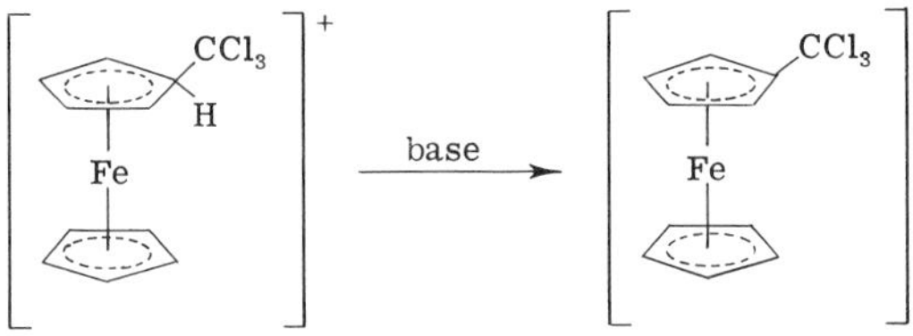

$$(5\text{-}17)$$

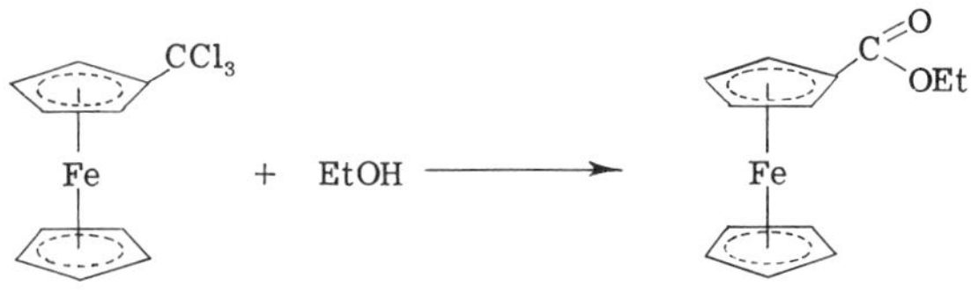

$$(5\text{-}18)$$

TABLE 5-4

Quantum Yields of the Photoreaction of Ferrocene in a
CCl_4–Ethanol Solution at Various Wavelengths[a]

Wavelength (nm)	Quantum yield of ethyl ferrocenecarboxylate	Quantum yield of decomposition of ferrocene
254	0.13	0.28
313	0.13	0.18
366	0.11	0.14
436	0.00	0.00

[a] Reprinted with permission from Akiyama *et al.* [34].

formation of the ferricenium ion and a ·CCl_3 radical. Presumably the CCl_3 radical then attacks the ferricenium ion to generate, after loss of H^+, $[Fe(\eta^5\text{-}C_5H_5)(\eta^5\text{-}C_5H_4CCl_3)]^+$. Rapid hydrolysis by ethanol would then generate the substituted products. The effect of substituents on the cyclopentadienyl ligand in the photoinduced ethoxycarbonylation was subsequently studied [34a], and it was noted that the relative quantum yields support ferrocene-to-CCl_4 charge transfer as the primary photoprocess.

At first glance, the results of Traverso and Scandola [28] and Sugimori and co-workers [33–34a] seem quite contradictory. However, their experimental observations, taken together, allow an overall explanation of the photochemistry of ferrocene to be derived, and this is shown in Fig. 5-9 for photolysis of neat CCl_4 solutions. The primary step in the photoprocess is dissociation of the CTTS state to give $[Fe(\eta^5\text{-}C_5H_5)_2]^+$, Cl^-, and ·CCl_3. In the absence of either EtOH or acrylamide, photolysis proceeds through

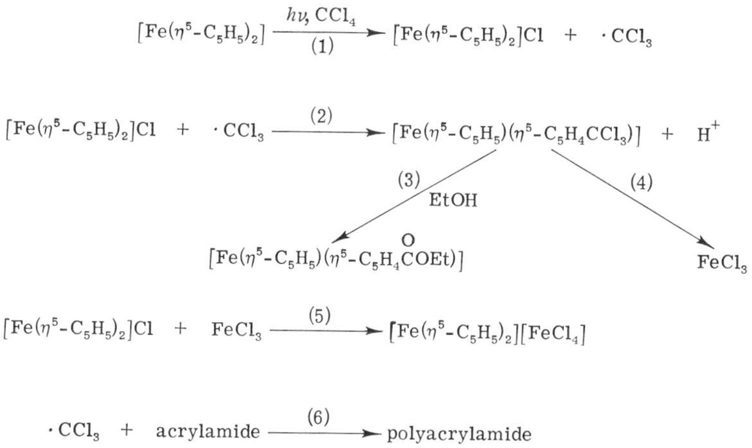

$$[Fe(\eta^5\text{-}C_5H_5)_2] \xrightarrow[(1)]{h\nu,\ CCl_4} [Fe(\eta^5\text{-}C_5H_5)_2]Cl\ +\ \cdot CCl_3$$

$$[Fe(\eta^5\text{-}C_5H_5)_2]Cl\ +\ \cdot CCl_3 \xrightarrow{(2)} [Fe(\eta^5\text{-}C_5H_5)(\eta^5\text{-}C_5H_4CCl_3)]\ +\ H^+$$

(3) EtOH   (4)

$$[Fe(\eta^5\text{-}C_5H_5)(\eta^5\text{-}C_5H_4\overset{O}{C}OEt)] \qquad\qquad FeCl_3$$

$$[Fe(\eta^5\text{-}C_5H_5)_2]Cl\ +\ FeCl_3 \xrightarrow{(5)} [Fe(\eta^5\text{-}C_5H_5)_2][FeCl_4]$$

$$\cdot CCl_3\ +\ acrylamide \xrightarrow{(6)} polyacrylamide$$

FIG. 5-9.   Proposed mechanism for ferrocene photoreaction.

reactions *1*, *2*, *4*, and *5* (Fig. 5-9) to give $[Fe(\eta^5\text{-}C_5H_5)_2][FeCl_4]$ as originally observed by Brand and Snedden [27]. In the presence of acrylamide, the $\cdot CCl_3$ radicals are efficiently trapped, reaction *2* does not occur, and high-yield conversion to $[Fe(\eta^5\text{-}C_5H_5)_2]Cl$ obtains. In the absence of acrylamide but in the presence of EtOH, substitution of ferrocene occurs by the sequence of reactions *1*, *2*, and *3* in Fig. 5-9.

In a somewhat related study, Horvath *et al.* [36] were able to isolate a stable ferrocene–lithium chloride complex $[Fe(\eta^5\text{-}C_5H_5)_2 \cdot LiCl]$ which upon irradiation decomposed to $[Fe(\eta^5\text{-}C_5H_5)_2]Cl$ and iron(III) chloride complexes. The absorption spectrum of this adduct was similar to that of ferrocene, except with a new band at 340–380 nm, similar to the CTTS band of Brand and Snedden [27,35], and it was only this band which was photoactive. Irradiation of the complex in the presence of 2,2′-bipyridine produced only red $[Fe(bipy)_3]^{2+}$, and the authors claimed that the primary photoreaction is that outlined in Eq. 5-19. The Fe(III) products were said to arise through

$$[Fe(\eta^5\text{-}C_5H_5)_2 \cdot LiCl] \xrightarrow{h\nu} Fe^{2+} + 2C_5H_5^- + LiCl \tag{5-19}$$

secondary thermal reactions. It is not clear how this study meshes with those described, except that a CTTS-like state is the photoactive excited state.

The photochemistry of ferrocene has also been examined in the presence of N_2O. Powell and Logan [37] observed that 254 nm excitation of ethanol or cyclohexane solutions of ferrocene under N_2O gave production of N_2 and ferricenium ion according to Eq. (5-20). The quantum yield was linearly

$$[Fe(\eta^5\text{-}C_5H_5)_2] + N_2O \xrightarrow{h\nu} [Fe(\eta^5\text{-}C_5H_5)_2]^+ + N_2 + [O]^- \tag{5-20}$$

dependent on the N_2O concentration but was significantly less than unity at all concentrations examined (e.g., 0.01 at $[N_2O] = 0.08 \ M$). Only 254 nm irradiation induced the reaction, and it was proposed that the low quantum yields obtained because the photoactive state was not the state directly populated by the light absorption. Finally, in a very early study, Thrush observed that photolysis of ferrocene in the gas phase gave cleavage to $[Fe(\eta^5\text{-}C_5H_5)]$ and C_5H_5 [38].

B. DERIVATIVES OF FERROCENE

Numerous compounds which derive from ferrocene by substitution of organic functional groups on the cyclopentadienyl rings have been prepared and characterized. Many of these organoferrocene derivatives have had their photochemical properties examined, and in most cases the photochemistry is that of the organic functional group. These studies fall outside the scope of this book and will not be discussed in detail here. Nesmeyanov and his co-workers have made many important contributions to this field [39–43], and the various studies have been adequately reviewed by several groups of workers [3,44,45].

One notable derivative which has been much studied in recent years is benzoylferrocene (XLIV), in some respects an organometallic analog to benzophenone. The lowest excited state in (XLIV) is probably LF, not $n\pi^*$.

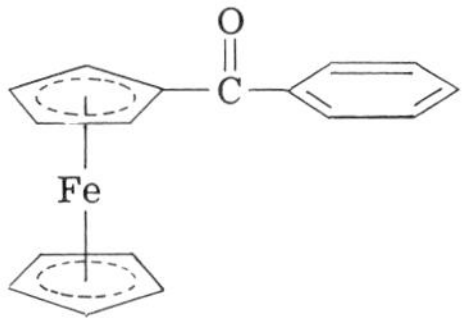

(XLIV)

And, unlike benzophenone which undergoes clean conversion to benz-pinacol when irradiated in alcohol solutions, benzoylferrocene shows a complex and somewhat confusing behavior. In 1973, Kemp and co-workers reported [46] that photolysis of benzoylferrocene and a series of related acylferrocenes in the polar solvents DMSO, DMF, and pyridine, gave photo-solvation via cleavage of a ring–metal bond (Eq. 5-21). The photoreaction

$$[Fe(\eta^5\text{-}C_5H_5)(\eta^5\text{-}C_5H_4\overset{\overset{O}{\|}}{\text{---}C}\text{---}C_6H_5)] \xrightarrow[\substack{DMSO/H_2O. \\ 280 \text{ nm}}]{h\nu} [Fe(\eta^5\text{-}C_5H_5)(\text{solvent})_n]PhCO_2 + C_5H_6$$

(XLV)

$$(5\text{-}21)$$

only occurred when water was present as an impurity in the solvents, and no reaction was observed in thoroughly dried solvents. NMR evidence verified the production of free C_5H_6 and indicated that the product (XLV) had one bound $\eta^5\text{-}C_5H_5$ ring. Infrared and UV evidence suggests the salt formulation given, and this is consistent with the observation that treatment with HCl gave benzoic acid. The benzoate was shown to derive from the originally substituted ring, and water was the source of oxygen in the carboxalate. Although a detailed mechanistic discussion was not presented, the results of Traverso et al. [47] suggest that a hydrogen-bonded intermediate such as (XLVI) may play a key role. When the photolysis was conducted in the

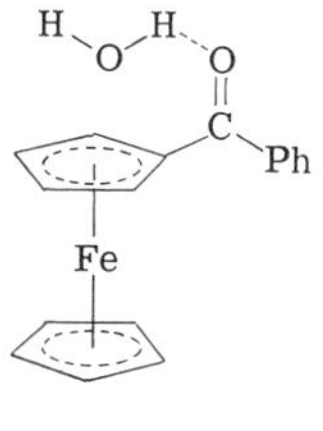

(XLVI)

presence of oxygen, a secondary reaction occurred which appeared to give rise to an Fe(III) species.

Bozak and Javaheripour also reported [48] in 1973 that photolysis of benzoyl- and other acylferrocenes in acidic alcohol solution (isopropanol + 12 N HCl) gave metal–ring cleavage, but they observed no debenzoylation. Protonation of acylferrocenes presumably occurs in these highly acidic solutions and the mechanism outlined in Eq. (5-22) was proposed [48].

$$\text{(5-22)}$$

Following neutralization, the keto form of benzoylcyclopentadienyl was isolated in 52% yield. It was proposed that the photoreaction occurs via the states populated by irradiation into the previously assigned [49] dissociative charge transfer bands at 370 and 480 nm. Examination of the reported spectra, however, reveals a strong similarity to the ferrocene spectrum for which analogous bands are clearly LF in character.

In still another report in 1973, Traverso *et al.* [47] described the photochemistry of benzoylferrocene in alcohol solution. Slight shifts and broadening in the infrared and electronic absorption spectra of benzoylferrocene in 10% methanol/CCl_4 and 10% ethanol/CCl_4 solutions suggested interaction of solvent with the complex, presumably through hydrogen bonding with the carbonyl group. *tert*-Butanol, isobutanol, or neat CCl_4 solutions showed no unusual spectral shifts, and prolonged photolysis of the complex in these apparent non-hydrogen-bonding solvents gave no observable reaction. Photolysis of neat ROH or ROH/CCl_4 solutions (R = CH_3, C_2H_5, or n-C_3H_7), however, gave rapid spectral changes and the formation of benzoylferricenium ion. In the presence of acrylamide, photolysis produced polymerization, suggesting a radical path. However, irradiation in the presence of $C(NO_2)_4$, a powerful scavenger of solvated electrons, gave reduction of $C(NO_2)_4$. The authors proposed that the reaction proceeds through production of solvated electrons according to Eq. (5-23). It was suggested that photoelectron ejection occurs from CTTS states lying under the benzoyl-

$$[Fe(\eta^5\text{-}C_5H_5)(\eta^5\text{-}C_5H_4COC_6H_5)] \xrightarrow{h\nu} [Fe(\eta^5\text{-}C_5H_5)(\eta^5\text{-}C_5H_4COC_6H_5)]^+ + e^-(ROH)$$
$$\text{(5-23)}$$

ferrocene intramolecular bands. No spectral evidence was obtained for these states, however.

C. $[Fe(\eta^5\text{-}C_5H_5)_2]^+$

The ferricenium cation has not been studied in great detail although several interesting reports have appeared. Koerner von Gustorf and Grevels [1] have reported that photolysis of $[Fe(\eta^5\text{-}C_5H_5)_2][FeX_4]$ in $CCl_4/EtOH$ solutions gave ferrocene and a large amount of C_2Cl_6, resulting from coupling of $\cdot CCl_3$ radicals.

The 254 nm induced decomposition of $[Fe(\eta^5\text{-}C_5H_5)_2]Cl$ was studied by Borrell and Henderson [50]. In aqueous HCl solution, decomposition occurred with a quantum yield of 0.029 independent of $[H^+]$ from 0.11 M to 2.76 M, and ferrocene was the only product observed. In $H_2O/MeOH/HCl$ solution, photoreduction to ferrocene occurred if $[H^+] = 1$ M, but in dilute HCl solution decomposition to $[Fe(OMe)]^{2+}$ obtained.

D. $[Fe(\eta^5\text{-}C_5H_5)(CO)_2X]$

Although the $[Fe(\eta^5\text{-}C_5H_5)(CO)_2X]$ (X = Cl^-, Br^-) complexes are inactive when irradiated in cyclohexane or diethyl ether solution, photolysis in DMSO or pyridine has been reported [51] to lead to production of $[Fe(\eta^5\text{-}C_5H_5)(CO)_2]_2$. The reaction apparently occurs through heterolysis of an Fe—X bond according to the reactions outlined in Eqs. (5-24)–(5-27).

$$[Fe(\eta^5\text{-}C_5H_5)(CO)_2X] \xrightarrow{h\nu} [Fe(\eta^5\text{-}C_5H_5)(CO)_2]^+ + X^- \tag{5-24}$$

$$[Fe(\eta^5\text{-}C_5H_5)(CO)_2]^+ + [Fe(\eta^5\text{-}C_5H_5)(CO)_2X] \longrightarrow$$
$$[Fe(\eta^5\text{-}C_5H_5)(CO)_2] + [Fe(\eta^5\text{-}C_5H_5)(CO)_2X]^+ \tag{5-25}$$

$$2[Fe(\eta^5\text{-}C_5H_5)(CO)_2] \longrightarrow [Fe(\eta^5\text{-}C_5H_5)(CO)_2]_2 \tag{5-26}$$

$$[Fe(\eta^5\text{-}C_5H_5)(CO)_2X]^+ \xrightarrow{H^+} \tfrac{1}{2}[C_5H_6]_2 + 2CO + FeCl^{2+} \tag{5-27}$$

Photohomolysis to produce chlorine radicals was ruled out because irradiation did not induce polymerization and Cl_2 was not detected. Heterolysis is also suggested by the observation that the reaction only proceeds in high-dielectric solvents. Production of $[Fe(\eta^5\text{-}C_5H_5)(CO)_2]_2$ occurs smoothly with maintenance of isosbestic points in the spectral changes only if $\lambda > 400$ nm is used. Irradiation of samples with $\lambda > 280$ nm leads to a secondary process in which substitution of $[Fe(\eta^5\text{-}C_5H_5)(CO)_2]_2$ occurs, apparently through carbonyl loss.

In a subsequent study [52], it was reported that photolysis of the substituted derivatives $[Fe(\eta^5\text{-}C_5H_5)(CO)(L)Br]$ gave production of $[Fe(\eta^5\text{-}C_5H_5)(CO)_2]_2$ when L = $P(OPh)_3$ and $[Fe(\eta^5\text{-}C_5H_5)(CO)_2(PPh_3)]Br$ when L = PPh_3. These two reactions must proceed through complicated mechanisms and their overall synthetic yields were around 10%.

E. $[Fe(\eta^5\text{-}C_5H_5)CO]_4$

Bock and Wrighton [53] examined the photochemistry of the tetranuclear cluster $[Fe(\eta^5\text{-}C_5H_5)CO]_4$ (XLVII) and found it to be essentially photoinert

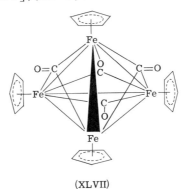

(XLVII)

to ligand substitution. Irradiation in halocarbon solution, however, led to oxidation of the cluster according to Eq. (5-28). Electronic absorption spectra

$$[Fe(\eta^5\text{-}C_5H_5)CO]_4 \xrightarrow[CCl_4]{h\nu} [Fe(\eta^5\text{-}C_5H_5)CO]_4{}^+ \qquad (5\text{-}28)$$

measured in different solvents showed the presence of charge transfer to

TABLE 5-5

Solvent Dependence of Quantum
Yields for Photooxidation of
$[Fe(\eta^5\text{-}C_5H_5)CO]_4{}^{a,b}$

S–CCl$_4$ (1:1 v/v)	$\Phi_{313\,nm}$
CH$_3$CN–CCl$_4$	0.21
MeOH–CCl$_4$	0.19
EtOH–CCl$_4$	0.21
i-PrOH–CCl$_4$	0.19
CH$_2$Cl$_2$–CCl$_4$	0.07
C$_6$H$_6$–CCl$_4$	0.10
CCl$_4$	0.05

[a] Reprinted with permission from Bock and Wrighton [53], *Inorg. Chem.* **16**, 1309 (1977). Copyright by the American Chemical Society.
[b] Degassed   S–CCl$_4$ (1:1 v/v) solutions   containing $2 \times 10^{-4}\,M$ $[Fe(\eta^5\text{-}C_5H_5)CO]_4$ irradiated   at 313 nm.

halocarbon solvent transitions in the range 290–320 nm, with the band position dependent on the solvent employed. As discussed earlier, a similar transition is shown by ferrocene in halocarbon solution [27]. Wavelength dependence studies showed that only the CTTS state was photoactive, and the quantum yields obtained in various solvent mixtures are given in Table 5-5.

F. RUTHENOCENE $[Ru(\eta^5\text{-}C_5H_5)_2]$

Ruthenocene has been found to be emissive at low temperature. Two groups have reported emission spectral properties as a function of temperature [24,25]. The emission has been attributed to the $^3E_1$ LF state. Lifetimes and quantum yields of the emission are temperature-dependent such that the ratio of lifetimes at two different temperatures is equal to the ratio of quantum yields at those same two temperatures, in accord with a temperature-independent radiative decay constant [24]. The emission at low temperatures is highly structured; the vibrational spacings are indicative of an association with the a_{1g} ring–metal stretch [54]. The emission exhibits a 0–0 band at $\sim 19,800$ cm^{-1}, and the triplet quencher properties of ruthenocene are consistent with such a triplet energy [24].

The photochemistry of ruthenocene has been independently reported by two groups of workers [29,55], with their articles received at their respective journals within 1 month of each other. The two reports are remarkably consistent, with the photochemistry of ruthenocene paralleling that of ferrocene. In cyclohexane solution ruthenocene is photoinert to 254 nm irradiation [55], but photolysis into the CTTS band which appears in chlorocarbon solvents (Fig. 5-10) leads to smooth oxidation to the ruthenicenium cation (Eq. 5-29). The ruthenicenium ion itself is too unstable to isolate in

$$[Ru(\eta^5\text{-}C_5H_5)_2] + R\text{—}Cl \xrightarrow{h\nu} [Ru(\eta^5\text{-}C_5H_5)_2]^+ + Cl^- + R \qquad (5\text{-}29)$$

pure form although it was characterized in solution by both groups of workers. Prolonged photolysis results in clouding of the solutions and precipitation of yellow $[Ru(\eta^5\text{-}C_5H_5)_2][RuCl_4]$, similar to the formation of $[Fe(\eta^5\text{-}C_5H_5)_2][FeCl_4]$ from ferrocene. Photolysis in the presence of acrylamide yielded polymerization, confirming the production of radicals.

The only significant difference between the two reports was in the quantum yields. Traverso et al. [55], with no details, reported a primary quantum yield of 1.0 with no effect shown by acrylamide. Borrell and Henderson [29], however, observed a slight concentration dependence and a marked wavelength and oxygen dependence. In deaerated solutions, 313 and 366 nm yields

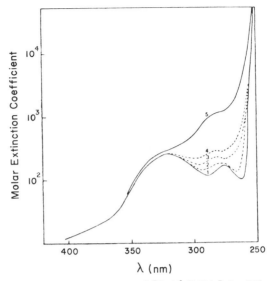

FIG. 5-10. Electronic absorption spectra of $[Ru(\eta^5\text{-}C_5H_5)_2]$ in CCl_4–ethanol solvent mixtures: (1) pure ethanol; (2) 5% CCl_4; (3) 10% CCl_4; (4) 15% CCl_4; (5) pure CCl_4. Reprinted with permission from Traverso et al. [55].

of 0.76 and 0.49, respectively, were obtained, whereas the corresponding yields in aerated solutions were 0.51 and 0.33. The 0.76 yield decreased to 0.63 when acrylamide was added, and this led the authors to suggest that a small portion of the ruthenocene is oxidized through secondary thermal reactions with radicals produced.

The mechanism of the photooxidation is clearly similar to that of ferrocene, although it has not been as thoroughly studied. Also, it should be noted that $[Ru(\eta^5\text{-}C_5H_5)_2]^+$ is not at all well characterized; it has never been made electrochemically, for example.

V. COBALT AND NICKEL COMPLEXES

In a comparison of the photoinduced decomposition of several metallocenes in cyclohexane or acidic methanol solutions, Borrell and Henderson [50] observed that cobaltocene $[Co(\eta^5\text{-}C_5H_5)_2]$ decomposed when irradiated with 254 nm, but a quantum yield could not be obtained because of a competing dark decomposition reaction. Cobalticenium ion $[Co(\eta^5\text{-}C_5H_5)_2]^+$ was stable to 254 nm irradiation.

Nickelocene $[Ni(\eta^5\text{-}C_5H_5)_2]$ slowly decomposed upon 254 nm irradiation giving a brown "polymer" and a quantum yield of 0.007 was estimated The product $[Ni(\eta^5\text{-}C_5H_5)_2]Cl$ was quite photosensitive and decomposed to Ni^{2+} and a yellow polymeric solid with a 254 nm quantum yield of 0.44. Borrell and Henderson [50] concluded from their series of studies that a correlation could be made with the photostability of the metallocenes and their number of valence electrons. Those complexes with an even number of electrons (18,20) were stable, but those with an odd number (17,19) were photoactive.

VI. SUMMARY

The several studies which have been conducted point to two common photoreactions of cyclopentadienyl complexes. The first is photoinduced oxidation proceeding through some type of CTTS state resulting from interaction of electron-accepting solvents with the complexes. We specifically propose that this interaction is of the type shown in (XLII) in which the principal interaction is through the cyclopentadienyl ligand. The second reaction mode appears to involve either complete breakage of the metal–cyclopentadienyl bond or a $\eta^5 \to \eta^{5-n}$ partial loosening of the bond which results in substitution of another ligand for C_5H_5. Much research remains to be conducted on cyclopentadienyl complexes, and it will be particularly important to characterize fully the nature of the excited states which give rise to these various reactions.

REFERENCES

1. E. Koerner von Gustorf and F.-W. Grevels, *Fortschr. Chem. Forsch.* **13**, 366 (1969).
2. R. E. Bozak, *Adv. Photochem.* **8**, 227 (1971).
3. C. R. Bock and E. A. Koerner von Gustorf, *Adv. Photochem.* **10**, 221 (1977).
4. T. C. Mckenzie, R. D. Sanner, and J. E. Bercaw, *J. Organomet. Chem.* **102**, 457 (1975).
5. R. L. Martin and G. Winter, *J. Chem. Soc.* p. 4709 (1965).
6. J. C. Green, M. L. H. Green, and C. K. Prout, *Chem. Commun.* p. 421 (1972).
7. J. L. Petersen, D. L. Lichtenberger, R. F. Fenske, and L. F. Dahl, *J. Am. Chem. Soc.* **97**, 6433 (1975).
8. C. J. Ballhausen and J. P. Dahl, *Acta Chem. Scand.* **15**, 1333 (1961).
9. R. W. Harrigan, G. S. Hammond, and H. B. Gray, *J. Organomet. Chem.* **81**, 79 (1974).

10. G. Condorelli, I. Fragala, A. Centineo, and E. Tondello, *J. Organomet. Chem.* **87**, 311 (1975).
11. E. Vitz and C. H. Brubaker, Jr., *J. Organomet. Chem.* **104**, C33 (1976).
12. E. Vitz, P. J. Wagner, and C. H. Brubaker, Jr., *J. Organomet. Chem.* **107**, 301 (1976).
13. E. Vitz and C. H. Brubaker, Jr., *J. Organomet. Chem.* **82**, C16 (1974).
13a. Z.-T. Tsai and C. H. Brubaker, Jr., *J. Organomet Chem.* **166**, 199 (1979).
14. M. H. Peng and C. H. Brubaker, Jr., *J. Organomet. Chem.* **135**, 333 (1977).
14a. J. G. S. Lee and C. H. Brubaker, Jr., *Inorg. Chim. Acta* **25**, 181 (1977).
15. Y. S. Sohn, D. N. Hendrickson, and H. B. Gray, *J. Am. Chem. Soc.* **93**, 3603 (1971).
16. See Sohn *et al.* [15] and Rösch and Johnson [17] for the pertinent references.
17. N. Rösch and K. H. Johnson, *Chem. Phys. Lett.* **24**, 179 (1974).
18. J. W. Rabalais, L. O. Werme, T. Bergmark, L. Karlsson, M. Hussain, and K. Siegbahn, *J. Chem. Phys.* **57**, 1185 (1972).
19. S. Evans, M. L. H. Green, B. Jewitt, A. F. Orchard, and C. F. Pygall, *J. Chem. Soc., Faraday Trans. 2* **68**, 1847 (1972).
20. D. R. Scott and R. S. Becker, *J. Chem. Phys.* **35**, 516 (1961).
21. A. N. Nesmeyanov, E. G. Perevalova, and O. A. Nesmeyanova, *Dokl. Akad. Nauk SSSR* **100**, 1099 (1955).
22. M. Kikuchi, K. Kikuchi, and H. Kokubun, *Bull. Chem. Soc. Jpn.* **47**, 1331 (1974).
23. A. Gilbert, J. M. Kelly, and E. Koerner von Gustorf, *Mol. Photochem.* **6**, 225 (1974).
24. M. S. Wrighton, L. Pdungsap, and D. L. Morse, *J. Phys. Chem.* **79**, 66 (1975).
25. W. G. Herkstroeter, *J. Am. Chem. Soc.* **97**, 4161 (1975).
26. K. Kikuchi, H. Kokubun, and M. Kikuchi, *Bull. Chem. Soc. Jpn.* **48**, 1378 (1975).
27. J. C. D. Brand and W. Snedden, *Trans. Faraday Soc.* **53**, 894 (1957).
28. O. Traverso and F. Scandola, *Inorg. Chim. Acta* **4**, 493 (1970).
29. P. Borrell and E. Henderson, *J. Chem. Soc., Dalton Trans.* p. 432 (1975).
30. H. Koller, Doctoral Dissertation, Göttingen University (1962).
31. A. M. Tarr and D. M. Wiles, *Can. J. Chem.* **46**, 2725 (1968).
32. E. Koerner von Gustorf, H. Koeller, M.-J. Jun, and G. O. Schenck, *Chem.-Ing.-Tech.* **35**, 591 (1963).
33. T. Akiyama, Y. Hoshi, S. Goto, and A. Sugimori, *Bull. Chem. Soc. Jpn.* **46**, 1851 (1973).
34. T. Akiyama, A. Sugimori, and H. Hermann, *Bull. Chem. Soc. Jpn.* **46**, 1855 (1973).
34a. T. Akiyama, P. Kitamura, T. Kato, H. Watanabe, T. Serizawa, and A. Sugimori, *Bull. Chem. Soc. Jpn.* **50**, 1137 (1977).
35. O. Traverso, R. Rossi, and V. Carassiti, *Synth. React. Inorg. Met.-Org. Chem.* **4**, 309 (1974).
36. E. Horvath, S. Sostero, O. Traverso, and V. Carassiti, *Gazz. Chim. Ital.* **104**, 1003 (1974).
37. J. A. Powell and S. R. Logan, *J. Photochem.* **3**, 189 (1974).
38. B. A. Thrush, *Nature (London)* **178**, 155 (1956).
39. A. N. Nesmeyanov, V. A. Sazonova, V. I. Ramanenko, V. N. Postnov, G. N. Zol'nikova, V. A. Blinova, and R. M. Kalyanova, *Dokl. Akad. Nauk SSSR* **173**, 589 (1967).
40. A. N. Nesmeyanov, A. Sazonova, V. I. Romanenko, and G. P. Zol'nikova, *Izv. Akad. Nauk SSSR, Ser. Khim.* p. 1694 (1965).
41. A. N. Nesmeyanov, A. Sazonova, V. I. Romanenko, N. A. Rodionova, and G. P. Zol'-nikova, *Dokl. Akad. Nauk SSSR* **155**, 1130 (1964).
42. A. N. Nesmayanov, A. Sazonova, and V. I. Romanenko, *Dokl. Akad. Nauk SSSR* **152**, 1358 (1963).
43. A. N. Nesmeyanov, V. A. Sazonova, A. V. Gerasumenko, and N. S. Sazonova, *Dokl. Akad. Nauk SSSR* **149**, 1354 (1963).
44. R. E. Bozak, *Adv. Photochem.* **8**, 227 (1971).

45. E. G. Perevalova and T. V. Nikitina, *Organomet. React.* **4**, 163 (1972).
46. L. H. Ali, A. Cox, and T. J. Kemp, *J. Chem. Soc., Dalton Trans.* p. 1468 (1973).
47. O. Traverso, R. Rossi, S. Sostero, and V. Carassiti, *Mol. Photochem.* **5**, 457 (1973).
48. R. E. Bozak and H. Javaheripour, *Chem. Ind. (London)* p. 696 (1973).
49. R. T. Lundquist and M. Cais, *J. Org. Chem.* **27**, 1167 (1962).
50. P. Borrell and E. Henderson, *Inorg. Chim. Acta* **12**, 215 (1975).
51. L. H. Ali, A. Cox, and T. J. Kemp, *J. Chem. Soc., Dalton Trans.* p. 1475 (1973).
52. D. M. Allen, A. Cox, T. J. Kemp, and L. H. Ali, *J. Chem. Soc., Dalton Trans.* p. 1899 (1973).
53. C. R. Bock and M. S. Wrighton, *Inorg. Chem.* **16**, 1309 (1977).
54. G. A. Crosby, G. D. Hager, K. W. Hipps, and M. L. Stone, *Chem. Phys. Lett.* **28**, 497 (1974).
55. O. Traverso, S. Sostero, and G A. Mazzocchin, *Inorg. Chim. Acta* **11**, 237 (1974).

Isocyanide Complexes

6

I. INTRODUCTION

A very large number of transition-metal isocyanide complexes have been prepared and characterized. Malatesta and Bonati's book "Isocyanide Complexes of Metals" [1] surveys the literature through 1968, and two excellent review articles [2,3] highlight some of the recent advances in the chemistry of this class of compounds.

Isocyanide ligands are particularly interesting because of their unique electronic properties. Isocyanides are good σ donors and also excellent π acceptors, in effect combining the bonding properties of CO and CN$^-$. Like carbon monoxide and cyanide, isocyanides lie high in the spectrochemical series, but their electronic properties vary somewhat with the nature of the organic portion of the ligand (R in RNC). Alkyl isocyanides are, in general, better σ donors than aryl isocyanides, whereas aryl isocyanides are superior in π-accepting ability. Further, aryl isocyanides can be tuned by varying the nature of the aryl substituents through the range of electron-withdrawing and electron-donating groups. For example, an approximate π-acceptor ordering is $p\text{-}CNC_6H_4NO_2 > p\text{-}CNC_6H_4Cl > CNC_6H_5 > p\text{-}CNC_6H_4CH_3$.

The superiority which aryl isocyanides show in π-accepting ability arises primarily because the energy of one of the CN π^* orbitals is lowered through conjugation with an arene π^* orbital. This interaction is illustrated in

Fig. 6-1. Conjugation of the out-of-plane CN π^*_v with an arene π^* orbital lowers its energy with respect to the in-plane CN π^*_h. Such conjugation, of course, cannot occur with alkyl isocyanides. The lowering of CN π^*_v relative to the alkyl isocyanide π^* orbitals allows increased overlap with the metal d orbitals and hence increased π acceptor ability. A similar conclusion has been reached by Bursten and Fenske [4] whose detailed calculations on a series of substituted phenyl isocyanides $CN—C_6H_4X$ show that the CN π^*_v orbitals possess a large percentage of C_6H_4X ring character.

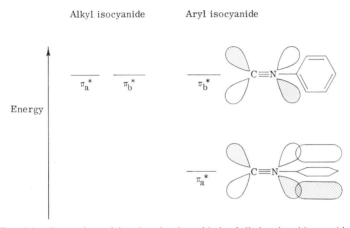

Fig. 6-1. Comparison of the π^*-molecular orbitals of alkyl and aryl isocyanides.

Intraligand, ligand-field, metal-to-isocyanide charge transfer, and isocyanide-to-metal charge transfer transitions have all been observed in isocyanide complexes of various types. Since the N—C portion of the ligand exhibits no electronic transitions below $50,000$ cm^{-1}, any IL absorption observed below this energy must arise from the R group. Alkyl isocyanides show virtually featureless electronic spectra below $50,000$ cm^{-1}, whereas aryl isocyanides show the characteristic aryl absorptions. Absorption spectra [5] of several aryl isocyanides are shown in Fig. 6-2, and the observed spectral shifts accord well with the nature of the R group.

Metal-to-isocyanide charge transfer transitions are often observed in isocyanide complexes, particularly those containing aryl isocyanides. As illustrated in Fig. 6-1, CN π^*_v is often sufficiently low in energy to be populated in low-lying excited states. Indeed, an excellent test for determining if a particular absorption band is MLCT is to replace the aryl isocyanides with alkyl isocyanides. If the band moves to much higher energy, then an MLCT assignment is suggested. The series of $[M(CNPh)_6]$ (M = Mo(0), Cr(0), W(0)) complexes, discussed in detail in the following section, have

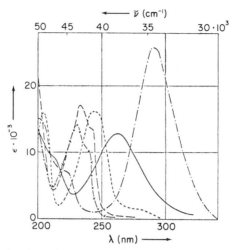

FIG. 6-2.   Electronic absorption spectra of p-CNC$_6$H$_4$CH$_3$ ($-\times-\times-$), p-CNC$_6$H$_4$Cl
($--$), p-CNC$_6$H$_4$OCH$_3$ ($----$), p-CHC$_6$H$_4$N(C$_2$H$_5$)$_2$ ($-\cdot-\cdot-$), and p-CNC$_6$H$_4$NO$_2$ ($———$)
in ethanol solution. Reprinted with permission from Ugi and Meyr [5].

been shown to exhibit rich MLCT spectra [6]. Likewise, square-planar
$[M(CNR)_4]^{n+}$ (M = Rh(I), Ir(I), Pt(II)) complexes show MLCT transitions
lowest in energy [7,8], and in many respects their spectra closely resemble
the spectra displayed by the series of $[M(diphos)_2]^+$ and $[MX(CO)L_2]$
(M = Rh(I), Ir(I); X = Cl$^-$, Br$^-$; L = phosphorus donor) complexes for
which the absorption bands have clearly been shown to be of MLCT origin
[9,10].

Isocyanide-to-metal charge transfer is less common than MLCT and is
observed only in those complexes which have low-lying empty d orbitals.
An excellent illustration comes from octahedral $[Mn(CNPh)_6]^{2+}$ with a
$(t_{2g})^5$ ground-state electronic configuration. Promotion of an electron
from an isocyanide ligand to the empty t_{2g} orbital can readily occur, and
LMCT transitions are observed within the visible spectral region [6].
Interestingly, the visible spectrum of $[Mn(CNPh)_6]^{2+}$ is quite similar to
that displayed by $[Mn(CN)_6]^{4-}$ for which analogous LMCT assignments
have been made [11].

MLCT transitions are normally shifted to high energy in alkyl isocyanide
complexes and LF transitions can often be observed. An example of an
isocyanide complex which shows LF bands is $[Fe(CNCH_3)_6]Cl_2$. The
spectrum [12] of this complex (Fig. 6-5) shows two bands at 315 nm ($\varepsilon = 320$)
and 263 nm ($\varepsilon = 340$) which can logically be assigned as the $^1A_{1g} \rightarrow {}^1T_{1g}$
and $^1A_{1g} \rightarrow {}^1T_{2g}$ d^6-octahedral transitions.

The photochemical properties of the isocyanide *ligands* have not been extensively studied. This is somewhat unfortunate since an investigator interested in the photochemistry of a complex would always like to know the possible ligand photoreactions. One report [13] does describe the photoisomerization of methyl isocyanide to acetonitrile (Eq. 6-1). The photolysis was conducted on gaseous samples using 254 nm irradiation.

$$CNCH_3 \xrightarrow{\ h\nu\ } NCCH_3 \tag{6-1}$$

The extinction coefficient of $CNCH_3$ at 254 nm was measured as 0.3 ± 0.1, and the quantum yield of reaction (6-1) estimated as about 2. The photoisomerization is pressure- and temperature-dependent, but no evidence suggesting a radical pathway was obtained.

In spite of the very large number of isocyanide complexes that are known, only a very few have had their photochemical properties examined. This is somewhat surprising since the variety of possible electronic excited states suggests opportunities for interesting comparative studies and especially for investigations of charge-transfer photochemistry. The few studies that have been conducted are described in the following paragraphs.

II.   CHROMIUM, MOLYBDENUM,
AND TUNGSTEN COMPLEXES

$[M(CNPh)_6]$ AND $[M(CN-2,6-(i-Pr)_2Ph)_6]$

The hexakis(aryl isocyanide) complexes of Cr(0), Mo(0), and W(0) are the only isocyanides of the Group VI metals which have been examined, but these complexes show a very rich photochemistry [14–16]. Detailed electronic absorption spectral studies have clearly shown [6] that the least energetic electronic transitions in these $[M(CNR)_6]$ complexes are of MLCT character, and the photochemistry presumably derives from these charge-transfer states.

The electronic absorption spectra of the $[M(CNPh)_6]$ complexes are shown in Fig. 6-3, and the spectral data are summarized in Table 6-1 [6]. Each of the three complexes shows intense bands in the visible spectral region which have been attributed to metal-to-isocyanide charge transfer transitions. The least energetic of these transitions represents population of the out-of-plane π^*_v orbital for which calculations [4] show substantial

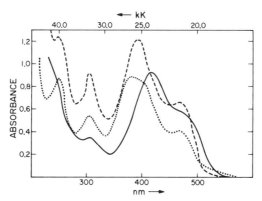

FIG. 6-3. Electronic absorption spectra of $[Cr(CNPh)_6]$ in 1:1 isopentane/diethyl ether (———), $[Mo(CNPh)_6]$ in 8:2:1 ethanol/methanol/diethyl ether (----), and $[W(CNPh)_6]$ in EPA ($\cdots$) at $77°K$. Reprinted with permission from Mann *et al.* [6].

C_6H_5 ring character. A comparison of the spectral values for these $[M(CNPh)_6]$ complexes with those of the corresponding hexacarbonyls $[M(CO)_6]$ (Table 6-2) clearly illustrates the effect of stabilization of the π^*_v-CN orbital due to its C_6H_5-π^* conjugation (Fig. 6-1). The lower energies of the $d\pi \rightarrow \pi^*$ bands in the isocyanide complexes can also be partially attributed to increased electron density on the metal due to an enhanced basicity of CNPh compared to CO.

Photolysis of the $[M(CNR)_6]$ (M = Cr, Mo, W; R = phenyl, 2,6-diisopropylphenyl) complexes in degassed pyridine solution results in photosubstitution of pyridine for isocyanide (Eq. 6-2). The electronic absorption spectral changes which obtain during photolysis of $[Cr(CN-2,6-(i-Pr)_2Ph)_6]$

$$[M(CNR)_6] \xrightarrow[\text{pyridine}]{h\nu} [M(CNR)_5py] + CNR \qquad (6\text{-}2)$$

in degassed pyridine solution are shown in Fig. 6-4 [16]. The isosbestic points shown in Fig. 6-4 and seen in the infrared and electronic absorption spectral changes for all the complexes studied suggest a clean conversion. Although pure $[M(CNR)_5py]$ products could not be isolated because of their air and solvent sensitivity, they were accurately identified by their infrared spectra. The $\nu_{C\equiv N}$ IR absorption pattern observed in each case is consistent with a C_{4v} $[M(CNR)_5L]$ configuration. The appearance of $\nu_{C\equiv N}$ corresponding to uncoordinated CNR in the IR during photolysis provided further support for the overall reaction expressed in Eq. (6-2). The photosubstitution reactions must be conducted in the absence of O_2, as otherwise oxidation occurs. A similar photoreactivity of $[Cr(CNPh)_6]$,

TABLE 6-1

Electronic Absorption Spectra of $[M(CNPh)_6]^z$ Complexes[a,b]

Complex	300 K			77 K			Assignment
	λ_{max}, nm	ν_{max}, kK	$\varepsilon_{max} \times 10^{-3}$	λ_{max}, nm	ν_{max}, kK	$\varepsilon_{max} \times 10^{-3}$	
$[Cr(CNPh)_6]^c$	458 sh	21.8	46.0	480	20.8	36.0	$d\pi \rightarrow \pi^*_v$ (CNPh)
	294	25.4	73.0	419	23.9	61.0	$d\pi \rightarrow \pi^*_v$ (CNPh)
	310	32.3	36.0	307	32.6	23.0	$d\pi \rightarrow \pi^*_v$ (CNPh)
$[Mo(CNPh)_6]^d$	453 sh	22.1	e	470	21.3	42.0	$d\pi \rightarrow \pi^*_v$ (CNPh)
	378	26.5	e	396	25.3	78.0	$d\pi \rightarrow \pi^*_v$ (CNPh)
	313 sh	31.9	e	307	32.6	53.0	$d\pi \rightarrow \pi^*_h$ (CNPh)
	255 sh	39.2	e	250	40.0	81.0	intraligand
$[W(CNPh)_6]^f$	446	22.4	44.0	467	21.4	35.0	$d\pi \rightarrow \pi^*_v$ (CNPh)
				405	24.7		
	367	27.2	58.0	380	26.3		$d\pi \rightarrow \pi^*_v$ (CNPh)
	320	81.3	41.0	306	32.7	46.0	$d\pi \rightarrow \pi^*_h$ (CNPh)
	250	40.0	—	250	40.0		intraligand
$[Mn(CNPh)_6]Cl^d$	340 sh	29.4	61.0	345	29.0	46.0	$d\pi \rightarrow \pi^*_v$ (CNPh)
	322	31.1	66.0	325	30.8	46.0	$d\pi \rightarrow \pi^*_v$ (CNPh)
	249 sh	40.2	51.0	250	40.0	—	intraligand
	234 sh	42.7	71.0	235	42.6	—	
	225	44.4	75.0				
$[Mn(CNPh)_6] \cdot [PF_6]_2^{d,g}$	—	—	—	549	18.2	5.3	$\sigma(CNPh) \rightarrow d\pi$
	490	20.4	4.6	490	20.4	6.4	$\sigma(CNPh) \rightarrow d\pi$

[a] Reprinted with permission from Mann *et al.* [6].
[b] Spectral corrected for solvent contraction.
[c] 1:1 isopentane/diethyl ether solution.
[d] 8:2:1 ethanol/methanol/diethyl ether solution.
[e] Solutions were too photosensitive to allow measurement.
[f] EPA solution.
[g] Ultraviolet spectra are not reported, owing to interfering absorption of the nitric acid added to the solutions.

TABLE 6-2

TABLE 6-2

Energies of MLCT Transitions in $[M(CO)_6]^z$ and
$[M(CNPh)_6]^z$ Complexes[a]

M	$d\pi \to \pi^*$ (CO) $(cm^{-1} \times 10^{-3})$	$d\pi \to \pi^*$ (CNPh) $(cm^{-1} \times 10^{-3})$
Cr(0)	35.7	21.8
	43.6	25.4
		32.3
Mo(0)	34.6	22.1
	42.8	26.5
		31.9
W(0)	34.7	22.4
	43.8	27.2
		31.3
Mn(I)	44.5	29.4
	49.9	31.1

[a] Reprinted with permission from Mann et al. [6].

$[Cr(p\text{-}CNC_6H_4OCH_3)_6]$, and $[Cr(p\text{-}CNC_6H_4CH_3)_6]$ had been previously observed by Sugimori and co-workers [14] who reported photoinduced substitution of fumaronitrile, dimethylfumarate, and maleic anhydride for RNC.

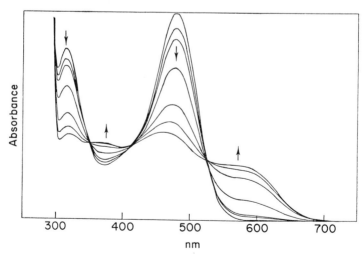

FIG. 6-4. Electronic absorption spectral changes which occur upon photolysis of a degassed pyridine solution of $[Cr(CN\text{-}2,6\text{-}(i\text{-}Pr)_2Ph]$ with fluorescent room lights. Spectra were recorded following periods of about 10 sec irradiations. Reprinted with permission from Mann [16].

Quantum yield values measured [15] for 313 and 436 nm irradiation into the MLCT bands of the complexes in pyridine solution are presented in Table 6-3. Comparison of data for the $[M(CNPh)_6]$ and $[M(CN\text{-}2,6\text{-}(i\text{-}Pr)_2Ph)_6]$ complexes of the same metal yields important mechanistic information. The 2,6-diisopropylphenyl isocyanide ligand induces severe steric hindrance. Yet, $[Cr(CNPh)_6]$ and $[Cr(CN\text{-}2,6\text{-}(i\text{-}Pr)_2Ph)_6]$ show identical quantum yields and wavelength dependences. These data led the authors to suggest [15] for these two complexes a *dissociative* mechanism which is independent of steric constraints [Eqs. (6-3) and (6-4)].

$$[Cr(CNR)_6] \xrightarrow{h\nu} [Cr(CNR)_5] + CNR \qquad (6\text{-}3)$$

$$[Cr(CNR)_5] + py \longrightarrow [Cr(CNR)_5py] \qquad (6\text{-}4)$$

TABLE 6-3

Quantum Yields for Photosubstitution of
Pyridine for CNR in $[M(CNR)_6]^a$

Complex	313 nm	426 nm
$[Cr(CNPh)_6]$	0.54	0.23
$[Mo(CNPh)_6]$	0.11	0.06
$[W(CNPh)_6]$	0.01	0.01
$[Cr(CN\text{-}2,6\text{-}(i\text{-}Pr)_2Ph)_6]$	0.55	0.23
$[Mo(CN\text{-}2,6\text{-}(i\text{-}Pr)_2Ph)_6]$	0.02	0.02
$[W(CN\text{-}2,6\text{-}(i\text{-}Pr)_2Ph)_6]$	$<1 \times 10^{-4}$	$<3 \times 10^{-4}$

^{a} Reprinted with permission from Mann [16].

A different interpretation was given [15] for photosubstitution of the analogous tungsten complexes. The quantum yields in Table 6-3 show a marked decrease in moving from $[W(CNPh)_6]$ to $[W(CN\text{-}2,6\text{-}(i\text{-}Pr)_2Ph)]$. This suggests an *associative* mechanism involving direct nucleophilic attack of pyridine on the positively charged metal center which is obtained in an MLCT state [Eqs. (6-5)–(6-7)]. The very low quantum yield observed for $[W(CN\text{-}2,6\text{-}(i\text{-}Pr)_2Ph)_6]$ is presumably due to the inability of the pyridine

$$[W(CNR)_6] \xrightarrow{h\nu} [W^+(CN^-R)_6]^* \qquad (6\text{-}5)$$

$$[W^+(CN^-R)_6]^* + py \longrightarrow [W^+(CN^-R)_6py] \qquad (6\text{-}6)$$

$$[W^+(CN^-R)_6py] \longrightarrow [W(CNR)_5py] + CNR \qquad (6\text{-}7)$$

to penetrate the bulky ligand barrier. The Mo complexes appear intermediate in properties, but the decreased quantum yield observed for $[Mo(CN\text{-}2,6\text{-}(i\text{-}Pr)_2Ph)_6]$ suggests some associative character in the photosubstitution.

Photolysis of the complexes in halocarbon solvents leads to a completely different photochemistry [15]. In degassed $CHCl_3$ solutions, 436 nm irradiation of the $[M(CN-2,6-(i-Pr)_2Ph)_6]$ complexes produced the one-electron oxidation products $[M(CN-2,6-(i-Pr)_2Ph)_6]^+$ (Eq. 6-8). The reaction products were identified by their IR spectra and by isolation and elemental analysis

$$[M(CN-2,6-(i-Pr)_2Ph)_6] \xrightarrow[CHCl_3]{hv} [M(CN-2,6-(i-Pr)_2Ph)_6]Cl \qquad (6-8)$$

of $[Cr(CN-2,6-(i-Pr)_2Ph)_6]Cl$. In degassed solution the measured 436 nm quantum yields were 0.19 for each of the three metals, but in the presence of oxygen the quantum yields increased markedly. With $[Cr(CN-2,6-(i-Pr)_2Ph)_6]$, for example, a quantum yield of 0.70 was obtained in oxygen-saturated $CHCl_3$.

Photolysis of the less hindered $[M(CNPh)_6]$ complexes in $CHCl_3$ solution produced different results. For M = Mo, W the seven-coordinate, two-electron oxidation products $[M(CNPh)_6Cl]^+$ were formed with 436 nm quantum yields of 0.11 and 0.28, respectively. Photolysis of $[Cr(CNPh)_6]$ in degassed $CHCl_3$ gave oxidation, but the nature of the final product was not ascertained.

The mechanism proposed [15] to explain the observed photooxidation is set out in Eqs. (6-9)–(6-13). The initially formed MLCT excited state is

$$[ML_6] \xrightarrow{436\ nm} [ML_6]^* \qquad (6-9)$$

$$[ML_6]^* + CHCl_3 \longrightarrow [ML_6^+ --- CHCl_3^{\cdot -}] \qquad (6-10)$$

$$[ML_6^+ --- CHCl_3^{\cdot -}] \longrightarrow [ML_6]^+ + CHCl_3^{\cdot -} \qquad (6-11)$$

$$[ML_6^+ --- CHCl_3^{\cdot -}] \longrightarrow [ML_6Cl^+ --- CHCl_2^{\cdot -}] \qquad (6-12)$$

$$[ML_6Cl^+ --- CHCl_2^{\cdot -}] \longrightarrow [ML_6Cl]^+ + CHCl_2^{\cdot -} \qquad (6-13)$$

believed to be quenched by electron transfer to $CHCl_3$, forming the $[ML_6^+ --- CHCl_3^{\cdot -}]$ radical pair. This species can then decompose along two pathways, either by simple dissociation to give $[ML_6]^+$ or by transfer of Cl and then dissociation to produce $[ML_6Cl]^+$. The preferred pathway appears to be the latter. In the sterically hindered $[M(CN-2,6-(i-Pr)_2Ph)_6]$ complexes, however, this route is blocked and reaction (6-11) predominates. The constant quantum yield of 0.19 for production of the $[M(CN-2,6-(i-Pr)_2Ph)_6]^+$ complexes led the authors to propose [15] that the rate-limiting step in the oxidation is diffusion of $CHCl_3^{\cdot -}$ away from the complex. The high quantum yields observed in the presence of O_2 presumably arise from formation of $O_2^{\cdot -}$ which can diffuse away from the metal faster than can $CHCl_3^{\cdot -}$

The $[M(CNR)_6]$ system is particularly interesting from a photochemical standpoint because it illustrates the variety of reactions which can occur from charge-transfer excited states. Electronic absorption spectral studies

[6] have clearly shown that the least energetic transitions in these complexes are of metal-to-isocyanide charge transfer character and in essence give a state with a one-electron oxidation of the metal and one-electron reduction of the ligand system. The electron residing out on the ligand system is available for transfer to an appropriate oxidant, if one is available. In halocarbon solvents, oxidation is observed, but the nature of the final products depends on the ability of the metal to achieve seven-coordination. In the absence of oxidant, simple photosubstitution occurs, but both associative and dissociative pathways obtain.

III. MANGANESE COMPLEXES

$[Mn(p\text{-}CNC_6H_4CH_3)_6]ClO_4$

The only manganese isocyanide complex which has been examined is $[Mn(p\text{-}CNC_6H_4CH_3)_6]^+$. Electronic absorption and IR spectral changes obtained [16] upon irradiation of neat pyridine solutions of the complex are similar to those observed [15,16] during photolysis of the $[M(CNR)_6]$ (M = Cr, Mo, W) complexes. In the IR, as the photolysis proceeds, a band at 2125 cm^{-1} due to photoreleased $p\text{-}CNC_6H_4CH_3$ appears, and bands at 2160 and 2075 cm^{-1} assignable to $[Mn(p\text{-}CNC_6H_4CH_3)_5py]^+$ appear The quantum yields measured at 313 and 366 nm for disappearance of $[Mn(p\text{-}CNC_6H_4CH_3)_6]^+$ are 0.22 and 0.21, respectively. It is interesting to note the similarity of these yields to the 0.19 quantum yield obtained for the $[Cr(CNAr)_6]$ complexes. Similar excited-state decay paths were proposed [16] for these two first-row d^6 complexes, but no mechanistic experiments were conducted on this system.

IV. IRON COMPLEXES

A. $[Fe(CNCH_3)_6]Cl_2$

The electronic absorption spectrum of $[Fe(CNCH_3)_6]Cl_2$ in H_2O, shown as the dashed curve in Fig. 6-5, displays bands at 315 nm ($\varepsilon = 320$) and 263 nm ($\varepsilon = 340$) [12] which can be assigned as the LF transitions $^1A_{1g} \rightarrow {}^1T_{1g}$ and $^1A_{1g} \rightarrow {}^1T_{2g}$, respectively. The spectrum is essentially identical

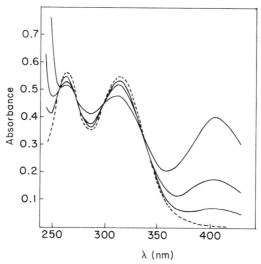

FIG. 6-5. Electronic absorption spectral changes which occur during irradiation of an aqueous solution of $[Fe(CNCH_3)_6]Cl_2$. The dashed curve is the spectrum of $[Fe(CNCH_3)_6]Cl_2$ before photolysis. Reprinted with permission from Carassitti *et al.* [12].

to that of $[Fe(CN)_6]^{4-}$ which has been interpreted in a like manner [11]. Irradiation of aqueous solutions of the complex with either 365 or 254 nm leads to the spectral changes shown in Fig. 6-5. Prolonged photolysis is reported to lead to replacement of all the isocyanides by water. The only intermediate positively identified was $[Fe(CNCH_3)_4(H_2O)_2]^{2+}$, which presumably is the species responsible for the 410 nm absorption feature which grows in during photolysis. The primary photoproduct must be $[Fe(CNCH_3)_5(H_2O)]^{2+}$, and the apparent sequence of reactions is given in Eqs. (6-14) and (6-15). The quantum yields measured at 254 and 365 nm are

$$[Fe(CNCH_3)_6]^{2+} \xrightarrow[H_2O]{hv} [Fe(CNCH_3)_5(H_2O)]^{2+} + CNCH_3 \qquad (6\text{-}14)$$

$$[Fe(CNCH_3)_5(H_2O)]^{2+} \xrightarrow[H_2O]{hv \text{ or } \Delta} [Fe(CNCH_3)_4(H_2O)_2]^{2+} + CNCH_3 \qquad (6\text{-}15)$$

0.09 and 0.06, respectively, and are independent of pH in the range 1.7–6.0. Complex behavior was observed in alkaline solutions.

B. $[Fe(CNCH_3)_4(AA)]^{2+}$ AND $[Fe(CNCH_3)_2(AA)_2]^{2+}$

The series of complexes $[Fe(CNCH_3)_4(AA)]^{2+}$ and $[Fe(CNCH_3)_2 \cdot (AA)_2]^{2+}$ (AA = 1,10-phenanthroline, 2,2′-bipyridine) undergo slow thermal decomposition in aqueous solution, yielding Fe^{2+}, $CNCH_3$, and AA

[17,18]. The decomposition is photoaccelerated by irradiation with 254 and 365 nm into apparent IL $\pi \to \pi^*$ and MLCT bands, respectively. The decomposition presumably occurs through intermediates achieved by successive substitution of $CNCH_3$ by H_2O. The quantum yields were reported to be in the range $10^{-2}-10^{-3}$.

C.   cis- AND trans-$[Fe(CN)_2(CNCH_3)_4]$

The electronic absorption spectra of the four complexes $[Fe(CN_6)]^{4-}$, $[Fe(CNCH_3)_6]^{2+}$, cis-$[Fe(CN)_2(CNCH_3)_4]$, and trans-$[Fe(CN)_2(CNCH_3)_4]$ are remarkably similar, as illustrated by the data presented in Table 6-4. The observed bands are all due to LF transitions, and the spectral similarities clearly indicate that CN^- and $CNCH_3$ are nearly identical with respect to their LF strengths. Acidic solutions of the dicyano complexes are thermally stable in the dark. When irradiated, however, rapid photosubstitution occurs [19]. The electronic absorption spectral changes obtained during photolysis for both the cis and trans complexes are similar to those shown in Fig. 6-5 for photolysis of $[Fe(CNCH_3)_6]^{2+}$ and suggest an identical photoreaction. Production of $CNCH_3$ was observed during irradiation, and addition of excess $CNCH_3$ to irradiated solutions completely reversed the spectral changes. An absorption band at 413 nm grows in during photolysis, and this spectral feature was assigned to $[Fe(CN)_2(CNCH_3)_2(H_2O)_2]$. This formulation was derived by interpolating the spectra of $[Fe(CN)_6]^{4-}$, $[Fe(CNCH_3)_6]^{2+}$, and $[Fe(H_2O)_6]^{2+}$ and applying the rule of average LF environment. The proposed mechanism is given in Eqs. (6-16) and (6-17)

TABLE 6-4

Comparison of Electronic Absorption Spectra of
Cyanide and Isocyanide Complexes

Complex	λ_{max} (nm)	ε (M^{-1} cm^{-1})	Reference
$[Fe(CN)_6]^{4-}$	323	304	[9]
	273	—	
$[Fe(CNCH_3)_6]^{2+}$	315	320	[10]
	263	340	
cis-$[Fe(CN)_2(CNCH_3)_4]$	315	360	[17]
	265 sh	460	
trans-$[Fe(CN)_2(CNCH_3)_4]$	315	330	[17]
	265 sh	410	

with $[Fe(CN)_2(CNCH_3)_3(H_2O)]$ as the initial primary photoproduct. The

$$[Fe(CN)_2(CNCH_3)_4] \xrightarrow[H_2O]{hv} [Fe(CN)_2(CNCH_3)_3(H_2O)] + CNCH_3 \qquad (6\text{-}16)$$

$$[Fe(CN)_2(CNCH_3)_3(H_2O)] \underset{}{\overset{fast, H_2O}{\rightleftharpoons}} [Fe(CN)_2(CNCH_3)_2(H_2O)_2] + CNCH_3 \qquad (6\text{-}17)$$

measured quantum yields were 0.14 and 0.13 for the cis and trans complexes, respectively, and were independent of irradiation wavelength.

Spectral measurements suggest that CN^- and $CNCH_3$ are similar in their overall LF strengths and in their π-acceptor properties. The various photochemical rules [20,21] would thus predict that aquation of either ligand could occur, but it is interesting that only photoaquation of $CNCH_3$ was observed. Loss of CN^-, however, involves charge separation, whereas loss of $CNCH_3$ does not. This suggests that whenever LF and π-acceptor arguments predict equal reaction probabilities, charge separation could become the controlling factor.

D. [Fe(PHTHALOCYANINE)(CNCH$_2$C$_6$H$_5$)L]

In 1974, Stynes reported [22] that photolysis of $[Fe(phthalocyanine)\cdot(CNCH_2C_6H_5)L]$ (L = py, piperidine, methylimidazole) (XLVIII) with

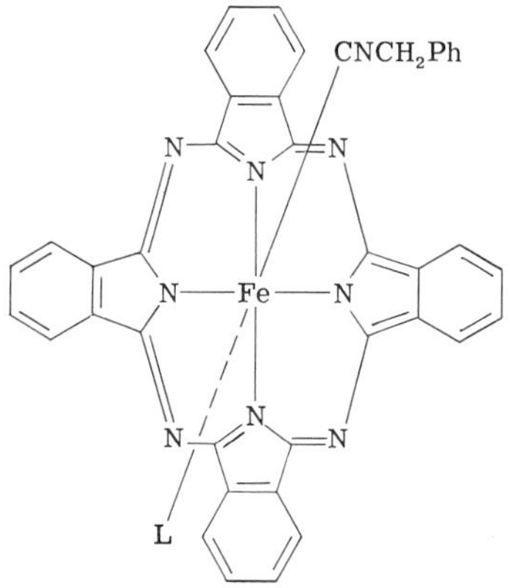

(XLVIII)

fluorescent room light led to a thousandfold increase in the rate of dissociation of $CNCH_2C_6H_5$ over the same dark reaction. The observed rate increase shifted the equilibrium shown in Eq. (6-18). For example, solutions which contained sufficient concentrations of CNR to shift the equilibrium

$$L + \underset{\text{blue}}{[Fe(phthalocyanine)(CNCH_2C_6H_5)L]} \rightleftharpoons \underset{\text{green}}{[Fe(phthalocyanine)L_2]} + CNCH_2C_6H_5$$

$$(6\text{-}18)$$

to the left were blue in color. When irradiated, the solution turned green, the color of $[Fe(phthalocyanine)L_2]$; but storage in the dark restored the original blue color. The color changes were cycled repeatedly with no loss of reversibility, and it was claimed [22] that the system represents a novel approach to solar energy storage in biological systems.

It should be mentioned that numerous reports [23–25] have cited the observation that irradiation increases the rate of dissociation of CNR from myoglobin and hemoglobin.

E. $[Fe(\eta^5\text{-}C_5H_5)(CO)(CN\text{---}CH(CH_3)(C_6H_5))]I$

Brunner and Vogel [26] observed that, whereas solutions containing the $(+)$ and $(-)$ diastereoisomers of $[Fe^*(\eta^5\text{-}C_5H_5)(CO)(CN\text{---}CH(CH_3)\cdot(C_6H_5))]I$ are stable in the dark, racemization rapidly occurred when solutions were irradiated with sunlight. No mechanistic or photochemical details were presented.

V.   RHODIUM AND IRIDIUM COMPLEXES

A. $[Rh_2(1,3\text{-}D\text{IISOCYANOPROPANE})_4]^{2+}$

Gray and co-workers have reported the preparation and spectral properties of $[Rh_2(1,3\text{-diisocyanopropane})_4]^{2+}$ [27,28]. This isocyanide bridged dimer has the structure shown in (XLIX). It shows a low-energy band at

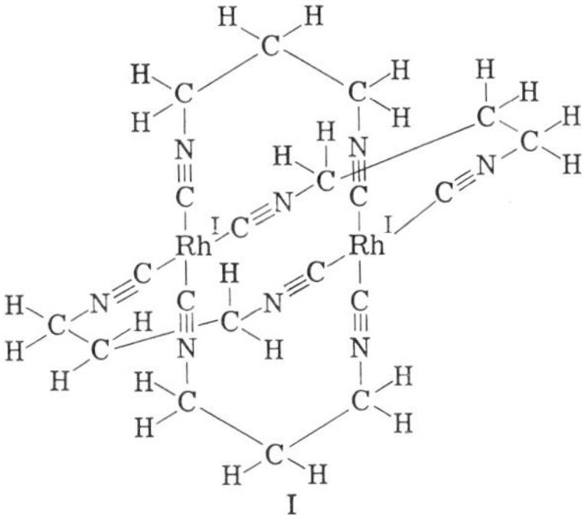

(XLIX)

553 nm ($\varepsilon = 14,500$) attributed to a $^1A_{1g} \rightarrow {}^1A_{2u}$ transition between orbitals which result from the rhodium–rhodium interaction [27]. When the dimer

is dissolved in 12 M HCl, a blue solution results which shows an absorption maximum at 578 nm ($\varepsilon = 52,700$). Photolysis of this 12 M HCl solution gives the electronic absorption spectral changes shown in Fig. 6-6. The final spectrum shown in Fig. 6-6 is identical to that of $[Rh_2(1,3\text{-diisocyano-propane})_4Cl_2]^{2+}$, and this complex was isolated as the product of the photolysis. Hydrogen gas was evolved during irradiation, and the overall reaction that occurred is shown in Eq. (6-19). The appearance quantum yield of $[Rh_2(1,3\text{-diisocyanopropane})_4Cl_2]^{2+}$ varies with $[H^+]$ but lies

$$[Rh_2(\text{bridge})_4]^{2+} + 2HCl \xrightarrow{\;h\nu\;} [Rh_2(\text{bridge})_4Cl_2]^{2+} + H_2 \qquad (6\text{-}19)$$

bridge = 1,3-diisocyanopropane

in the range 0.002–0.008. Although the detailed mechanism of this photoreaction is still under investigation, it has been suggested that it may proceed through the sequence of reactions (6-20)–(6-22) [28a]. Hydrogen is ap-

$$2[Rh_2(\text{bridge})_4]^{2+} \xrightarrow{\;\Delta\;} [Rh_2(\text{bridge})_4]_2^{6+} + H_2 \qquad (6\text{-}20)$$

$$[Rh_2(\text{bridge})_4]_2^{6+} \xrightarrow{\;h\nu\;} [Rh_2(\text{bridge})_4]^{4+} + [Rh_2(\text{bridge})_4]^{2+} \qquad (6\text{-}21)$$

$$[Rh_2(\text{bridge})_4]^{4+} + 2Cl^- \longrightarrow [Rh_2(\text{bridge})_4Cl_2]^{2+} \qquad (6\text{-}22)$$

parently produced in a thermal process which yields the $Rh^I\!-\!Rh^{II}\!-\!Rh^{II}\!-\!Rh^I$ tetramer shown in Eq. (6-20). This tetramer can be photochemically cleaved to regenerate one molecule of the starting $Rh^I\!-\!Rh^I$ dimer and one molecule

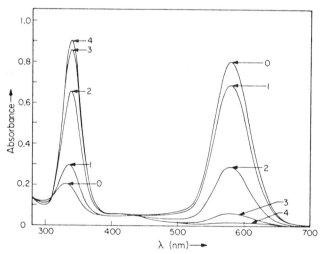

FIG. 6-6.   Irradiation (546 nm) of $1.6 \times 10^{-4}\ M$ $[Rh_2(\text{bridge})_4](BF_4)_2$ in degassed 12 M HCl solution. The four scans after $\tau = 0$ were taken at intervals of 2 min. Reprinted with permission from Mann *et al.* [28], *J. Am. Chem. Soc.* **99**, 5525 (1977). Copyright by the American Chemical Society.

of a Rh^{II}–Rh^{II} dimer which subsequently picks up Cl^- to give the observed product. The tetramer is apparently responsible for the blue color observed in acid solution. The production of H_2 from this system led the authors to argue that by properly adjusting the photochemical and thermal kinetic properties of this system it could be used to produce H_2 in solar energy conversion schemes [28].

B. $[Ir(CNR)_4]^+$

Bedford and Rouschias [29] have briefly noted that photolysis of deoxygenated solutions of $[Ir(CNR)_4]X$ induced a rapid color change from blue to red-orange. The color changes were observed in MeOH ($X = Cl^-$), acetone ($X = BPh_4^-$), CH_3CN, Me_2SO, and $HCONMe_2$ ($X = BPh_4^-$, PF_6^-). In several cases the changes were reversible; for example, when red irradiated methanol solutions of $[Ir(CNCH_3)_4]Cl$ were stored in the dark for 6 days at $25°C$, they regained their original blue color. The photoreaction was monitored in the ^1H NMR, and the absence of the characteristic triplet of free $CNCH_3$ ($\tau = 6.92$) seemed to rule out photosubstitution and photodissociation. Rather, the slightly different methyl resonances observed before and after photolysis led the authors to propose that photolysis gave association of a solvent molecule with $[Ir(CNCH_3)_4]^+$ to produce pentacoordinate $[Ir(CNCH_3)_4S]X$ (Eq. 6-23). Irradiation in the presence of excess $CNCH_3$

$$[Ir(CNCH_3)_4]Cl \xrightarrow[S]{h\nu} [Ir(CNCH_3)_4S]Cl \qquad (6\text{-}23)$$

produced pentacoordinate $[Ir(CNCH_3)_5]^+$ (Eq. 6-24), and these were cited

$$[Ir(CNCH_3)_4]Cl + CNCH_3 \xrightarrow{h\nu} [Ir(CNCH_3)_5]Cl \qquad (6\text{-}24)$$

as the first examples of a true photoinduced association reaction.

In a later study, Kawakami et al. [30] prepared a series of aryl isocyanide complexes $[Ir(CN\text{-aryl})_4]X$ and studied their chemistry. They likewise observed that these complexes were deep blue-purple in color, but molecular weight measurements indicated that the complexes existed in the form of oligomers $[Ir(CNR)_4]_n^{n+}$, even in solution. Such solution oligomerization has been recently shown to occur for $[Rh(CNR)_4]^+$ [8], and it is expected to be even more extensive for the iridium complexes. Like Bedford and Rouschias [29], Kawakami et al. [30] observed that when the blue solutions of $[Ir(CNR)_4]^+$ were irradiated, they turned orange. NMR data, however, did not indicate formation of $[Ir(CNR)_4S]^+$. Although the photolysis products were not identified and the photoreaction not investigated in detail, it was noted that the aryl isocyanide complexes appear less photoreactive than does $[Ir(CNCH_3)_4]_n^{n+}$.

The photochemistry of $[Ir(CNCH_3)_4]Cl$ was subsequently examined by Geoffroy and co-workers [31]. Molecular weight measurements and electronic absorption spectral studies did indeed show that $[Ir(CNCH_3)_4]^+$ exists as oligomers in solution and that the deep blue-purple colors are due to metal–metal interaction in the oligomers. For example, the electronic absorption spectrum of oligomeric $[Ir(CNCH_3)_4]_n^{n+}Cl_n$ in degassed MeOH solution is shown as the initial spectrum in Fig. 6-7. Using the molecular orbital diagram of Mann *et al.* [8], the intense band at 620 nm was assigned as a spin-allowed transition from a metal–metal orbital of principally d_{z^2} character to an oligomer orbital of principally isocyanide a_{2u} character. When blue solutions of $[Ir(CNCH_3)_4]_n^{n+}$ were irradiated in the presence of O_2, very complicated spectral changes obtained. However, when *degassed* solutions were photolyzed with $\lambda \leq 525$ nm, the solutions turned orange and smooth spectral changes were observed (Fig. 6-7). The orange solution displayed the final electronic absorption spectrum shown in Fig. 6-7. This spectrum is *identical* to that of $[Ir(CN-t-Bu)_4]^+$, which apparently does not oligomerize in the solid state or in solution [31] and is very similar to the spectra of an extensive series of $[IrX(CO)L_2]$ [10] and $[Ir(diphos)_2]^+$ [9] complexes. It is precisely the kind of spectrum expected for a planar Ir(I) complex containing π-acceptor ligands. The photoreaction which occurs is thus not photoassociation but rather photoinduced dissociation of the iridium–isocyanide oligomers (Eq. 6-25). The monomeric $[Ir(CNR)_4]^+$ complexes are

$$[Ir(CNCH_3)_4]_n^{n+} \underset{\Delta}{\overset{h\nu}{\rightleftarrows}} n[Ir(CNR)_4]^+ \qquad (6\text{-}25)$$

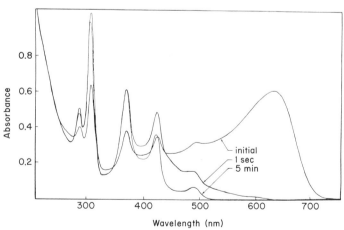

FIG. 6-7. Electronic absorption spectral changes during $\lambda \leq 525$ nm photolysis of a 4.02×10^{-5} M methanol solution of $[Ir(CNCH_3)_4]Cl$. Reprinted with permission from Geoffroy *et al.* [31], *Inorg. Chem.* **17**, 777 (1978). Copyright by the American Chemical Society.

extremely reactive toward oxidative addition and rapidly decompose in the presence of O_2. Photolysis in the presence of excess $CNCH_3$ does indeed lead to formation of $[Ir(CNCH_3)_5]^+$ but only through *thermal* addition of $CNCH_3$ to the photoproduced monomeric $[Ir(CNCH_3)_4]^+$. The 436 nm quantum yield for disappearance of the oligomers was 0.3, calculated per monomer unit [31].

VI. GOLD COMPLEXES

The complex $[Au(CNCH_3)_2]ClO_4$ has been reported [32] to decompose in light, yielding metallic gold, but no further details were given.

VII. SUMMARY

Relatively few isocyanide complexes have been examined, but those which have been studied show a very rich photochemistry. Photosubstitution by both dissociative and associative pathways occurs, and redox chemistry has been observed from charge transfer states. The electronic properties of isocyanides are particularly interesting from a photochemical standpoint in that excited states of their complexes can often be "tuned" by varying the nature of the isocyanide. Such tuning should allow unique opportunities for probing excited-state processes and for detailed examination of charge transfer photochemistry. The study of isocyanide complexes is an especially fertile area, and we eagerly look forward to future developments.

References

1. L. Malatesta and F. Bonati, "Isocyanide Complexes of Metals." Wiley, New York, 1969.
2. P. M. Treichel, *Adv. Organomet. Chem.* **11**, 21 (1973).
3. F. Bonati and G. Minghetti, *Inorg. Chim. Acta* **9**, 95 (1974).
4. B. E. Bursten and R. F. Fenske, *Inorg. Chem.* **16**, 963 (1977).
5. I. Ugi and R. Meyr, *Chem. Ber.* **93**, 239 (1960).
6. K. R. Mann, M. Cimolino, G. L. Geoffroy, G. S. Hammond, A. A. Orio, G. Albertin, and H. B. Gray, *Inorg. Chim. Acta* **16**, 97 (1976).
7. H. Isci and W. R. Mason, *Inorg. Chem.* **14**, 913 (1975).

8. K. R. Mann, J. G. Gordon, II, and H. B. Gray, *J. Am. Chem. Soc.* **97**, 3553 (1975).

9. G. L. Geoffroy, M. S. Wrighton, G. S. Hammond, and H. B. Gray, *J. Am. Chem. Soc.* **96**, 3105 (1974).

10. R. Brady, B. R. Flynn, G. L. Geoffroy, H. B. Gray, J. Peone, Jr., and L. Vaska, *Inorg. Chem.* **15**, 1485 (1976).

11. J. J. Alexander and H. B. Gray, *J. Am. Chem. Soc.* **90**, 4260 (1968).

12. V. Carassitti, G. Condorelli, L. L. Condorelli-Costanzo, *Ann. Chim.* (*Rome*) **55**, 329 (1965).

13. D. H. Shaw and H. O. Pritchard, *Can. J. Chem.* **45**, 2749 (1967).

14. K. Iuchi, S. Asada, and A. Sugimori, *Chem. Lett.* p. 801 (1974); K. Iuchi, S. Asada, T. Kinugasa, K. Kanamori, and A. Sugimori, *Bull. Chem. Soc. Jpn.* **49**, 577 (1976).

15. K. R. Mann, H. B. Gray, and G. S. Hammond, *J. Am. Chem. Soc.* **99**, 306 (1977).

16. K. R. Mann, Ph.D. Thesis, California Institute of Technology, Pasadena (1976).

17. G. Condorelli and L. L. Condorelli-Costanzo, *Ann. Chim.* (*Rome*) **56**, 1140 (1966).

18. G. Condorelli and L. L. Condorelli-Costanzo, *Ann. Chim.* (*Rome*) **56**, 1159 (1966).

19. G. Condorelli, L. Giallongo, A. Guiffrida, and G. Romeo, *Inorg. Chim. Acta* **7**, 7 (1973).

20. A. W. Adamson, *J. Phys. Chem.* **71**, 798 (1967).

21. M. Wrighton, H. B. Gray, and G. S. Hammond, *Mol. Photochem.* **5**, 165 (1973).

22. D. V. Stynes, *J. Am. Chem. Soc.* **96**, 5942 (1974); see also C. Irwin and D. V. Stynes, *Inorg. Chem.* **17**, 2682 (1978).

23. E. Antonini and M. Brunori, "Hemoglobin and Myoglobin in Their Reactions with Ligands." North-Holland Publ., Amsterdam, 1971.

24. B. Talbot, M. Brunori, E. Antonini, and J. Wyman, *J. Mol. Biol.* **58**, 261 (1971).

25. F. X. Cole and Q. H. Gibson, *J. Biol. Chem.* **248**, 4998 (1973).

26. H. Brunner and M. Vogel, *J. Organomet. Chem.* **35**, 169 (1972).

27. N. S. Lewis, K. R. Mann, J. G. Gordon, II, and H. B. Gray, *J. Am. Chem. Soc.* **98**, 7461 (1976).

28. K. R. Mann, N. S. Lewis, V. M. Miskowski, D. K. Erwin, G. S. Hammond, and H. B. Gray, *J. Am. Chem. Soc.* **99**, 5525 (1977).

28a. H. B. Gray, *Natl. Meet. Am. Chem. Soc.* *177* (Abstr. Inorg. 166), Honolulu, Hawaii, April 1–6, 1979.

29. W. M. Bedford and G. Rouschias, *Chem. Commun.* p. 1224 (1972).

30. K. Kawakami, M. A. Haga, and T. Tanaka, *J. Organomet. Chem.* **60**, 363 (1973).

31. G. L. Geoffroy, M. G. Bradley, and M. E. Keeney, *Inorg. Chem.* **17**, 777 (1978).

32. G. Bergerhoff, *Z. Anorg. Allg. Chem.* **327**, 139 (1964).

Hydride Complexes

7

I. INTRODUCTION

Although transition-metal hydride complexes do not strictly fit the definition of organometallic compounds, they are discussed here because the metal–hydrogen bond is of common occurrence in organometallic complexes and because hydride complexes are of critical importance in many transition-metal-assisted or -catalyzed organic transformations. Furthermore, the physical and chemical properties of metal hydrides more closely resemble those of organometallics than they do those of the classical coordination compounds. The field of hydride chemistry has expanded tremendously since the isolation and characterization of the first stable transition-metal hydride, $[Re(\eta^5-C_5H_5)_2H]$ by Wilkinson and Birmingham [1] in 1955. Thousands of hydride complexes have been prepared and characterized, and many have had their thermal reactivity examined in detail [2–8]. However, in spite of the importance of hydride complexes in organometallic chemistry and catalysis, relatively few hydrides have had their photochemical properties examined. Those studies are summarized in the following sections, and the general conclusions which can be made concerning the photoreactivity of hydride complexes are given at the end of this chapter.

II. VANADIUM COMPLEXES

[VH₃(CO)₃(diars)]

Ellis and co-workers [9] have briefly reported the preparation and photolysis of $[VH_3(CO)_3(diars)]$ (diars = o-$(Me_2As)_2C_6H_4$). The complex is thermally stable with respect to replacement of hydride by CO, but irradiation under a CO atmosphere readily leads to the formation of $[VH(CO)_4\cdot(diars)]$ (Eq. 7-1). No mechanistic details were presented although the

$$[VH_3(CO)(diars)] \xrightarrow[CO]{hv} H_2 + [VH(CO)_4(diars)] \qquad (7\text{-}1)$$

reaction appears to proceed by simple expulsion of H_2 to generate transient $[VH(CO)_3(diars)]$ which in turn is scavenged by CO.

III. MOLYBDENUM AND TUNGSTEN COMPLEXES

A. [MoH₄(dppe)₂]

The tetrahydride $[MoH_4(dppe)_2]$ (dppe = $Ph_2PCH_2CH_2PPh_2$) is a well-defined and thermally stable complex possessing the $Mo_4H_4P_4$ core shown in (L) [10,11]. The complex is relatively inert to H_2 loss and can be

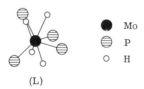

	● Mo
	⊖ P
	○ H

(L)

stored in solution under N_2 or Ar for several days without noticeable decomposition [12]. It has been noted, however, that refluxing the complex in toluene solution under N_2 leads to *trans*-$[Mo(N_2)_2(dppe)_2]$ in high yield [13]. Stirring a benzene solution of the complex at room temperature under an atmosphere of CO gives a mixture of *cis*- and *trans*-$[Mo(CO)_2(dppe)_2]$ [12].

The complex is quite photosensitive, and it has been demonstrated that irradiation leads to loss of 2 moles of H_2 per mole of complex irradiated [12].

In the absence of added substrate an orange solid can be isolated from the irradiated solutions, and spectral evidence suggested its formulation as $[Mo(dppe)_2]$, or possibly its dimer. The apparent photoreaction is that shown in Eq. (7-2). Further support for the formation of $[Mo(dppe)_2]$ as

$$[MoH_4(dppe)_2] \xrightarrow{hv} 2H_2 + [Mo(dppe)_2] \qquad (7\text{-}2)$$

the primary photoproduct comes from its reactivity. Irradiation of solutions of $[MoH_4(dppe)_2]$ under N_2 at room temperature gives $trans\text{-}[Mo(N_2)_2 \cdot (dppe)_2]$ in 93% yield after a few hours of photolysis, and irradiation under carbon monoxide gives a mixture of cis- and $trans\text{-}[Mo(CO)_2(dppe)_2]$. The analogous $[MoH_4(L)_4]$ (L = $PMePh_2$, $PEtPh_2$) complexes were also demonstrated to lose H_2 upon photolysis, and they give the corresponding bis(dinitrogen) complexes when irradiated under N_2 [12].

B. $[Mo(\eta^5\text{-}C_5H_5)_2H_2]$ AND $[W(\eta^5\text{-}C_5H_5)_2H_2]$

In a series of four communications M. L. H. Green and co-workers [14–17] reported that UV irradiation of solutions of $[W(\eta^5\text{-}C_5H_5)_2H_2]$ leads to apparent elimination of H_2 and generation of reactive $[W(\eta^5\text{-}C_5H_5)_2]$. The only products actually isolated from solutions were those of the general formula $[W(\eta^5\text{-}C_5H_5)_2H(R)]$ or $[W(\eta^5\text{-}C_5H_5)_2R_2]$, in which R derives from the solvent. Irradiation of the dihydride in benzene solution with a Pyrex-filtered 450 W high-pressure Hg lamp for 20 h, for example, gave $[W(\eta^5\text{-}C_5H_5)_2H(C_6H_5)]$ in yields of 40–80%, depending on conditions which, however, were not specified [15]. Yields were reproducibly greater than 60% if dilute solutions (0.005 M) were employed, and only small quantities of other decomposition products were obtained. No reaction was observed when the dihydride was heated in refluxing benzene solution over a period of several days. Photolysis in C_6D_6 solution gives formation of $[W(\eta^5\text{-}C_5H_5)_2D(C_6D_5)]$, strongly indicating that the reaction proceeds according to Eq. (7-3). Irradiation of the dihydride in toluene solution gives the analo-

$$[W(\eta^5\text{-}C_5H_5)_2H_2] \xrightarrow{hv} H_2 + [W(\eta^5\text{-}C_5H_5)_2] \xrightarrow{C_6H_6} [W(\eta^5\text{-}C_5H_5)_2H(C_6H_5)] \quad (7\text{-}3)$$

gous $[W(\eta^5\text{-}C_5H_5)_2H(p\text{-}CH_3C_6H_4)]$ in 45% yield. No quantum yield data or mechanistic details were presented although it was suggested that the reaction may proceed by initial formation of a tungstenocene–benzene π complex [15]. It was subsequently shown by other workers that the lower limit 313 nm quantum yield for elimination of H_2 is 0.02. [18,19].

Photolysis of $[W(\eta^5\text{-}C_5H_5)_2H_2]$ in mesitylene solution gives an orange product which was shown by x-ray crystallography to be $[W(\eta^5\text{-}C_5H_5)_2 \cdot$

$\{CH_2(3,5\text{-}Me_2C_6H_3)_2\}$] (LI) [14]. An analogous complex $[W(\eta^5\text{-}C_5H_5)_2 \cdot$

(LI)

$(CH_2C_6H_4CH_3)_2$] was obtained from photolysis of the dihydride in p-xylene solution. It was claimed [14] that formation of these two derivatives represented the first examples of direct insertion of a metal into saturated, uncoordinated C—H groups. Because of the reaction stoichiometry, it was proposed that the mechanism of the reaction leading to formation of the products must proceed by reversible transfer of hydrogen or an alkyl substituent between tungsten and a C_5H_5 ring [Eqs. (7-4)–(7-6)].

$$[W(\eta^5\text{-}C_5H_5)_2H(R)] \xrightleftharpoons{h\nu} [W(\eta^5\text{-}C_5H_5)(C_5H_6)R] \tag{7-4}$$

$$[W(\eta^5\text{-}C_5H_5)(C_5H_6)R] + RH \longrightarrow [W(\eta^5\text{-}C_5H_5)(C_5H_6)H(R)_2] \tag{7-5}$$

$$[W(\eta^5\text{-}C_5H_5)(C_5H_6)H(R)_2] \longrightarrow [W(\eta^5\text{-}C_5H_5)_2(R)_2] + H_2 \tag{7-6}$$

Photolysis of the dihydride in methanol solution leads to the formation of $[W(\eta^5\text{-}C_5H_5)_2H(OMe)]$ and $[W(\eta^5\text{-}C_5H_5)_2Me(OMe)]$ in a 1:5 ratio [16]. In a separate experiment it was shown that irradiation of $[W(\eta^5\text{-}C_5H_5)_2H(OMe)]$ in MeOH gave a 30% conversion to $[W(\eta^5\text{-}C_5H_5)_2 \cdot Me(OMe)]$, but irradiation of the latter gave no further reaction. Photolysis of $[W(\eta^5\text{-}C_5H_5)_2H_2]$ in ethanol solution gives only $[W(\eta^5\text{-}C_5H_5)_2H(OEt)]$. It was proposed that the formation of $[W(\eta^5\text{-}C_5H_5)_2Me(OMe)]$ proceeds according to the sequence of reactions (7-7)–(7-11).

$$[W(\eta^5\text{-}C_5H_5)_2H_2] \xrightleftharpoons{h\nu} [W(\eta^5\text{-}C_5H_5)_2] + H_2 \tag{7-7}$$

$$[W(\eta^5\text{-}C_5H_5)_2] + CH_3OH \longrightarrow [W(\eta^5\text{-}C_5H_5)_2H(\text{—}CH_2OH)] \tag{7-8}$$

$$[W(\eta^5\text{-}C_5H_5)_2H(\text{—}CH_2OH)] \longrightarrow OH^- + [W(\eta^5\text{-}C_5H_5)_2H(CH_2)]^+ \tag{7-9}$$

$$[W(\eta^5\text{-}C_5H_5)_2H(CH_2)]^+ \xrightleftharpoons{} [W(\eta^5\text{-}C_5H_5)_2CH_3]^+ \tag{7-10}$$

$$[W(\eta^5\text{-}C_5H_5)_2CH_3]^+ + CH_3OH \longrightarrow [W(\eta^5\text{-}C_5H_5)_2(CH_3)(OCH_3)] + H^+ \tag{7-11}$$

Green and co-workers later showed that the intermediate which results from photolysis of $[W(\eta^5\text{-}C_5H_5)_2H_2]$ is rather unselective to electronic effects in its reactions with a series of substituted benzenes. The product ratios instead appeared to be largely determined by steric effects. For example, fluorobenzene gave a 40:60 ratio of the meta and para isomers of

$[W(\eta^5\text{-}C_5H_5)_2H(C_6H_4F)]$, and methyl benzoate gave a 50:50 ratio of m- and p-$[W(\eta^5\text{-}C_5H_5)_2H(C_6H_4CO_2Me)]$. Likewise, the intermediate was shown to be unselective as to its attack on aliphatic or aromatic C—H bonds in toluene. In this case, a 60/40 ratio of $[W(\eta^5\text{-}C_5H_5)_2H(C_6H_4Me)]$ and $[W(\eta^5\text{-}C_5H_5)_2(CH_2Ph)(C_6H_4Me)]$ resulted. Photolysis of $[W(\eta^5\text{-}C_5H_5)_2H_2]$ in the presence of $SiMe_4$ was also reported [17] to give a dimeric product $[W_2(\eta^5\text{-}C_5H_5)_2(\eta^1,\eta^5\text{-}C_5H_4)H(CH_2SiMe_3)]$ which resulted from insertion into a C—H bond of $SiMe_4$ and which possesses a bridging $\eta^1,\eta^5\text{-}C_5H_4$ ligand.

Three different intermediates, each of which could be produced photochemically from $[W(\eta^5\text{-}C_5H_5)_2H_2]$ and each of which could undergo the observed insertion reactions, have been suggested [17]. Irradiation could lead to direct concerted elimination of H_2 to give tungstenocene (Eq. 7-7), or irradiation could induce hydrogen migration to one of the C_5H_5 rings to give a coordinated diene (Eq. 7-12). Alternatively, photolysis could induce slipping of one of the C_5H_5 ligands to give a species with an $\eta^3\text{-}C_5H_5$ ligand (Eq. 7-13). All of these possible intermediates are 16-valence-electron species,

$$[W(\eta^5\text{-}C_5H_5)_2H_2] \longrightarrow [W(\eta^5\text{-}C_5H_5)H(C_5H_6)] \qquad (7\text{-}12)$$

$$[W(\eta^5\text{-}C_5H_5)_2H_2] \longrightarrow [W(\eta^5\text{-}C_5H_5)(\eta^3\text{-}C_5H_5)H_2] \qquad (7\text{-}13)$$

and each could be reactive enough to undergo the observed insertion reactions. No detailed mechanistic discussion was presented by Green and co-workers although their initial papers [14-16] seemed to favor the first mechanism, namely concerted photoinduced loss of H_2. For example, their proposed reaction sequence [Eqs. (7-7)–(7-9)] for the reactions with methanol begin with this step. Likewise, their observation [15] of $[W(\eta^5\text{-}C_5H_5)_2 \cdot D(C_6D_5)]$ as the only product from photolysis in C_6D_6 solution appears to eliminate the second reaction as a possibility because hydrogen transfer to one of the C_5H_5 rings should lead to some degree of H/D scrambling and some $[W(\eta^5\text{-}C_5H_5)H(C_6D_5)]$ should appear as product. As described below, the mechanistic experiments which have been conducted for the analogous molybdenum complex strongly favor concerted elimination of H_2 as the primary photochemical step. Additional support for the intermediacy of $[W(\eta^5\text{-}C_5H_5)_2]$ in these reactions comes from the photochemical studies of $[W(\eta^5\text{-}C_5H_5)_2CO]$ discussed in Chapter 2 [Eq. (2-21)].

Bradley and Geoffroy [18,19] subsequently observed that irradiation of the corresponding molybdenocene dihydride also gives elimination of H_2 and generation of transient molybdenocene (Eq. 7-14). The reaction can be

$$[Mo(\eta^5\text{-}C_5H_5)_2H_2] \longrightarrow [Mo(\eta^5\text{-}C_5H_5)_2] + H_2 \qquad (7\text{-}14)$$

induced by irradiation with $\lambda \leq 366$ nm, and the quantum yield of disappearance of $[Mo(\eta^5\text{-}C_5H_5)_2H_2]$ measured at 313 nm was 0.02 [18,19].

Because of the air sensitivity of the system the quantum yield was measured in a sealed and degassed UV cell; hence back-reaction with the photo-generated H_2 was not prevented. Thus, the measured yield must be treated as a lower limit. At the same wavelength of irradiation, elimination of H_2 from $[Mo(\eta^5\text{-}C_5H_5)_2H_2]$ is an order of magnitude more efficient than elimination of H_2 from $[W(\eta^5\text{-}C_5H_5)_2H_2]$. Unlike $[W(\eta^5\text{-}C_5H_5)_2]$, photo-generated $[Mo(\eta^5\text{-}C_5H_5)_2]$ does not insert into solvent molecules, but instead a red-brown product which appeared identical to a previously described [20,21] "polymeric" $[Mo(\eta^5\text{-}C_5H_5)_2]_x$ was isolated. This material was later shown by Green and co-workers [22] to contain the dimer shown in (LII).

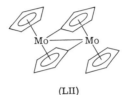

(LII)

The photogenerated molybdenocene can be readily trapped by added substrate, and photolysis in the presence of CO and C_2H_2 led to near-quantitative formation of the corresponding adducts $[Mo(\eta^5\text{-}C_5H_5)_2CO]$ and $[Mo(\eta^5\text{-}C_5H_5)_2C_2H_2]$ [18,19]. Irradiation in the presence of excess PEt_3 and PPh_3 led to formation of the previously unknown $[Mo(\eta^5\text{-}C_5H_5)_2PR_3]$ complexes. These appear useful as synthetic intermediates because of the lability of the Mo—PR_3 bond [18,19].

The mechanism of the photoelimination reaction is of obvious importance. Heterolytic cleavage of a Mo—H bond to generate H^+ or H^- appeared unlikely in the nonpolar aliphatic and aromatic solvents employed in the study of $[Mo(\eta^5\text{-}C_5H_5)_2H_2]$. Mass spectral analysis of the gases over irradiated toluene-d_8 solutions of $[Mo(\eta^5\text{-}C_5H_5)_2H_2]$ showed predominantly H_2 with less than 10% HD produced. Since toluene is an efficient scavenger for free hydrogen atoms, the lack of an appreciable amount of HD indicated that hydrogen atoms are not produced to a significant extent in the photolysis. Thus, elimination of hydrogen from the $[Mo(\eta^5\text{-}C_5H_5)_2H_2]$ complex directly as H_2 is implied [18,19]. Similar experiments point to direct loss of H_2 from $[W(\eta^5\text{-}C_5H_5)_2H_2]$ [18,19].

Elimination of H_2 can occur via three different routes: direct concerted elimination of H_2, forced concerted loss of H_2 from a photogenerated $[M(\eta^5\text{-}C_5H_5)(\eta^3\text{-}C_5H_5)H_2]$ intermediate (Eq. 7-13), and intramolecular hydrogen abstraction from a $\eta^5\text{-}C_5H_5$ ligand. Irradiation of $[Mo(\eta^5\text{-}C_5H_5)_2D_2]$ in C_6H_6 or C_6D_6 solution gave a mixture of D_2 and HD in an approximate ratio of 1.6:1 in each case [18,19]. The substantial production of D_2 indicated that *concerted elimination* of D_2 occurred since abstraction of hydrogen from a C_5H_5 ring would have yielded HD. The labeling experiments did not allow the authors to distinguish between direct concerted loss of H_2 from the photoexcited complex and forced concerted loss of H_2 caused by interaction of solvent or substrate with a $[Mo(\eta^5\text{-}C_5H_5)(\eta^3\text{-}C_5H_5)H_2]$ intermediate. However, it was argued [18,19] that a 16-valence-electron $[Mo(\eta^5\text{-}C_5H_5)(\eta^3\text{-}C_5H_5)H_2]$ intermediate should not spontaneously lose H_2, and it was viewed as extremely unlikely that a solvent such as hexane would interact with this species in such a way as to force H_2 loss.

The HD produced upon photolysis of $[Mo(\eta^5\text{-}C_5H_5)_2D_2]$ in C_6H_6 or C_6H_6 solution was suggested to arise through a subsequent thermal reaction of photogenerated $[Mo(\eta^5\text{-}C_5H_5)_2]$ with unreacted $[Mo(\eta^5\text{-}C_5H_5)_2D_2]$ [18,19]. This could proceed by hydrogen abstraction to give a dimeric intermediate with an $\eta^1,\eta^5\text{-}C_5H_4$ ligand, such as that shown in (LIII), which could then eliminate HD. Such an intermediate is consistent with the dimeric product (LII) isolated by Green and coworkers [22], and the overall reaction shown in Fig. 7-1 was suggested.

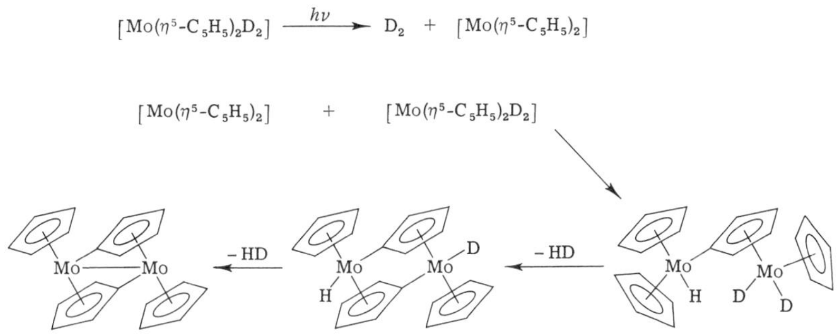

FIG. 7-1. Proposed mechanism for photoinduced D_2 elimination.

The nature of the photoactive excited state which leads to elimination of H_2 was also discussed by Bradley and Geoffroy [18,19]. The electronic absorption spectra of $[Mo(\eta^5\text{-}C_5H_5)_2H_2]$ and $[W(\eta^5\text{-}C_5H_5)_2H_2]$ are shown in Fig. 7-2, and an adaptation of molecular orbital diagrams which were independently calculated [23,24] for the molybdenum complex is shown in Fig. 7-3. The lowest unoccupied orbital was calculated [24] to be a strongly antibonding $d_{xy}\text{-}H_2\sigma^*$ combination of b_2 symmetry, and it was proposed that population of this orbital was responsible for H_2 elimination. The lowest energy shoulder at 230 nm seen in the spectra of both the dihydride complexes was assigned as the $^1A_1 \rightarrow {}^1B_1$ transition, and the intense band at 270 nm was tentatively assigned as a metal to cyclopentadienyl charge transfer transition [18,19].

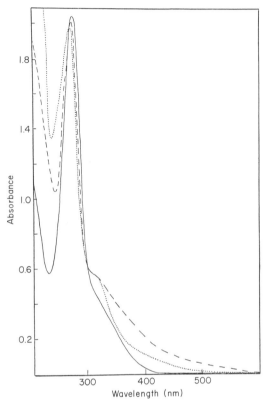

FIG. 7-2. Electronic absorption spectra of $[Mo(\eta^5\text{-}C_5H_5)_2H_2]$ (--), $[W(\eta^5\text{-}C_5H_5)_2H_2]$ (———), and $[Mo(\eta^5\text{-}C_5H_5)_2CO]$ (· · ·) in degassed hexane solution. Reprinted with permission from Geoffroy and Bradley [19], *Inorg. Chem.* **17**, 2410 (1978). Copyright by the American Chemical Society.

C. $[W(\eta^5\text{-}C_5H_5)H(CO)_3]$

In a study of the thermal and photochemical substitution reactions of $[W(\eta^5\text{-}C_5H_5)H(CO)_3]$, Hoffman and Brown [25] observed that 311 nm photolysis promoted substitution of PBu_3 for CO (Eq. 7-15). The quantum yield for the overall substitution reaction varied from 6 to 30 and suggested

$$[W(\eta^5\text{-}C_5H_5)H(CO)_3] + PBu_3 \xrightarrow{h\nu} [W(\eta^5\text{-}C_5H_5)H(CO)_2(PBu_3)] + CO \quad (7\text{-}15)$$

a radical chain mechanism. It was proposed that excitation initially gave homolysis of the W—H bond with a low quantum yield and that the sequence of reactions shown in Eqs. (7-16)–(7-19) obtained. From their concurrent

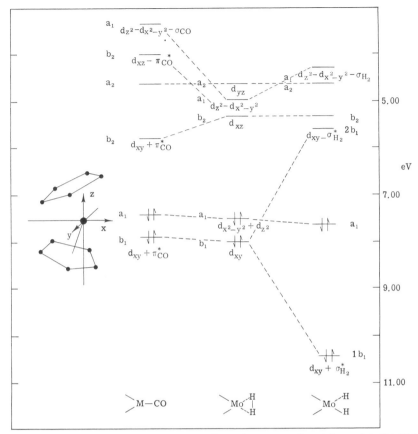

FIG. 7-3.   Molecular orbital diagrams for $[Mo(\eta^5\text{-}C_5H_5)_2CO]$ and $[Mo(\eta^5\text{-}C_5H_5)_2H_2]$. Reprinted with permission from Geoffroy and Bradley [19], *Inorg. Chem.* **17**, 2410 (1978). Copyright by the American Chemical Society.

thermal substitution studies, Hoffman and Brown [25] estimated a chain

$$[W(\eta^5\text{-}C_5H_5)H(CO)_3] \xrightarrow{h\nu} [\cdot W(\eta^5\text{-}C_5H_5)(CO)_3] + \cdot H \qquad (7\text{-}16)$$

$$[\cdot W(\eta^5\text{-}C_5H_5)(CO)_3] \longrightarrow [\cdot W(\eta^5\text{-}C_5H_5)(CO)_2] + CO \qquad (7\text{-}17)$$

$$[\cdot W(\eta^5\text{-}C_5H_5)(CO)_2] + PBu_3 \longrightarrow [\cdot W(\eta^5\text{-}C_5H_5)(CO)_2(PBu_3)] \qquad (7\text{-}18)$$

$$[\cdot W(\eta^5\text{-}C_5H_5)(CO)_2(PBu_3)] + [W(\eta^5\text{-}C_5H_5)H(CO)_3] \longrightarrow$$

$$[W(\eta^5\text{-}C_5H_5)H(CO)_2(PBu_3)] + [\cdot W(\eta^5\text{-}C_5H_5)(CO)_3] \qquad (7\text{-}19)$$

length of about 2000 for the sequence of reactions shown in Eqs. (7-16)–(7-19). With the overall quantum yield of 30, the quantum yield for the primary photochemical process is about 0.015. Support for the initial W—H homolysis reaction came from a trapping experiment with Ph_3CCl which gave complete conversion to $[W(\eta^5-C_5H_5)Cl(CO)_3]$ (Eq. 7-20).

$$[W(\eta^5-C_5H_5)H(CO)_3] + Ph_3CCl \xrightarrow{h\nu} [W(\eta^5-C_5H_5)Cl(CO)_3] \qquad (7\text{-}20)$$

While W—H bond cleavage does obtain here, it is likely that the primary photoprocesses are dominated by CO extrusion.

IV. RHENIUM COMPLEXES

A. $[ReH_3(Ph_2PCH_2CH_2PPh_2)]$

$[ReH_3(Ph_2PCH_2CH_2PPh_2)]$ is thermally quite stable and shows no tendency to lose H_2. It is quite photosensitive, however, and irradiation leads to smooth reductive elimination of H_2 to generate $[ReH(Ph_2PCH_2CH_2 \cdot PPh_2)_2]$ as an apparent intermediate [12]. The photogenerated monohydride complex is quite reactive, and the several reactions shown in Fig. 7-4 were observed.

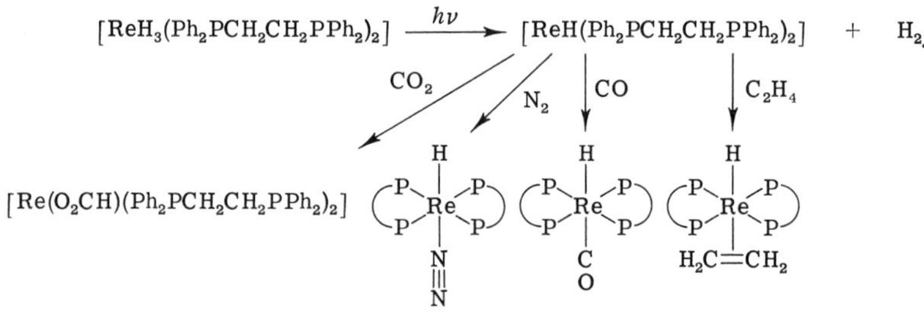

Fig. 7-4. Photoreactions of $[ReH_3(Ph_2PCH_2CH_2PPh_2)_2]$ [12].

The product actually isolated from photolysis of $[ReH_3(Ph_2PCH_2CH_2 \cdot PPh_2)_2]$ in degassed solution appeared to be an ortho-metallated derivative which results from the loss of another H_2 molecule. Photolysis in the presence of H_2, CO, and C_2H_4 leads to nearly quantitative yields of the known trans dinitrogen, carbon monoxide, and ethylene complexes. Irradiation under a carbon dioxide atmosphere led to CO_2 uptake and production of a formate complex. Interestingly, addition of CO_2 was shown to be reversible, and the formate complex can be readily converted to the N_2 complex by heating under an N_2 atmosphere. None of these reactions occur thermally, and light is essential to induce H_2 loss from the parent $[ReH_3(Ph_2PCH_2CH_2PPh_2)_2]$.

The $[ReH(N_2)(Ph_2PCH_2CH_2PPh_2)_2]$ complex was also shown to lose N_2 readily upon photolysis, as exemplified by reaction (7-21) [12].

$$[ReH(N_2)(Ph_2PCH_2CH_2PPh_2)_2] \xrightarrow[CO_2]{hv} N_2 + [Re(O_2CH)(Ph_2PCH_2CH_2PPh_2)_2] \quad (7\text{-}21)$$

B. $[MnH(CO)_5]$ AND $[ReH(CO)_5]$

In connection with a study of the mechanism of thermal substitution of PR_3 for CO in $[ReH(CO)_5]$, Byers and Brown [25a] observed that 311 nm photolysis of a hexane solution 10^{-3} M in $[ReH(CO)_5]$ and 10^{-2} M in PBu_3 gave slow formation of $[ReH(CO)_4PBu_3]$ and eventually $[ReH(CO)_3(PBu_3)_2]$. After 12 h irradiation the reaction was only about 66% complete. Photolysis presumably induces dissociation to CO to yield $[ReH(CO)_4]$ which captures PBu_3. As mentioned in Chapter 2, photolysis of $[MnH(CO)_5]$ in a frozen Ar matrix was shown to give CO loss to generate $[MnH(CO)_4]$ [26].

V.   IRON AND RUTHENIUM COMPLEXES

A. $[FeH_2(N_2)(PEtPh_2)_3]$

In 1968, Sacco and Aresta [27] prepared and characterized $[FeH_2(N_2)\cdot(PEtPh_2)_3]$ and reported that it underwent the reversible reaction shown in Eq. (7-22) when the solid material was exposed to sunlight. Although few

$$[FeH_2(N_2)(PEtPh_2)_3] \xrightleftharpoons{hv} H_2 + [\overline{FeH(N_2)(C_6H_4PEtPh)}(PEtPh_2)_2] \quad (7\text{-}22)$$

details were given, the complex apparently loses H_2 and the resultant $[Fe(N_2)(PEtPh_2)_3]$ complex undergoes an ortho-metallation reaction with iron inserting into a C—H bond of one of the $PEtPh_2$ ligands. No experimental data were presented to support their claim for reaction (7-22), and no measure of the efficiency was presented.

In 1972, Koerner von Gustorf and co-workers [28] reexamined the complex and also noted that the solid shows a strong photochromism, changing from yellow-orange to brown within minutes upon exposure to sunlight. The original color returns after storage of the material for several days in the dark. These workers noted that the reverse reaction is about 60 times faster under an N_2 atmosphere than under an Ar atmosphere, and they concluded that N_2 rather than H_2 was lost upon photolysis. This would generate

$[FeH_2(PEtPh_2)_3]$, and it was reasoned that such a complex should be capable of hydrogenating unsaturated hydrocarbons. In confirmation of this hypothesis, photolysis of a benzene solution of $[FeH_2(N_2)(PEtPh_2)_3]$ in the presence of methyl acrylate gave 60–65% methyl propionate, 35–40% H_2, and the theoretical amount of N_2. Similar results were obtained using 2,3-dimethylbutadiene as the substrate.

Darensbourg [29] has reported that photolysis of $[FeH_2(N_2)(PEtPh_2)_3]$ under a CO atmosphere for 30 min gave the carbonyl derivatives $[Fe(CO)_4(PEtPh_2)]$ and *trans*-$[Fe(CO)_3(PEtPh_2)_2]$ in a 1:1 ratio. It was proposed that $[FeH_2(CO)(PEtPh_2)_3]$ was initially formed, and they observed that irradiation of this complex under CO gave products identical to the final products obtained from $[FeH_2(N_2)(PEtPh_2)_3]$.

None of these three studies represented a thorough examination of the photochemistry of $[FeH_2(N_2)(PEtPh_2)_3]$. The available evidence seems to indicate that both H_2 and N_2 can be lost upon photolysis, but presumably N_2 loss occurs first. Elimination of H_2 is not surprising in view of the photochemistry of the other hydrides discussed in this chapter which lose H_2 upon photolysis.

B. $[RuClH(CO)(PPh_3)_3]$, $[RuH_2(CO)(PPh_3)_3]$, AND $[RuClH(CO)_2(PPh_3)_2]$

A thorough examination of the photochemical properties of these three closely related Ru(II) complexes has been reported [30]. Interestingly, each complex was found to show completely different photochemical behavior. Irradiation of $[RuClH(CO)(PPh_3)_3]$ (LIV) gives elimination of CO, photolysis of $[RuH_2(CO)(PPh_3)_3]$ (LV) induces elimination of H_2, and $[RuClH(CO)_2(PPh_3)_2]$ (LVI) undergoes photoisomerization.

(LIV)   (LV)   (LVI)

Irradiation of a thoroughly degassed CH_2Cl_2 solution of $[RuClH(CO) \cdot (PPh_3)_3]$ with 366 nm gives the electronic absorption spectral changes shown in Fig. 7-5 [30]. The solution turns deep purple during photolysis and the IR and electronic absorption spectra indicate that $[RuClH(PPh_3)_3]$ is formed (Eq. 7-23). The reaction proceeds with a 313 nm quantum yield of 0.02,

$$[RuClH(CO)(PPh_3)_3] \xrightarrow{h\nu} CO + [RuClH(PPh_3)_3] \qquad (7\text{-}23)$$

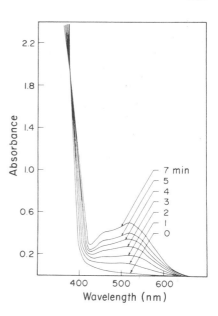

FIG. 7-5. Electronic absorption spectral changes during 366 nm photolysis of a 1×10^{-3} M CH$_2$Cl$_2$ solution of [RuClH(CO)·(PPh$_3$)$_3$]. Reprinted with permission from Geoffroy and Bradley [30], *Inorg. Chem.* **16**, 744 (1977). Copyright by the American Chemical Society.

and it was proposed that the active excited state is LF in nature. The [RuClH(PPh$_3$)$_3$] product is extremely air-sensitive, and the photoreaction must be conducted under N$_2$ or vacuum. The measured quantum yield is only a lower limit since it was determined using degassed and sealed cells from which the photoreleased CO could not escape, and back-reaction of CO with [RuClH(PPh$_3$)$_3$] was not prevented.

Perhaps the most important aspect of the photochemical reaction expressed in Eq. (7-23) is the product. [RuClH(PPh$_3$)$_3$] has been reported [31] to be one of the most active of all known homogeneous hydrogenation catalysts. Irradiation of [RuClH(CO)(PPh$_3$)$_3$] in the presence of 1-hexene and H$_2$ led to rapid hydrogenation of the olefin, confirming the generation of [RuClH(PPh$_3$)$_3$] [31]. One of the problems which has hampered use of [RuClH(PPh$_3$)$_3$] is its relative difficulty of preparation and its extreme air sensitivity. Solutions, for example, change color from purple to green upon exposure to air and deactivate completely. The photochemical reaction expressed in Eq. (7-23) thus offers a convenient means of generating solutions of the catalyst from an air-stable and very easily prepared precursor.

Unfortunately, the overall reaction resulting from photolysis of [RuClH(CO)(PPh$_3$)$_3$] is not as simple as that expressed in Eq. (7-23) since the initial complex has a high affinity for carbon monoxide, and the reaction shown in Eq. (7-24) reduces the overall yield of [RuClH(PPh$_3$)$_3$] [30]. The dicarbonyl

$$[\text{RuClH(CO)(PPh}_3)_3] + \text{CO} \longrightarrow [\text{RuClH(CO)}_2(\text{PPh}_3)_2] + \text{PPh}_3 \qquad (7\text{-}24)$$

derivative once formed, does not lose carbon monoxide thermally or upon photolysis (see the following discussion), and the maximum yield of $[RuClH(PPh_3)_3]$ that was obtained was approximately 85%.

Replacement of chloride by hydride in $[RuClH(CO)(PPh_3)_3]$ gives $[RuH_2(CO)(PPh_3)_3]$ of structure (LV). This complex is also photosensitive, but the spectral data indicate that molecular hydrogen rather than carbon monoxide is eliminated upon photolysis (Eq. 7-25). The $[Ru(CO)(PPh_3)_3]$ photoproduct produced from H_2 loss should be extremely reactive and the

$$[RuH_2(CO)(PPh_3)_3] \xrightarrow{h\nu} H_2 + [Ru(CO)(PPh_3)_3] \qquad (7\text{-}25)$$

IR spectrum of the material isolated after photolysis suggested an ortho-metallated derivative. The photoproduct was trapped, however, by irradiation under a carbon monoxide atmosphere, and the well-characterized $[Ru(CO)_3(PPh_3)_2]$ was formed (Eq. 7-26).

$$[Ru(CO)(PPh_3)_3] + 2CO \longrightarrow [Ru(CO)_3(PPh_3)_2] + PPh_3 \qquad (7\text{-}26)$$

The dicarbonyl complex $[RuClH(CO)_2(PPh_3)_2]$ also proved to be photosensitive [30]. The IR and electronic absorption spectral changes showed that 366 nm irradiation apparently induces photoisomerization of the octahedral complex into another octahedral isomer, much as has been observed [32] for a number of $[RuX_2(CO)_2(PR_3)_2]$ derivatives. The initial complex has the *cis*-dicarbonyl structure shown in (LVI). The exact configuration of the photoproduct was not determined although the presence of two carbonyl vibrations in its IR spectrum indicate another *cis*-dicarbonyl structure. Although no excited-state or mechanistic details were presented, it was argued that an LF state would most likely be responsible for the observed reaction.

VI. COBALT AND IRIDIUM COMPLEXES

A. $[CoH(PF_3)_4]$ AND $[IrH(PF_3)_4]$

Kruck and co-workers [33] have reported that UV photolysis of Et_2O solutions of $[CoH(PF_3)_4]$ gave the bridged complex (LVII). The product was

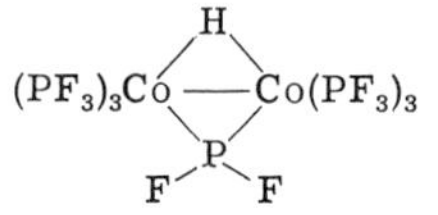

$$(PF_3)_3Co \overset{\displaystyle H}{\underset{\displaystyle F{\diagdown}F}{\diagup\diagdown}} Co(PF_3)_3$$

(LVII)

isolated by sublimation and characterized by its mass spectrum. Photolysis of a mixture of $[CoH(PF_3)_4]$ and $[IrH(PF_3)_4]$ gave the corresponding mixed-metal derivative $[(PF_3)_3Co(H)(PF_2)Ir(PF_3)_3]$.

Photolysis of $[IrH(PF_3)_4]$ alone in Et_2O solution leads to elimination of H_2 and to a nonbridged dimer with an Ir—Ir bond (Eq. 7-27) [34]. Approximately 20% yield was obtained after 17 h irradiation, and the product was

$$[IrH(PF_3)_4] \xrightarrow{h\nu} H_2 + [(PF_3)_4Ir—Ir(PF_3)_4] \qquad (7\text{-}27)$$

well-characterized by its IR, ^{19}F-NMR, and mass spectra.

B. $[CoH_2(phen)(PR_3)_2]^+$ AND $[CoH_2(bipy)(PR_3)_2]^+$

Camus *et al.* [35] have reported that photolysis of degassed methanol solutions of the title complexes with $PR_3 = PEt_3$, PPr_3, PBu_3, and PEt_2Ph, gave a darkening of the solution which was reversed upon addition of H_2. They proposed that the reversible reaction shown in Eq. (7-28) occurred with photoinduced elimination of H_2 as the primary photochemical step. In

$$[CoH_2(phen)(PR_3)_2]^+ \underset{H_2}{\overset{h\nu}{\rightleftarrows}} H_2 + [Co(phen)(PR_3)_2]^+ \qquad (7\text{-}28)$$

agreement with the proposed reaction, no color changes were observed when the photolysis was conducted under an H_2 atmosphere, and irradiation of $[CoD_2(bipy)(PR_3)_2]^+$ under H_2 gave the corresponding dihydride complexes.

C. $[IrClH_2(CO)(PPh_3)_2]$, $[IrCl_2H(CO)(PPh_3)_2]$, $[IrH_2(Ph_2PCHCHPPh_2)_2]Cl$, AND $[IrH_2(Ph_2PCH_2CH_2PPh_2)_2]Cl$

The three complexes $[IrClH_2(CO)(PPh_3)_2]$, $[IrH_2(Ph_2PCHCHPPh_2)_2]^+$, and $[IrH_2(Ph_2PCH_2CH_2PPh_2)_2]^+$ have all been shown to undergo elimination of molecular hydrogen upon photolysis [36] (Eq. 7-29). The carbonyl complex will lose H_2 thermally, but it was shown that the rate of elimination

$$[IrH_2(Ph_2PCH_2CH_2PPh_2)_2]^+ \xrightarrow{h\nu} H_2 + [Ir(Ph_2PCH_2CH_2PPh_2)_2]^+ \qquad (7\text{-}29)$$

was about 2.6 times faster when irradiated. The diphosphine complexes, however, do not lose H_2 thermally, and photolysis is the only known means by which H_2 can be removed to form $[Ir(diphos)_2]^+$. These reactions are reversible, and the sequence of photoinduced elimination and thermal addition of H_2 can be cycled repeatedly without noticeable decomposition. Similar reactions were also observed for the analogous dioxygen adducts

$[IrO_2(Ph_2PCHCHPPh_2)_2]^+$ and $[IrO_2(Ph_2PCH_2CH_2PPh_2)_2]^+$, which also do not lose O_2 thermally [36]. The electronic absorption spectral changes accompanying irradiation of the dihydride and O_2 complexes are virtually identical, and they are shown in Fig. 7-6 for photolysis of $[IrO_2(Ph_2PCH_2CH_2PPh_2)_2]^+$ and $[IrO_2(Ph_2PCHCHPPh_2)_2]^+$. Although quantum yields were not measured, it was noted that 366 nm light was sufficient to induce elimination of H_2 from $[IrClH_2(CO)(PPh_3)_2]$ and $[IrH_2(Ph_2PCH_2CH_2PPh_2)_2]^+$ but that 313 nm was required for $[IrH_2 \cdot (Ph_2PCHCHPPh_2)_2]^+$. The analogous $[IrIH_2(CO)(PPh_3)_2]$ complex also undergoes photoinduced elimination of H_2, but the product $[IrI(CO) \cdot (PPh_3)_2]$ is photosensitive and undergoes subsequent photodecomposition.

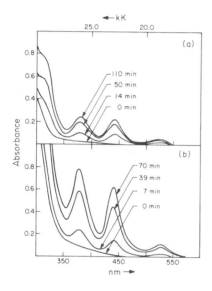

Fig. 7-6. Electronic absorption spectral changes accompanying 366 nm photolysis ethanol solutions of
(a) $[IrO_2(Ph_2PCH_2CH_2PPh_2)_2]Cl$ and
(b) $[IrO_2(Ph_2PCHCHPPh_2)_2]Cl$.
Reprinted with permission from Geoffroy et al. [36], J. Am. Chem. Soc. **97**, 3933 (1975). Copyright by the American Chemical Society.

Irradiation of argon-purged solutions of $[IrCl_2H(CO)(PPh_3)_2]$ gives elimination of HCl and generation of $[IrCl(CO)(PPh_3)_2]$ (Eq. 7-30) [36].

$$[IrCl_2H(CO)(PPh_3)_2] \xrightarrow[\text{Ar purge}]{h\nu} HCl + [IrCl(CO)(PPh_3)_2] \qquad (7\text{-}30)$$

Under a vigorous inert-gas purge the reaction was forced to completion, and it was noted that the rate of dehydrochlorination was about $\frac{1}{5}$ that of dehydrogenation of $[IrClH_2(CO)(PPh_3)_2]$ under similar conditions. $[IrClH(Ph_2PCHCHPPh_2)_2]^+$ and $[IrClH(PPh_2CH_2CH_2PPh_2)_2]^+$ were photosensitive, but apparently elimination of HCl did not occur since the $[Ir(diphos)_2]^+$ complexes were not generated [36].

The electronic absorption spectral data for the complexes examined are summarized in Table 7-1. The spectra are not at all well resolved and it is

TABLE 7-1

Lowest Electronic Absorption Bands in Iridium Adduct Complexes[a]

Complex	ν_{max}, kK	ε_{max}, M^{-1} cm^{-1}
$[H_2IrBr(CO)(P(C_6H_{11})_3)_2]$	27.4	38.6
$[H_2IrI(CO)(P(C_6H_{11})_3)_2]$	27.0	34.2
$[H_2IrCl(CO)(P(i\text{-}C_3H_7)_3)_2]$	27.6	35.2
$[H_2IrBr(CO)(P(i\text{-}C_3H_7)_3)_2]$	27.4	6.6
$[H_2IrI(CO)(P(i\text{-}C_3H_7)_3)_2]$	27.0	26.1
$[H_2Ir(Ph_2PCH_2CH_2PPh_2)_2]Cl$	27.0	26
$[H_2Ir(Ph_2PCHCHPPh_2)_2]Cl$	28.6	40
$[H_2IrCl(CO)(PPh_3)_2]$	26.3	25
$[O_2Ir(Ph_2PCH_2CH_2PPh_2)_2]Cl$	27.4	200
$[O_2Ir(Ph_2PCHCHPPh_2)_2]Cl$	26.0	130
$[H(Cl)Ir(Ph_2PCH_2CH_2PPh_2)_2]Cl$	27.8	60
$[H(Cl)Ir(Ph_2PCHCHPPh_2)_2]Cl$	27.8	135
$[H(Cl)IrCl(CO)(PPh_3)_2]$	26.7	140

[a] Reprinted with permission from Geoffroy *et al.* [36], *J. Chem. Soc.* **97**, 3933 (1975). Copyright by the American Chemical Society.

difficult to make any definitive assignments, although it was proposed [36] that the active excited states arise from metal-to-phosphine charge transfer. Such a transition would leave the metal electron-deficient [Ir(IV)] in the excited state, and it was suggested that reduction of the Ir(IV) nucleus could occur by the electron-rich $O_2{}^{2-}$ and $H_2{}^{2-}$ fragments giving simultaneous elimination of the small molecules [36]. No information concerning the mechanism of elimination of H_2 was given, although photolysis in toluene solution gave no evidence for the production of bibenzyl, suggesting that hydrogen atom formation did not occur to a significant extent.

D. $[IrClH_2(PPh_3)_3]$ AND *fac*-$[IrH_3(PPh_3)_3]$

Perhaps the most definitive study yet performed on a metal hydride complex was that of $[IrClH_2(PPh_3)_3]$ [37]. This complex is thermally resistant to loss of H_2 and has the structure shown in (LVIII). No elimination of H_2

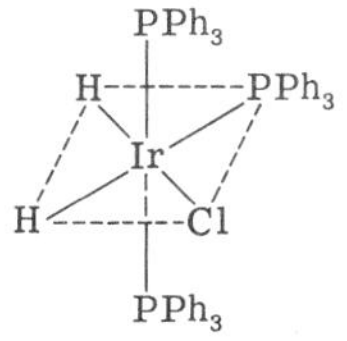

(LVIII)

occurs, for example, when samples are heated to 150°C for 24 h under vacuum. The complex is extremely photosensitive, and the white solid turns orange within minutes upon exposure to sunlight or to fluorescent room light. Solutions similarly turn orange when irradiated, and the electronic absorption spectral changes obtained during 366 nm photolysis of a degassed CH_2Cl_2 solution are shown in Fig. 7-7. Monitoring the photolysis in the IR revealed that the two metal–hydride vibrations at 2215 and 2110 cm^{-1} smoothly decreased in intensity during photolysis and no new bands appeared in the ν_{M-H} region. Mass spectral analysis of the gases above irradiated solutions showed substantial amounts of H_2, and the final spectrum shown in Fig. 7-7 is identical to that of an authentic sample of $[IrCl(PPh_3)_3]$. The photoreaction rapidly reverses under an H_2 atmosphere although the formation of the orange color was not suppressed even by a vigorous H_2 purge. If the photolysis was not prolonged (<30 min), an orange solid identical in properties to $[IrCl(PPh_3)_3]$ was isolated. These various results show that the primary process is photoinduced elimination of H_2 to generate $[IrCl(PPh_3)_3]$. If the photolysis is prolonged, however, or if the photoreleased H_2 is removed from the solution, the originally generated orange color slowly bleaches. Evaporation of solvent from these latter solutions gave a cream-colored solid which showed IR bands characteristic of an

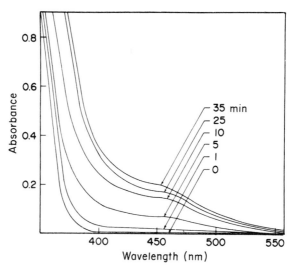

FIG. 7-7. Electronic absorption spectral changes accompanying 366 nm photolysis of a 2.3×10^{-2} M degassed CH_2Cl_2 solution of $[IrClH_2(PPh_3)_3]$. Reprinted with permission from Geoffroy and Pierantozzi [37], *J. Am. Chem. Soc.* **98**, 8054 (1976). Copyright by the American Chemical Society.

ortho-metallated derivative. The overall photochemical process is that summarized in Eq. (7-31).

$$[IrClH_2(PPh_3)_3] \underset{H_2}{\overset{h\nu}{\rightleftharpoons}} H_2 + [IrCl(PPh_3)_3]$$

(7-31)

The ortho-metallation reaction is also reversible, and the sequence of reactions shown in Eq. (7-31) can be cycled repeatedly if the photoreleased H_2 is not allowed to escape. Elimination of H_2 could be induced by irradiation with $\lambda < 400$ nm, and the quantum yield of formation of $[IrCl(PPh_3)_3]$ measured for 254 nm irradiation is 0.56. This must be treated as a lower limit, however, since the measurements were made in sealed vessels from which the photoreleased H_2 could not escape, and back-reaction with H_2 was not prevented.

Elimination of H_2 was shown to proceed in a concerted fashion since irradiation of an equimolar mixture of $[IrClH_2(PPh_3)_3]$ and $[IrClD_2(PPh_3)_3]$ gave only H_2 and D_2 with no HD detected [37]. If the reaction had proceeded stepwise by elimination of $H^+(D^+)$, $H^{\cdot}(D^{\cdot})$, or $H^-(D^-)$, some HD would have formed, especially from photolysis of $[IrClD_2(PPh_3)_3]$. This result accords well with the previous report [38] that *thermal* elimination of H_2 from $[IrH_2(CO)_2(PEtPh_2)_2]^+$ occurs via a concerted path.

The electronic absorption spectrum of the complex revealed little as to the nature of the photoactive excited state since no well-resolved bands were observed. A molecular orbital diagram was drawn on the basis of symmetry considerations, and it is shown in Fig. 7-8 [37]. The highest occupied orbital $\sigma_{x^2-y^2}$ is the principal bonding orbital between H_2 and iridium, and the lowest unoccupied orbital $\sigma^*_{x^2-y^2}$ is strongly antibonding between the metal and H_2. It was suggested that the active excited state involved either depopulation of $\sigma_{x^2-y^2}$ or population of $\sigma^*_{x^2-y^2}$, since either should greatly weaken the Ir—H_2 bonding.

It was suggested that the $[IrClH_2(PPh_3)_3]-[IrCl(PPh_3)_3]$ system could serve as a model system for hydrogen storage and energy storage. $[IrCl(PPh_3)_3]$ readily takes up H_2 to store it as $[IrClH_2(PPh_3)_3]$ and then easily releases it on demand by irradiation with UV light or with sunlight. Furthermore, when hydrogen adds to $[IrCl(PPh_3)_3]$, approximately 15–20

296   7. Hydride Complexes

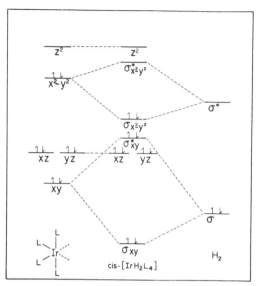

Fig. 7-8. Molecular orbital energy level diagram for six-coordinate *cis*-dihydride complexes of iridium. Reprinted with permission from Geoffroy and Pierantozzi [37], *J. Am. Chem. Soc.* **98**, 8054 (1976). Copyright by the American Chemical Society.

kcal/mol of energy is released [39] (Eq. 7-32). Irradiation of $[IrClH_2(PPh_3)_3]$ with 366 nm (78 kcal/mol) gives H_2 and $[IrCl(PPh_3)_3]$, which can be

$$[IrCl(PPh_3)_3] + H_2 \longrightarrow [IrClH_2(PPh_3)_3] + 15\text{–}20 \text{ kcal/mol} \qquad (7\text{-}32)$$

separately stored. When the components are later allowed to react to generate $[IrClH_2(PPh_3)_3]$, approximately 15–20 kcal/mol of energy is given off, and this amount of energy was then stored. Because of the high cost of iridium, however, a system like this can only serve as a model.

$[IrH_3(PPh_3)_3]$ exists as meridional and facial isomers with structures (LIX) and (LX), respectively. Neither isomer will lose H_2 under thermal conditions.

(LIX)   (LX)

Irradiation of a degassed benzene solution of the synthetic mixture of the two isomers was reported [37] to give elimination of H_2. Infrared and high-pressure liquid chromatography data indicated that the elimination was more efficient from the meridional isomer. The expected primary photo-product $[IrH(PPh_3)_3]$ was not isolated but rather an ortho-metallated derivative $[Ir(C_6H_4PPh_2)(PPh_3)_2]$ was obtained. This complex presumably resulted by initial ortho-metallation followed by photoelimination of a second molecule of H_2 (Eq. 7-33). The reactive intermediate $[IrH(PPh_3)_3]$

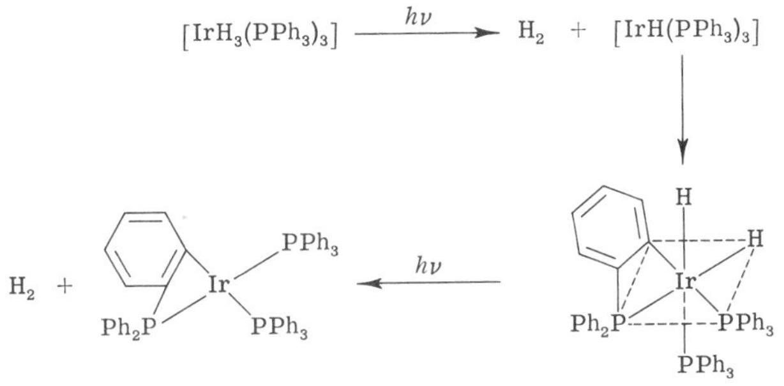

$$[IrH_3(PPh_3)_3] \xrightarrow{h\nu} H_2 + [IrH(PPh_3)_3]$$

$$(7\text{-}33)$$

was trapped, however, by irradiation under a carbon monoxide atmosphere to produce the known $[IrH(CO)(PPh_3)_3]$ in quantitative yield (Eq. 7-34).

$$[IrH(PPh_3)_3] + CO \longrightarrow [IrH(CO)(PPh_3)_3] \qquad (7\text{-}34)$$

Photolysis of $[IrH_3(PPh_3)_3]$ under a hydrogen atmosphere affords a different product. Prolonged photolysis of a benzene solution gives an insoluble white precipitate which was identified as $[IrH_5(PPh_3)_2]$. Irradiation under H_2 apparently allows the detection of a small quantum yield process involving the loss of PPh_3 (Eq. 7-35).

$$[IrH_3(PPh_3)_3] \xrightarrow[H_2]{h\nu} [IrH_5(PPh_3)_2] + PPh_3 \qquad (7\text{-}35)$$

VII.   SUMMARY

Although it is true that only a relatively small number of hydride complexes have been examined, we believe that several aspects concerning their photoreactivity are becoming clear. First, it appears that irradiation of complexes containing only one hydride ligand will often lead to "typical" photochemical reactions. For example, photolysis of $[RuClH(CO)(PPh_3)_3]$ and $[MH(CO)_5]$ (M = M, Re) induces CO loss as with many other metal carbonyls, and irradiation of $[RuClH(CO)_2(PPh_3)_2]$ gives photoisomerization analogous to that observed [32] for a series of $[RuX_2(CO)_2L_2]$ (X = halide; L = tertiary phosphine) complexes.

Strong evidence has been provided that if a complex has two or more hydrogens it will lose H_2 when irradiated regardless of how thermally resistant is the complex to H_2 loss. This now seems to be a general reaction for di- and polyhydrides of all the transition elements, and indeed it has already been observed for complexes of V, Mo, W, Fe, Ru, Co, and Ir. More studies are clearly necessary, however, before these general conclusions can be accepted with confidence.

REFERENCES

1. G. Wilkinson and J. M. Birmingham, *J. Am. Chem. Soc.* **77**, 3421 (1955).
2. J. Chatt, *Science* **160**, 723 (1968).
3. A. P. Ginsberg, *Transition Met. Chem.* **1**, 111 (1965).
4. M. L. H. Green, *Endeavour* **26**, 129 (1967).
5. M. L. H. Green and D. J. Jones, *Adv. Inorg. Chem. Radiochem.* **7**, 115 (1965).
6. H. D. Kaesz and R. B. Saillant, *Chem. Rev.* **72**, 231 (1972).
7. D. M. Roundhill, *Adv. Organomet. Chem.* **13**, 273 (1975).
8. G. L. Geoffroy and J. R. Lehman, *Adv. Inorg. Chem. Radiochem.* **20**, 189 (1977).
9. J. E. Ellis, R. A. Faltynek, and S. G. Hentges, *J. Am. Chem. Soc.* **99**, 626 (1977).
10. F. Pennella, *Inorg. Synth.* **15**, 42 (1976).
11. P. Meaken, L. J. Guggenberger, W. C. Peet, E. L. Muetterties, and J. P. Jesson, *J. Am. Chem. Soc.* **95**, 1467 (1973).
12. G. L. Geoffroy, M. G. Bradley, and R. Pierantozzi, to be submitted.
13. B. L. Haymore, private communication.
14. K. Elmitt, M. L. H. Green, R. A. Forder, I. Jefferson, and K. Prout, *Chem. Commun.* p. 474 (1974).
15. C. Giannotti and M. L. H. Green, *Chem. Commun.* p. 1114 (1972).
16. L. Farrugia and M. L. H. Green, *Chem. Commun.* p. 416 (1975).
17. M. L. H. Green, M. Berry, C. Couldwell, and K. Prout, *Nouv. J. Chim.* **1**, 187 (1977).
18. G. L. Geoffroy and M. G. Bradley, *J. Organomet. Chem.* **134**, C27 (1977).
19. G. L. Geoffroy and M. G. Bradley, *Inorg. Chem.* **17**, 2410 (1978).
20. J. L. Thomas, *J. Am. Chem. Soc.* **95**, 1838 (1973).
21. J. L. Thomas and H. H. Brintzinger, *J. Am. Chem. Soc.* **94**, 1386 (1972).
22. M. Berry, S. G. Davies, and M. L. H. Green, *Chem. Commun.* p. 99 (1978).
23. J. L. Petersen, D. L. Lichtenberger, R. F. Fenske, and L. F. Dahl, *J. Am. Chem. Soc.* **97**, 6433 (1975).
24. H. H. Brintzinger, L. L. Lohr, Jr., and K. L. Tang Wong, *J. Am. Chem. Soc.* **97**, 5146 (1975).
25. N. W. Hoffman and T. L. Brown, *Inorg. Chem.* **17**, 613 (1978).
25a. B. H. Byers and T. L. Brown, *J. Am. Chem. Soc.* **99**, 2527 (1977).
26. A. J. Rest and J. J. Turner, *Chem. Commun.* p. 375 (1969).
27. A. Sacco and M. Aresta, *Chem. Commun.* p. 1223 (1968).
28. E. Koerner von Gustorf, I. Fischler, J. Leitich, and H. Dreeskamp, *Angew. Chem., Int. Ed. Engl.* **11**, 1088 (1972).
29. D. J. Darensbourg, *Inorg. Nucl. Chem. Lett.* **8**, 529 (1972).
30. G. L. Geoffroy and M. G. Bradley, *Inorg. Chem.* **16**, 744 (1977).
31. P. S. Hallman, B. R. McGarvey, and G. Wilkinson, *J. Chem. Soc. A* p. 3143 (1968).

32. C. F. J. Barnard, J. A. Daniels, J. Jeffery, and R. J. Mawby, *J. Chem. Soc., Dalton Trans.* p. 953 (1976).

33. T. Kruck, G. Sylvester, and I. P. Kunau, *Z. Naturforsch., Teil B* **28**, 38 (1973).

34. T. Kruck, G. Sylvester, and I. P. Kunau, *Angew. Chem., Int. Ed. Engl.* **10**, 725 (1971).

35. A. Camus, C. Cocevar, and G. Mestroni, *J. Organomet. Chem.* **39**, 355 (1972).

36. G. L. Geoffroy, H. B. Gray, and G. S. Hammond, *J. Am. Chem. Soc.* **97**, 3933 (1975).

37. G. L. Geoffroy and R. Pierantozzi, *J. Am. Chem. Soc.* **98**, 8054 (1976).

38. M. J. Mays, R. N. F. Simpson, and F. P. Stefanini, *J. Chem. Soc. A* p. 3000 (1970).

39. Estimated from data given by L. Vaska and M. F. Werneke, *Trans. N.Y. Acad. Sci.* [2] **33**, 70 (1971).

Alkyl Complexes

8

I. INTRODUCTION

Like transition-metal hydride complexes, studies of the preparation, characterization, and chemistry of metal alkyls have increased dramatically in recent years. This has arisen primarily through the recognition that the metal–carbon bond in metal alkyls is inherently strong and that decomposition does not readily occur by its simple homolytic cleavage. A principal mode of decomposition of metal alkyls is β-hydride elimination (Eq. 8-1), and

$$L_nM—CH_2CH_2R \longrightarrow L_nMH + CH_2\!=\!CHR \qquad (8\text{-}1)$$

consequently much attention has been given to alkyl ligands such as CH_3, $CH_2Si(CH_3)_3$, $CH_2C(CH_3)_3$, and $CH_2C_6H_5$ which do not have β-hydrogens and cannot decompose by this low-energy pathway.

A number of metal alkyl complexes have had their photochemical properties examined, although the clear majority of studies have been of a qualitative nature. Few quantitative and mechanistic studies have been conducted, and perhaps the most thorough study is that of the cobalamins and related cobaloximes discussed in Section VI,B. Several metal–carbonyl–alkyl complexes have been reported to undergo photosubstitution of carbon monoxide. These include the cyclopentadienyl complexes $[M(\eta^5\text{-}C_5H_5)(CH_3)(CO)_3]$ (M = Cr, Mo, W) [1–3] and $[Fe(\eta^5\text{-}C_5H_5)R(CO)_2]$ (R = CH_3 [4], C_6H_5 [5]), and the η^1-allyl complexes $[M(\eta^5\text{-}C_5H_5)(\eta^1\text{-}C_3H_5)(CO)_3]$ (M = Mo, W) which rearrange to η^3-allyl complexes upon photolysis (Eq. 8-2) [6,7]. The

$$[M(\eta^5\text{-}C_5H_5)(\eta^1\text{-}C_3H_5)(CO)_3] \xrightarrow{h\nu} [M(\eta^5\text{-}C_5H_5)(\eta^3\text{-}C_3H_5)(CO)_2] + CO \quad (8\text{-}2)$$

norbornyl complexes $[M(1\text{-norbornyl})_4]$ (M = Ti, V, Cr, Mn, Fe, Co, Zr, Hf) (LXI) decompose when irradiated in pentane solution and give quantitative yields of norbornane or 1,1'-binorbornane [8]. The complex $[VO(CH_2\text{-}SiMe_3)_3]$ has been reported [9] to be light-sensitive, but no further details were given.

(LXI)

II. TITANIUM, ZIRCONIUM, AND HAFNIUM COMPLEXES

A. $[M(\eta^5\text{-}C_5H_5)_2(CH_3)_2]$ (M = Ti, Zr, Hf)

The first report of the photochemical properties of $[Ti(\eta^5\text{-}C_5H_5)_2(CH_3)_2]$ and $[Ti(\eta^5\text{-}C_5Me_5)_2(CH_3)_2]$ was in 1974 when Harrigan *et al.* [10] noted that these complexes were photosensitive. When $[Ti(\eta^5\text{-}C_5H_5)_2(CH_3)_2]$ was irradiated in degassed $CHCl_3$ solution, a complicated product mixture of $[Ti(\eta^5\text{-}C_5H_5)_2(CH_3)Cl]$, $[Ti(\eta^5\text{-}C_5H_5)_2Cl_2]$, and $[Ti(\eta^5\text{-}C_5H_5)Cl_3]$ was obtained. It was proposed that $[Ti(\eta^5\text{-}C_5H_5)_2(CH_3)Cl]$ was the primary photoproduct and that the other two complexes arose through secondary photolysis [10].

Rausch, Alt, and co-workers [11,12] have studied $[Ti(\eta^5\text{-}C_5H_5)_2(CH_3)_2]$ and its Zr and Hf analogs in somewhat more detail. Photolysis of each of these in hydrocarbon solution was initially reported [11] to produce methane and compounds formulated as metallocenes (Eq. 8-3). These "metallocenes"

$$[M(\eta^5\text{-}C_5H_5)_2(CH_3)_2] \xrightarrow{h\nu} \text{``}[M(\eta^5\text{-}C_5H_5)_2]\text{''} + 2CH_4 \quad (8\text{-}3)$$

were not identical to those characterized by other workers, although in each case elemental analysis was consistent with the empirical formula $C_{10}H_{9,10}M$. Their reactivity was somewhat consistent with a metallocene formulation since their reactions with CO, NO, N_2, H_2, olefins, and acetylenes led to known metallocene adducts. The titanocene species reacted with

CO, for example, to give near-quantitative formation of $[Ti(\eta^5\text{-}C_5H_5)_2(CO)_2]$. However, cryoscopic molecular weight measurements of the photogenerated titanocene species gave values of 680–710, which suggested an oligomeric structure [11a]. Yields of the "titanocene" produced by this procedure were greater than 90% following 2–3 h photolysis. Lower yields of "zirconocene" and "hafnocene" were obtained, however, and longer irradiation periods were required [11].

The mechanism by which these reactions proceed is of obvious importance. A series of deuteration studies showed that in benzene solution the methane produced did not arise by abstraction of hydrogen from the solvent but predominantly by abstraction from an $\eta^5\text{-}C_5H_5$ ligand [11a,12]. This was independently confirmed by Bamford et al. [12a], who showed that photolysis of $[Ti(\eta^5\text{-}C_5H_5)(CD_3)_2]$ in toluene-d_8 solution gave only formation of CD_3H with no CD_4 detected, again implicating hydrogen abstraction from one of the C_5H_5 ligands. Such a process would probably yield compounds such as (LXII) with $\eta^1,\eta^5\text{-}C_5H_4$ bridging ligands as the metallocene

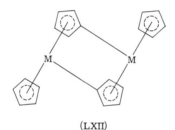

(LXII)

products. Such a structure is consistent with the apparent oligomeric nature of the metallocenes. These results appear to rule out simple photoinduced homolysis of the titanium–methyl bonds to generate *free* methyl radicals since the latter should readily abstract hydrogen from the toluene solvent to give the more stable benzyl radicals. On the other hand, recent studies by Samuel et al. [13] employing the spin traps nitrosodurene and 5,5-dimethyl-1-pyrroline-1-oxide led to the observation of ESR signals characteristic of their methyl adducts, and the authors proposed that homolysis of the titanium–methyl bonds did obtain upon photolysis as shown in Eq. (8-4). However, the deuteration studies cited previously [11a,12,12a]

$$[M(\eta^5\text{-}C_5H_5)_2(CH_3)_2] \xrightarrow{h\nu} [M(\eta^5\text{-}C_5H_5)_2CH_3] + \,^{\cdot}CH_3 \qquad (8\text{-}4)$$

apparently unambiguously rule out production of *free* radicals, and one must surmise that the spin-trapping agents employed in the study of Samuel et al. [13] somehow interact either with the excited complex or with some intermediate to form the spin adducts detected by ESR spectroscopy.

Photolysis of the dimethyl derivatives in the presence of alkynes was initially reported to give metallocycles as the major product [11]. However, in a recent communication [14] it was noted that insertion products could also be obtained. For example, irradiation of a dilute solution containing $[Ti(\eta^5\text{-}C_5H_5)_2(CH_3)_2]$ and diphenylacetylene for 4 h gave 18% yield of the metallocycle (LXIII) and 11% yield of the insertion product (LXIV). If bis(pentafluorophenyl)acetylene was employed in the reaction, only the insertion product was obtained and no metallocycle was isolated. The extensive chromatography that was required to separate the products (LXIII) and (LXIV) was the apparent source of error in the initial report [11].

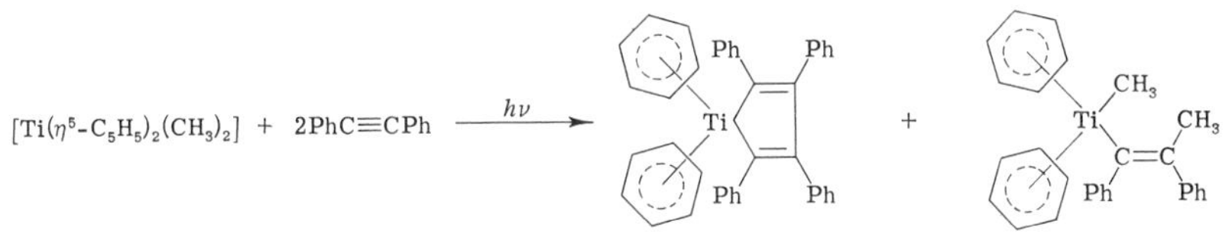

$$[Ti(\eta^5\text{-}C_5H_5)_2(CH_3)_2] + 2PhC{\equiv}CPh \xrightarrow{h\nu}$$

(LXIII)   (LXIV)

$$(8\text{-}5)$$

Bis(η^5-indenyl)dimethyl derivatives of Ti, Zr, and Hf (LXV) have been

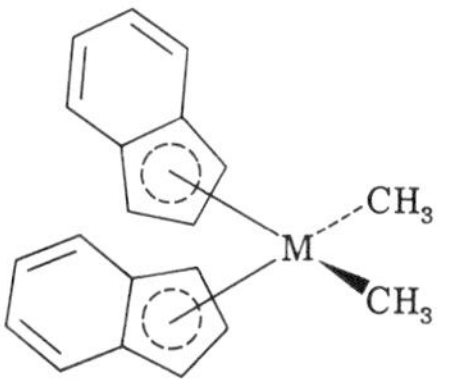

(LXV)

reported [11a,14a] to show a photochemical behavior similar to that of the $[M(\eta^5\text{-}C_5H_5)_2(CH_3)_2]$ derivatives described. Photolysis in degassed pentane solution gave methane and bis(indenyl) metal complexes. In the presence of alkynes metallocycles were produced, and photolysis under a CO atmosphere gave high yields of the dicarbonyls $[M(\eta^5\text{-}C_9H_7)_2(CO)_2]$.

Irradiation of the corresponding zirconium fluorenyl complex $[Zr(\eta^5\text{-}C_{13}H_9)_2(CH_3)_2]$ (LXVI) with unfiltered light from a 450 W medium-pressure

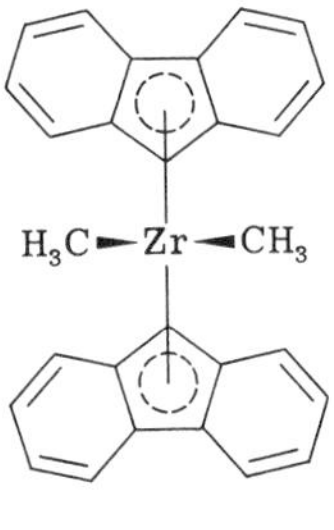

$$H_3C \longrightarrow Zr \longleftarrow CH_3$$

(LXVI)

Hg lamp gave cleavage of both zirconium–methyl bonds and a pyrophoric product analyzing for $[Zr(C_{13}H_9)_2]$ [12]. The gas released was mainly methane plus smaller amounts of ethane and ethylene' The NMR spectrum of the product did not show the resonance expected for the single proton on the 5-membered ring, and it was proposed that its abstraction could have occurred via the photogenerated methyl radicals to form methane and a carbon–zirconium σ bond. This argument was supported by IR analysis and the studies [11a] which showed ring proton abstraction during photolysis of $[Ti(\eta^5\text{-}C_5H_5)_2(CH_3)_2]$. It was suggested [12] that this may be a general process in the photolysis of cyclopentadienylalkyl complexes.

Samuel and Giannotti [14b] have recently reported that photolysis of the dialkytitanocenes $[Ti(\eta^5\text{-}C_5H_5)_2R_2]$ ($R = CH_3$, CH_2Ph) in the presence of elemental sulfur produced titanocene pentasulfide and other polysulfides according to Eq. (8-6). Photolysis of $[Zr(\eta^5\text{-}C_5H_5)_2(CH_3)_2]$ under the same conditions did not produce zirconium pentasulfide.

$$[Ti(\eta^5\text{-}C_5H_5)_2R_2] + S_8 \xrightarrow[C_6H_6]{h\nu} \quad + \quad RS_nR \quad (8\text{-}6)$$

B. $[M(\eta^5\text{-}C_5H_5)_2(Ar)_2](M = Ti, Zr)$

Peng and Brubaker [15] examined diphenyltitanocene $[Ti(\eta^5\text{-}C_5H_5)_2 \cdot (C_6H_5)_2]$ and suggested that its photoreactivity parallels that of the corresponding dimethyl complex in apparently giving homolysis of a Ti phenyl bond in the primary event. Photolysis in benzene solutions was stated to produce biphenyl, benzene, and an oligomeric material formulated as $[Ti(\eta^5\text{-}C_5H_5)_2H]_x$ (Eq. 8-7). The evidence for benzene formation was obtained by

$$[Ti(\eta^5\text{-}C_5H_5)_2(C_6H_5)_2] \xrightarrow{h\nu} [Ti(\eta^5\text{-}C_5H_5)_2H]_x + C_6H_6 + C_6H_5\text{—}C_6H_5 \quad (8\text{-}7)$$

photolysis of the complex in benzene-d_6 and observing growth of the benzene NMR resonance over a period of several hours. Stable titanocene adducts were formed by photolysis of diphenyltitanocene in the presence of carbon monoxide and diphenylacetylene.

However, Rausch et al. [15a] have obtained evidence that indicates that *both* Ti–aryl bond homolysis and reductive coupling of the two aryl ligands obtain upon photolysis of $[Ti(\eta^5\text{-}C_5H_5)_2(aryl)_2]$. For example, irradiation of $[Ti(\eta^5\text{-}C_5H_5)_2(C_6H_5)_2]$ in benzene-d_6 solution produces biphenyl-d_0 and biphenyl-d_5 in about equal amounts. With $[Ti(\eta^5\text{-}C_5H_5)_2(p\text{-}C_6H_4CH_3)_2]$, a mixture of toluene, 4-methylbiphenyl, and 4,4'-dimethylbiphenyl obtains upon photolysis. Irradiation of $[Ti(\eta^5-C_5H_5)_2(C_6H_5)_2]$ in the presence of CO leads to moderate yields of $[Ti(\eta^5-C_5H_5)_2]$ [15a].

In contrast to the chemistry observed with $[Ti(\eta^5-C_5H_5)_2(C_6H_5)_2]$, Erker [16] has suggested that photolysis of bis(aryl)zirconocene complexes leads only to coupling of the aryl ligands and formation of substituted biphenyls. For example, photolysis of bis(p-tolyl)zirconocene in benzene solution gave almost quantitative yield of 4,4′-dimethylbiphenyl (Eq. 8-8). The solution turned deep blue-brown upon photolysis, and it was suggested that the

$$[(\eta^5\text{-}C_5H_5)_2Zr(\!-\!\langle\bigcirc\rangle\!-\!CH_3)_2]$$

$$\downarrow h\nu$$

$$H_3C\!-\!\langle\bigcirc\rangle\!-\!\langle\bigcirc\rangle\!-\!CH_3 \;+\; "[Zr(\eta^5\text{-}C_5H_5)_2]"$$

(8-8)

color arose from production of zirconocene. Irradiation of the corresponding diphenyl complex gave biphenyl, and 3,3′-dimethylbiphenyl was obtained from photolysis of bis(m-tolyl)zirconocene. No mechanistic experiments were conducted although this system appears to be ideally suited for such studies. The biphenyl products probably arise through concerted reductive elimination and coupling of the aryl ligands. The near-quantitative yield of 4,4′-dimethylbiphenyl from photolysis of $[Zr(\eta^5-C_5H_5)_2(C_6H_4CH_3)_2]$ in benzene solution appears to rule out Zr–tolyl homolysis to produce tolyl radicals. Such a radical process in benzene solution should produce substantial amounts of C_6H_5—$C_6H_4CH_3$ which was apparently not detected.

The inconsistencies among the three studies [15,15a,16] clearly indicate the need for additional mechanistic studies on both of these systems, as well as further studies on the dimethyl complexes discussed. It should be noted that the benzene detected [15] upon photolysis of $[Ti(\eta^5-C_5H_5)_2(C_6H_5)_2]$ could have arisen by H/D exchange catalyzed by the "titanocene" intermediates.

C. $[Ti(CH_2Ph)_4]$ AND $[Zr(CH_2Ph)_4]$

Both of these tetrakis(benzyl) derivatives have been shown to be sensitive to light [17]. Irradiation of toluene solutions of $[Ti(CH_2Ph)_4]$ at $-78°$ gave nearly 50% reduction to Ti(III), but no mechanistic details were given. The zirconium derivative is more sensitive than $[Ti(CH_2Ph)_4]$, and its solutions gave a rapid color change from yellow to brown when irradiated. Reaction of the latter irradiated solutions with CH_3OD gave formation of HD, and this suggested the presence of a Zr—H bond in the photoproduct. A brown solid which decomposes at $0°$ was isolated by precipitation from the irradiated solutions. Analytical data were consistent with the formulation (LXVII), and the overall photoreaction shown in Eq. (8-9) was suggested.

$$[Zr(CH_2Ph)_4] \xrightarrow{h\nu}$$

(LXVII)

Photolysis was proposed to induce migration of a benzyl ligand to another coordinated ligand, with hydrogen transfer back to the vacant site on the metal [17].

Ballard and van Lienden [18] also noted the photosensitivity of $[Zr(CH_2Ph)_4]$ and studied its utility for the polymerization of vinyl monomers. $[Zr(CH_2Ph)_4]$ will thermally catalyze the polymerization of styrene, but Pyrex-filtered irradiation increased the polymerization rate 4-fold. When the light was turned off, the polymerization system returned to the thermal rate, and there seemed to be no restrictions on the number of times the system could be cycled between the thermal and photochemical rates. The electronic absorption spectrum of a mixture of styrene and $[Zr(CH_2Ph)_4]$ shows a strong absorption at 320 nm which tails considerably into the visible spectral region. The spectrum of this mixture was shown to be only the sum of the spectra of the individual components, and no spectral evidence for a Zr–styrene complex was obtained. Assignment of the spectral transitions was not attempted. Since Zr(IV) is d^0, however, the intense band at 320 nm is presumably alkyl-to-zirconium charge transfer in character.

Wavelength dependence experiments were conducted and two different photochemical processes were identified. For wavelengths between 450 and 600 nm, the photopolymerization rate was completely independent of the radiation intensity, but below 450 nm the rate was directly proportional to $I^{1/2}$. It was established that the thermal polymerization did not proceed through a radical path but more likely through coordination polymerization with a species such as (LXVIII) as the key intermediate. Experimental evi-

(LXVIII)

dence suggested that each metal center could only support one polymerization center. Rate measurements and the product distribution resulting from irradiation with wavelength less than 450 nm indicated that the photo-polymerization proceeds through a free-radical pathway, presumably initiated by photocleavage of a Zr–benzyl or Zr–polymer bond.

However, it was proposed [18] that the polymerization resulting from 450–600 nm irradiation was related to that occurring in the dark. This conclusion was supported by rate data and by the observation that the thermal and photochemical rates could be cycled repeatedly. It was suggested that the propagation center resembles (LXVIII) and that irradiation probably increases the rate of alkyl (or polymer) migration to the coordinated monomer. It was also noted that the molecular weight of the polymer formed from photopolymerization was lower than that obtained thermally, indicating that irradiation increases the rate of termination as well as propagation.

D. $[Ti(CH_3)Cl_3]$

De Vries [19] has reported that solutions of $[Ti(CH_3)Cl_3]$ can be photo-decomposed, but no evidence was obtained for production of methyl radicals. The gases evolved during photochemical decomposition of the complex in C_6D_{12} solution did not contain any deuterium, while the gaseous product from photodecomposition of $[Ti(CD_3)Cl_3]$ in n-heptane solution was principally CD_4. No further details were presented.

III.   CHROMIUM, MOLYBDENUM, AND TUNGSTEN COMPLEXES

A. $[M(\eta^5\text{-}C_5H_5)CH_3(CO)_3]$ (M = Cr, Mo, W)

The photosensitivity of these complexes has been examined by several groups of workers. Barnett and Treichel [2] first studied the photolysis of $[M(\eta^5\text{-}C_5H_5)CH_3(CO)_3]$ (M = Mo, W) in the presence of PPh_3. Thermal substitution of carbon monoxide by phosphine was observed, but the reaction proceeded much faster under UV irradiation. The resultant yield of products, however, was lower due to photodecomposition. For the molybdenum complex, the principal products were $[Mo(\eta^5\text{-}C_5H_5)CH_3(CO)_2PPh_3]$ and $[Mo(\eta^5\text{-}C_5H_5)(COCH_3)(CO)_2PPh_3]$ which were isolated in yields of 14% and 6%, respectively, following 3 h irradiation of a hexane solution of the complex. Both products were believed to arise through initial photo-substitution of CO by PPh_3 with the acetyl derivative coming from a secondary thermal reaction with carbon monoxide (Eq. 8-10). A similar process

$$[Mo(\eta^5\text{-}C_5H_5)CH_3(CO)_3] + PPh_3 \xrightarrow{\quad h\nu \quad} [Mo(\eta^5\text{-}C_5H_5)CH_3(CO)_2PPh_3] + CO \qquad (8\text{-}10)$$

$$\Big\downarrow \Delta, CO$$

$$[Mo(\eta^5\text{-}C_5H_5)(COCH_3)(CO)_2PPh_3]$$

had been previously shown to lead to the acetyl derivative of $[Fe(\eta^5\text{-}C_5H_5)\cdot CH_3(CO)_2]$ [4]. The tungsten complex gave only photosubstitution of CO by PPh_3 and no acetyl product could be isolated.

In subsequent reports, Alt [20] confirmed Barnett and Treichel's observation of photosubstitution of $[W(\eta^5\text{-}C_5H_5)CH_3(CO)_3]$ with phosphine and phosphite ligands and showed an analogous photoreactivity pattern for $[Cr(\eta^5\text{-}C_5H_5)CH_3(CO)_3]$. The chromium complex also gave the acetyl derivative $[Cr(\eta^5\text{-}C_5H_5)(COCH_3)(CO)_2PPh_3]$, presumably from a thermal reaction of the initially formed $[Cr(\eta^5\text{-}C_5H_5)(CH_3)(CO)_2PPh_3]$. Alt also observed that photolysis of the $[M(\eta^5\text{-}C_5H_5)CH_3(CO)_3]$ (M = Cr, Mo, W) complexes in pentane solution in the absence of excess ligand led to formation of the binuclear complexes $[M(\eta^5\text{-}C_5H_5)(CO)_3]_2$ and $[M(\eta^5\text{-}C_5H_5)(CO)_2]_2$, (Eq. 8-11). Methane was identified as the gaseous product, and

$$[M(\eta^5\text{-}C_5H_5)CH_3(CO)_3] \xrightarrow[300\ nm]{h\nu} [M(\eta^5\text{-}C_5H_5)(CO)_3]_2 + [M(\eta^5\text{-}C_5H_5)(CO)_2]_2 + CH_4$$

$$(8\text{-}11)$$

labeling studies [20a] showed that abstraction of hydrogen from a cyclo-pentadienyl ligand had occurred. It was suggested that homolysis of the M—CH_3 represented the primary photochemical event in these complexes. Photolysis of the $[M(\eta^5\text{-}C_5H_5)CH_3(CO)_3]$ complexes in $CHCl_3$ gave high yields of $[Cr(\eta^5\text{-}C_5H_5)Cl_2]_2$ for chromium and $[M(\eta^5\text{-}C_5)Cl(CO)_3]$ for Mo and W.

Severson *et al.* [21] have also examined the photoreactivity of $[W(\eta^5\text{-}C_5H_5)R(CO)_3]$ (R = CH_3, $CH_2C_6H_5$) and presented the most definitive mechanistic data. Their results essentially confirmed the observations of Treichel [2] and Alt [20], but their key result was in noting that the quantum yield of formation of $[W(\eta^5\text{-}C_5H_5)(CO)_3]_2$ from $[W(\eta^5\text{-}C_5H_5)CH_3(CO)_3]$ was greatly suppressed when the complex was irradiated under a CO atmosphere when compared to photolysis under Ar. This strongly implicates CO dissociation in the primary photochemical event, rather than M—CH_3 homolysis as suggested by Raush and Alt [20a]. Methane was suggested to arise from secondary thermal or photochemical reactions, but no further mechanistic experiments were reported. The overall mechanism shown in Eq. (8-12) was suggested [21]. Photolysis of the benzyl complexes in the presence of the PPh$_3$ did not lead to photosubstitution but rather to isomerization of the η^1-CH$_2$C$_6$H$_5$ ligand to η^3-CH$_2$C$_6$H$_5$ (Eq. 8-13).

$$[W(\eta^5\text{-C}_5H_5)CH_3(CO)_3] \xrightarrow[-CO]{h\nu} [W(\eta^5\text{-C}_5H_5)CH_3(CO)_2]$$

$$[W(\eta^5\text{-C}_5H_5)CH_3(CO)_2(PPh_3)] \qquad\qquad [W(\eta^5\text{-C}_5H_5)(CO)_2]$$

$$\Big\downarrow +CO \qquad\qquad (8\text{-}12)$$

$$[W(\eta^5\text{-C}_5H_5)(CO)_3]$$

$$\Big\downarrow \text{dimerization}$$

$$[W(\eta^5\text{-C}_5H_5)(CO)_3]_2$$

$$[W(\eta^5\text{-C}_5H_5)(\eta^1\text{-CH}_2C_6H_5)(CO)_3] \xrightarrow{h\nu} [W(\eta^5\text{-C}_5H_5)(\eta^3\text{-CH}_2C_6H_5)(CO)_2] + CO$$

$$(8\text{-}13)$$

Alt [3,20] later noted that irradiation of $[W(\eta^5\text{-C}_5H_5)CH_3(CO)_3]$ in the presence of acetylene gave formation of the metallocycle (LXIX) (Eq. 8-14). The primary product is probably $[W(\eta^5\text{-C}_5H_5)CH_3(CO)_2C_2H_2]$ arising from photosubstitution of CO which then rearranges with incorporation of CO to give (LXIX) [3].

$$[W(\eta^5\text{-C}_5H_5)CH_3(CO)_3] + C_2H_2 \xrightarrow{h\nu}$$

$$(8\text{-}14)$$

(LXIX)

B.   [M(CARBENE)$_2$(CO)$_4$](M = Cr, Mo, W)

Herberhold and co-workers [22,222] have described the photochemical cis–trans isomerization of a series of [M(carbene)$_2$(CO)$_4$] (M = Cr, Mo, W) complexes in which the carbene is an imidazole derivative. The cis isomers are thermodynamically more stable, but photolysis induces rapid conversion to the trans isomers (Eq. 8-15). The trans isomers thermally rearrange to the cis

$$cis\text{-} \quad \left[\begin{array}{c} H_3C \\ | \\ N \\ \bigcirc \\ N \\ | \\ H_3C \end{array} \text{Cr(CO)}_4 \right]_2 \xrightarrow{h\nu} trans\text{-} \quad \left[\begin{array}{c} H_3C \\ | \\ N \\ \bigcirc \\ N \\ | \\ H_3C \end{array} \text{Cr(CO)}_4 \right]_2 \qquad (8\text{-}15)$$

isomers, and this thermal isomerization was studied in great detail. However, little information was presented concerning the photochemical process. The cis isomers themselves were initially prepared by the photoinduced disproportionation reaction shown in Eq. (8-16) [22].

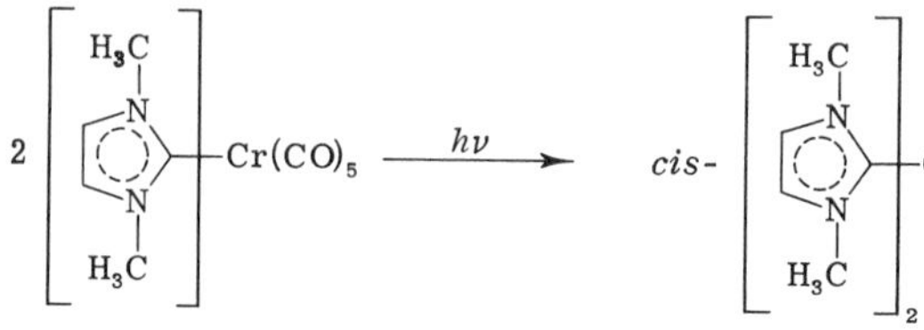

$$(8\text{-}16)$$

The trans isomers can be electrochemically converted to the cis isomers with no net current flow, and it has been suggested [23] that a photoelectric cell could be designed around the photoinduced cis → trans and electrochemical trans → cis isomerizations.

The carbene complexes $[M(CR_1R_2)(CO)_5]$ (M = Cr, W; R_1 = OMe, OEt; R_2 = Me, Et, i-Pr, Ph) have been reported [24] to undergo substitution of CO by phosphine ligands to give principally cis-$[M(CR_1R_2)(CO)_4PR_3]$.

C. $[W(\eta^5\text{-}C_5H_5)_2(-CHR_1CHR_2CH_2-)]$

Irradiation of the metallocyclobutane derivative $[W(\eta^5\text{-}C_5H_5)_2 \cdot (-CHR_1CHR_2CH_2-)]$ ($R_1 = R_2 = H$; $R_1 = H$, $R_2 = CH_3$; $R_1 = CH_3$, $R_2 = H$) (LXX) has been reported [25] to give evolution of olefin through the proposed sequence of reactions shown in Eq. (8-17). None of the complexes in brackets was isolated, and the only evidence for the mechanism written was the nature of the olefin products. No further photochemical details were presented.

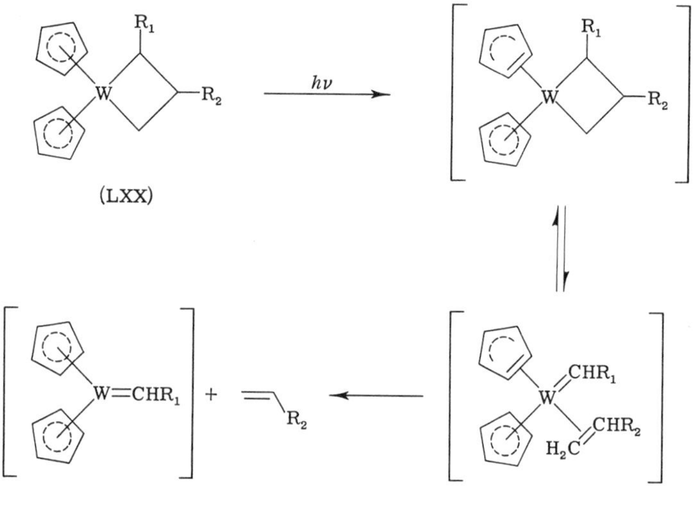

$$(8\text{-}17)$$

IV. MANGANESE COMPLEXES

$[MnR(CO)_5]$ $(R = CH_3, C_6H_5)$

Lappert and co-workers [26] have shown that irradiation of $[MnR(CO)_5]$ $(R = CH_3, C_6H_5)$ with UV light leads to homolytic cleavage of the Mn–carbon bond (Eq. 8-18). Evidence for homolytic cleavage was obtained by

$$[MnR(CO)_5] \xrightarrow[CHCl_3.]{hv} R^{\cdot} + [Mn(CO)_5] \qquad (8\text{-}18)$$

trapping both radicals ($R^{\cdot}$ and $^{\cdot}Mn(CO)_5$) with nitrosodurene 2,3,5,6-Me_4C_6H—NO. The nitroxides produced (LXXI) and (LXXII) were identi-

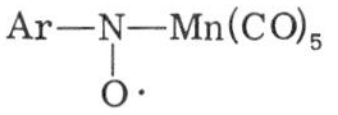

$$\underset{\underset{O^{\cdot}}{|}}{Ar-N-Mn(CO)_5} \qquad \underset{\underset{O^{\cdot}}{|}}{Ar-N-R}$$

(LXXI) 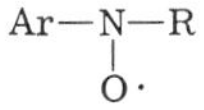   (LXXII)

fied by ESR spectroscopy. No further details concerning the photochemical conditions or reaction efficiencies were given.

V. IRON AND OSMIUM COMPLEXES

A. $[Fe(i\text{-}C_3H_7)_3]$

In a series of reports Fischer and Muller [27–29] described a method for the synthesis of olefin and arene complexes through photolysis of $[Fe(i\text{-}C_3H_7)_3]$. No information was given concerning the photochemical steps, but the overall reaction appears to be replacement of alkyl radicals by olefin or aromatic molecules (Eq. 8-19). The equimolar product mixture of propane

$$[Fe(i\text{-}C_3H_7)_3] + 2C_6H_8 \xrightarrow{hv} [Fe(C_6H_6)(C_6H_8)] + \tfrac{3}{2}C_3H_8 + \tfrac{3}{2}C_3H_6 + H_2 \quad (8\text{-}19)$$

and propylene suggests that the reaction may proceed through photoinduced β-hydride elimination to produce a metal hydride intermediate such as (LXXIII). The latter could then collapse to give propane and propylene.

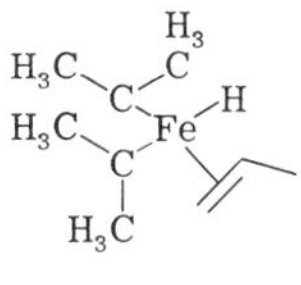

(LXXIII)

B. $[Fe(C_{16}H_{21}N_2O_8)(CO)_4]$

Irradiation of 1,1,1-tetracarbonyl-2,3,1-diazaferrole (LXXIV) in the pres-

(**LXXIV**)

ence of substrate leads to a variety of products, all apparently arising through initial photoelimination of CO [30]. With PPh_3, $[FeR(CO)_3PPh_3]$ results, but irradiation of (LXXIV) in the presence of diphenylacetylene and 2,3-dimethylbutadiene resulted in incorporation of the acetylene or diene in the organic portions of the molecule.

C. $[Fe(\eta^5\text{-}C_5H_5)(R)(CO)(PR_3)]$ AND $[Fe(\eta^5\text{-}C_5H_5)(R)(CO)_2]$

Treichel and co-workers [4] have reported that irradiation of $[Fe(\eta^5\text{-}C_5H_5)CH_3(CO)_2]$ in the presence of PPh_3 leads to substitution of CO by PPh_3 (Eq. 8-20). In addition, an acetyl derivative was produced which was

$$[Fe(\eta^5\text{-}C_5H_5)(CH_3)(CO)_2] + PPh_3 \xrightarrow{hv} [Fe(\eta^5\text{-}C_5H_5)(CH_3)(CO)(PPh_3)] + CO \quad (8\text{-}20)$$

shown to arise from a thermal reaction of the primary photoproduct with CO. In a subsequent report, Nesmeyanov and co-workers [5] demonstrated a similar photoreaction between $[Fe(\eta^5\text{-}C_5H_5)(C_6H_5)(CO)_2]$ and $P(OPh)_3$. Prolonged UV photolysis, however, produced the sequence of reactions shown in Eqs. (8-21) and (8-22). It is not clear how the dimeric product arises

$$[Fe(\eta^5\text{-}C_5H_5)(C_6H_5)(CO)_2] + P(OPh)_3 \xrightarrow{hv} [Fe(\eta^5\text{-}C_5H_5)(C_6H_5)(CO)(P(OPh)_3)] \quad (8\text{-}21)$$

$$2[Fe(\eta^5\text{-}C_5H_5)(C_6H_5)(CO)(P(OPh)_3)] + P(OPh)_3 \xrightarrow{hv}$$

$$[Fe(\eta^5\text{-}C_5H_5)(P(OPh)_3)_2]_2 + CO + C_6H_5\text{—}C_6H_5 \quad (8\text{-}22)$$

although it was proposed that photosubstitution of the second CO occurs to give $[Fe(\eta^5\text{-}C_5H_5)(C_6H_5)(P(OPh)_3)_2]$ which then decomposes with loss of C_6H_5 in a secondary thermal or photochemical step.

D. $[Os(CH_3)_2(CO)_4]$

Norton and co-workers [31] have briefly noted that photolysis of $[Os(CH_3)_2(CO)_4]$ in hexane solution led to the formation of methane but produced no ethane. Presumably, the methane derives from photohomolysis of an osmium–methyl bond and scavenging of hydrogen by the resultant methyl radical.

VI. COBALT COMPLEXES

A. $[Co(CN)_5(CH_2C_6H_5)]^{3-}$

Vogler and Hirschmann [32] observed that $[Co(CN)_5(CH_2C_6H_5)]^{3-}$ was quite photosensitive, producing $[Co(CN)_5]^{3-}$ and bibenzyl when irradiated with 313 nm in deaerated water solution (Eq. 8-23). The 313 nm quantum

$$2[Co(CN)_5(CH_2C_6H_5)]^{3-} \xrightarrow{h\nu} 2[Co(CN)_5]^{3-} + C_6H_5CH_2CH_2C_6H_5 \quad (8\text{-}23)$$

yield of disappearance of $[Co(CN)_5(CH_2C_6H_5)]^{3-}$ was 0.13. The electronic absorption spectrum of the complex shows an intense band at 295 nm ($\varepsilon = 18,000$), but LF bands were *not* observed at lower energy. The intense band was assigned as a benzyl-to-cobalt CT transition, and it was proposed that the photoreaction proceeds through population of such a CT state leading to homolytic cleavage of the cobalt–carbon bond. This may exemplify the notion suggested in Fig. 1-12 where the LMCT is actually a $\sigma_b \rightarrow \sigma^*$ type excitation.

Irradiation of $[Co(CN)_5(CH_2C_6H_5)]^{3-}$ in a strongly alkaline solution which was then subsequently saturated with oxygen gave formation of $[(CN)_5CoO_2Co(CN)_5]^{3-}$ through a thermal reaction between O_2 and $[Co(CN)_5]^{3-}$. Irradiation in the *presence* of oxygen gave different results. The superoxo complex was not formed, and benzaldehyde was the predominant organic product. It was suggested that under these conditions the reactions shown in Eqs. (8-24) and (8-25) occur. It was suggested that reaction

$$[Co(CN)_5(CH_2C_6H_5)]^{3-} + O_2 \xrightarrow{h\nu} [Co(CN)_5(O_2CH_2C_6H_5)]^{3-} \quad (8\text{-}24)$$

$$[Co(CN)_5(O_2CH_2C_6H_5)]^{3-} + H_2O \longrightarrow [Co(CN)_5H_2O]^{2-} + C_6H_5CHO + OH^-$$
$$(8\text{-}25)$$

(8-24) proceeds by initial generation of benzyl radicals which are scavenged by O_2 and then react with photogenerated $[Co(CN)_5]^{3-}$ to give the peroxo complex; the 313 nm quantum yield for disappearance of $[Co(CN)_5 \cdot (CH_2C_6H_5)]^{3-}$ under these conditions was 0.15. The similarity of the quantum yields in aerated and deaerated solutions provides strong support for the hypothesis that the same primary photoreaction, homolytic cleavage of a Co—C bond, occurs in both cases.

Sheats and McConnell [33] have recently demonstrated the use of this photochemical reaction as a biophysical tool to trap nitroxides. For example, they used reactions of the type shown in Eq. (8-26) to measure rates of lateral

$$[Co(CN)_5(CH_2CO_2^-)]^{4-} + HO{-}\!\!\!\!\!\underset{}{\bigcirc}\!\!\!\!\!{N}{-}O$$

$$h\nu \downarrow$$

$$[Co^{II}(CN)_5]^{3-} + HO{-}\!\!\!\!\!\underset{}{\bigcirc}\!\!\!\!\!{N}{-}O{-}CH_2{-}CO_2^-$$

(8-26)

diffusion in phospholipid bilayers and to determine the number of nitroxide labels on the outer surface of liposomes. This type of reaction promises to be of great utility in biophysical and biochemical studies.

B. METHYLCOBALAMIN, COENZYME B_{12}, COBALOXIMES, AND RELATED DERIVATIVES

These complexes, all of which contain cobalt within a macrocyclic ligand and with an axial cobalt–carbon bond, have all been found to be photosensitive and have been the subject of numerous photochemical studies. Space does not permit a detailed summary of all the studies which have been conducted on these classes of compounds, and we present here only a general summary of the various observations and discuss pertinent articles from the most recent literature. The reader is referred to an excellent review of the subject by Koerner von Gustorf et al. [34] which presents a detailed discussion of those reports appearing prior to 1975.

The various photochemical studies of these compounds have been conducted primarily because ot the interest of researchers in the synthesis, properties, and biological activity of vitamin B_{12} and its derivatives. The structure of vitamin B_{12}, as determined by Crowfoot-Hodgkin et al. [35], is shown in Fig. 8-1. It consists of cobalt in a corrin ring complexed axially by α-5,6-dimethylbenzimidazole nucleotide and by cyanide ion. Replacement

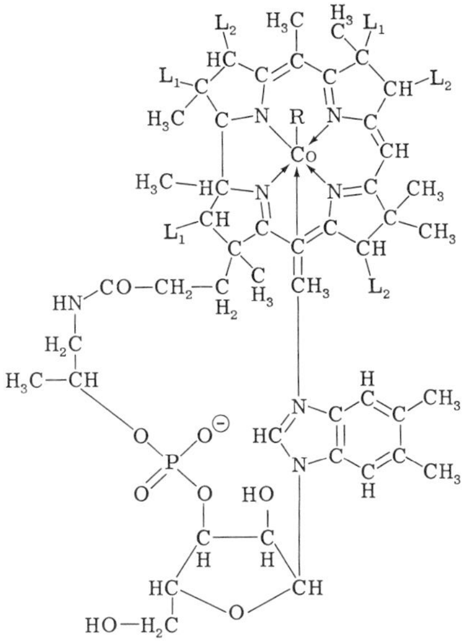

FIG. 8-1.   Structure of vitamin B_{12} [34,35].

$$L_1 = -CH_2-CO-NH_2$$
$$L_2 = -CH_2-CH_2-CO-NH_2$$
$$R = CN^{\ominus}$$

of the axial CN^- by a methyl group gives methylcobalamin and by 5'-deoxyadenosine gives coenzyme B_{12}. The formal oxidation state of cobalt is $3+$. The metal under certain conditions can be reduced to the $2+$ state, and these derivatives are labeled B_{12_r}. Numerous complexes have been prepared to model various aspects of B_{12} chemistry, and the majority of these have centered around the bis(dimethylglyoximato) complexes shown in (LXXV). These possess a donor ligand B and an alkyl ligand R on the axis

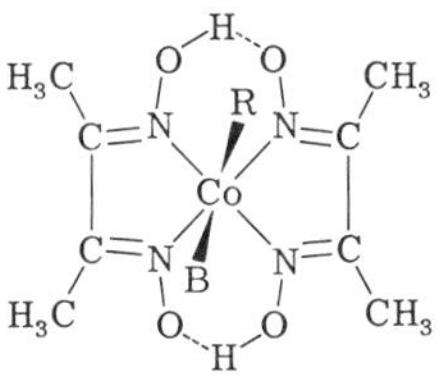

(LXXV)

and have been given the common name cobaloximes.

Methylcobalamin and its derivatives are thermally stable but are photosensitive. When irradiated in the presence of O_2, methylcobalamin gives formation of formaldehyde and a cobalt(III) aquo derivative (Eq. 8-27). The

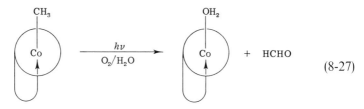

methylcobalamin

$$(8\text{-}27)$$

quantum yield varies with pH and with irradiation wavelength but ranges between 0.2 and 0.5 [34]. In the absence of oxygen, photolysis yields B_{12_r} and a mixture of methane and ethane (Eq. 8-28). Deuteration studies showed

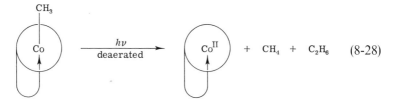

$$(8\text{-}28)$$

that the ethane arose primarily from coupling of two methyl radicals from methylcobalamin [36]. Quantum yields under anaerobic conditions are much smaller than those obtained in the presence of O_2, presumably due to rapid recombination of methyl radicals with B_{12_r}.

Both aerobic and anaerobic processes are consistent with the primary photochemical reaction being homolytic cleavage of the cobalt–carbon bond, initially producing B_{12_r} and methyl radicals. Endicott and Ferraudi [37] have recently observed this process in flash-photolysis studies. In all flash-photolyzed samples—oxygenated or deoxygenated, water or isopropanol solution—they observed generation of B_{12_r}. In deaerated solution, recombination of some of the methyl radicals with B_{12_r} to regenerate a portion of the methylcobalamin was demonstrated. Most of the methyl radicals, however, dimerized to ethane, giving a relatively large net generation of B_{12_r}. It was noted that in continuous photolysis experiments, in which the irradiation intensities would be several orders of magnitude lower than in the flash experiments, the stationary-state concentrations of methyl radicals would rarely be as high as 10^{-8} M. Under those conditions in deaerated solutions the B_{12_r} concentration would rapidly become high enough to scavenge efficiently most of the methyl radicals produced, and the resultant quantum yields would be low. Lappert and co-workers [38] have also demonstrated that homolysis of the cobalt–alkyl bond occurs upon photolysis of coenzyme B_{12} and ethylcobalamin by trapping the 5'-deoxyadenosyl and ethyl radicals produced with $(CH_3)_3CNO$. They were able to detect the spin-trapped $(CH_3)_3CN(\dot{O})R$ radicals by ESR spectroscopy.

The photochemistry of the cobaloximes is similar to that of the cobalamins except that the quantum yields are much lower [34]. The nature of the primary photochemical processes is also less clear. Giannotti and co-workers [39,40], for example, have proposed a mechanism different than that for methylcobalamin. They obtained evidence that photoaquation of the axial ligand B precedes oxygen insertion when methylcobaloxime is irradiated in aerated solution (Eq. 8-29). After the initial photoaquation, another photon

$$(8\text{-}29)$$

is necessary to form the peroxo complex, and it is likely that excitation induces homolytic cleavage of the $Co-CH_3$ bond. The resultant methyl radicals could scavenge O_2 to give $\cdot OOCH_3$ which could then recombine with cobaloxime to give the peroxo products. Giannotti and his co-workers [40] actually isolated a series of alkylperoxycobaloximes by photolysis in the presence of O_2 at temperatures ranging from -20 to $-120°C$. An interesting question is why photoaquation of the axial ligand B has to precede the peroxy formation. Perhaps, as has been suggested by Koerner von Gustorff *et al.* [34], a competition for the excitation energy exists. With a base such as pyridine, photosubstitution of the base must be the preferred reaction; but with water, cleavage of the Co–methyl bond is favored.

It would appear that homolytic cleavage of the cobalt–alkyl bond in the cobalamins and cobaloximes should proceed from an alkyl-to-cobalt CT state ($\sigma_b \rightarrow \sigma^*$). This has generally been accepted, and such transitions have been identified in the cobaloximes in the 400–450 nm region with $\varepsilon = 10^3$ [34]. In the cobalamins the corresponding LMCT transitions must be hidden underneath the intense $\pi \rightarrow \pi^*$ transitions of the corrin ring. It is likely that the photoactive LMCT states are reached by internal conversion from the initially populated $\pi \rightarrow \pi^*$ states. An LMCT state does not explain the apparent competition for the excitation energy discussed in connection with the photochemical reactions of methylcobaloxime. Such competition usually occurs from LF states, and there is an apparent inconsistency here which remains to be resolved.

In a related study, Mok and Endicott [41] reported that photolysis of the macrocyclic cobalt–alkyl complex $[CoCH_3([14]aneH_4)OH_2]^{2+}$ (LXXVI)

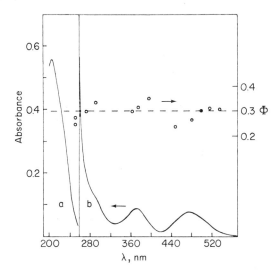

(LXXVI)

gave homolytic cleavage of the Co—CH$_3$ bond according to Eq. (8-30). The quantum yield of reaction (8-30) is 0.30 and is independent of excitation

$$[\text{CoCH}_3([14]\text{aneN}_4)\text{OH}_2]^{2+} \xrightarrow{hv} [\text{Co}([14]\text{aneN}_4)]^{2+} + {}^\cdot\text{CH}_3 + \text{H}_2\text{O} \quad (8\text{-}30)$$

wavelength, as illustrated by the data presented in Fig. 8-2. It was established that the energy threshold for the reaction is 18,000 cm^{-1}. Although the absorption spectrum of the complex shown in Fig. 8-2 is quite similar to that displayed by the series of $[\text{Co}(\text{NH}_3)_5\text{X}]^{3+}$ complexes in which the lower energy bands are assigned as LF transitions, the authors preferred not to make this traditional assignment. They proposed instead that the 478 nm band be assigned as a transition from the alkyl–cobalt bonding orbital to the cobalt $d_{x^2-y^2}$ orbital. Depopulation of the cobalt–carbon bonding orbital

FIG. 8-2.   Electronic absorption spectrum and wavelength dependence of homolysis quantum yield Φ for $[\text{Co}([14]\text{aneN}_4)(\text{OH}_2)\text{CH}_3]^{2+}$; (curve a) 2.35×10^{-5} M $[\text{Co}([14]\text{aneN}_4)$·$(\text{OH}_2)\text{CH}_3]$ in 10^{-3} M HClO$_4$; (curve b) 9.78×10^{-4} M $[\text{Co}([14]\text{aneN}_4)(\text{OH}_2)\text{CH}_3]^{2+}$ in 10^{-3} M HClO$_4$. Dashed line represents average value of quantum yields. Reprinted with permission from Mok and Endicott [41], *J. Am. Chem. Soc.* **99**, 1276 (1977). Copyright by the American Chemical Society.

is consistent with the observed homolytic cleavage of the Co—C bond. On the other hand, a normal LF transition could also lead to similar chemistry. The cobalt–alkyl bond is quite covalent and population of d_{z^2}, antibonding between Co—CH_3, should greatly weaken the Co—CH_3 bond. It seems reasonable that if the methyl group were eliminated, the electrons in the covalent cobalt–methyl bond would flow in the direction which would lead to the most stable products, and ·CH_3 would surely be expected to form preferentially to CH_3^-.

C. $[Rh(C_2H_5)(NH_3)_5]^{2+}$ AND *trans*-$[Rh(C_2H_5)(OH_2)(NH_3)_4]^{2+}$

Inoue and Endicott [42] have recently studied the photochemical properties of $[Rh(C_2H_5)(NH_3)_5]^{2+}$ and *trans*-$[Rh(C_2H_5)(OH_2)(NH_3)_4]^{2+}$ in aqueous solution. The *trans*-$[Rh(C_2H_5)(OH_2)(NH_3)_4]^{2+}$ complex was examined in most detail because of complicating thermal reactions with $[Rh(C_2H_5)(NH_3)_5]^{2+}$. The experimental evidence was consistent with photoinduced homolytic cleavage of the Rh—C_2H_5 bond (Eq. 8-31). The quantum yield varied with wavelength of irradiation and approached a limiting

$$\textit{trans-}[Rh(C_2H_5)(OH_2)(NH_3)_4]^{2+} \xrightarrow[H_2O]{h\nu} [Rh(OH_2)_2(NH_3)_4]^{2+} + \cdot C_2H_5 \quad (8\text{-}31)$$

value of 0.4 as the excitation energy increased. In aerated solution the final product was suggested to be $[Rh(O_2C_2H_5)(OH_2)(NH_3)_4]^{2+}$ formed either through the sequence of reactions (8-32) and (8-33) or through the reactions (8-34) and (8-35).

$$[Rh(OH_2)_2(NH_3)_4]^{2+} + O_2 \longrightarrow [Rh(O_2)(OH_2)(NH_3)_4]^{2+} \quad (8\text{-}32)$$

$$[Rh(O_2)(OH_2)(NH_3)_4]^{2+} + \cdot C_2H_5 \longrightarrow [Rh(O_2C_2H_5)(OH_2)(NH_3)_4]^{2+} \quad (8\text{-}33)$$

$$\cdot C_2H_5 + O_2 \longrightarrow \cdot O_2C_2H_5 \quad (8\text{-}34)$$

$$[Rh(OH_2)_2(NH_3)_4]^{2+} + \cdot O_2C_2H_5 \longrightarrow [Rh(O_2C_2H_5)(OH_2)(NH_3)_4]^{2+} \quad (8\text{-}35)$$

VII.   NICKEL, PALLADIUM, AND PLATINUM COMPLEXES

A. $[Ni(\eta^5\text{-}C_5H_5)(CH_2\text{—}\triangleleft)CO]$

Brown and co-workers [43] have reported that Pyrex-filtered irradiation of $[Ni(\eta^5\text{-}C_5H_5)(CH_2\text{—}\triangleleft)CO]$ leads to the transformation expressed in

Eq. (8-36). Although no quantitative or mechanistic details were presented,

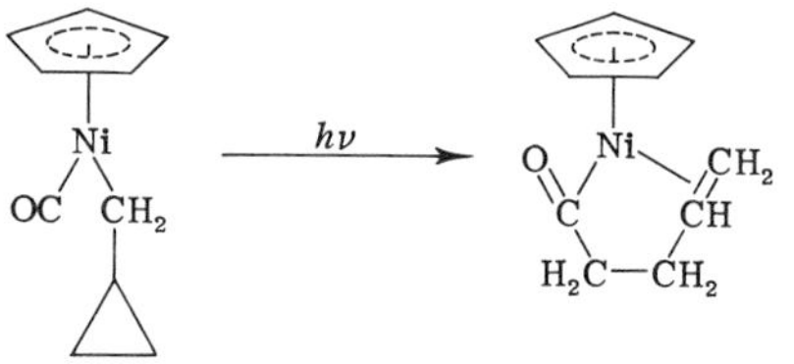

$$\tag{8-36}$$

it was proposed that the reaction proceeds through carbonyl insertion into the cyclopropane and then coordination of the olefin.

B. cis-$[MR_2L_2](M = Pd, Pt)$

Van Leeuwen *et al.* [44] have described the photochemical properties of several complexes of the general formula cis-$[MR_2L_2]$ (M = Pd, Pt; L = PPh_3, $L_2 = Ph_2PCH_2CH_2PPh_2$; R = CH_3, CH_2CH_3) in chloroform solution. The reactions were monitored using the chemically induced dynamic nuclear polarization (CIDNP) technique, and most attention was focused on $[Pt(CH_3)_2L_2]$. These authors suggested the sequence of events shown in Eq. (8-37). They did not speculate on the immediate details of the first step,

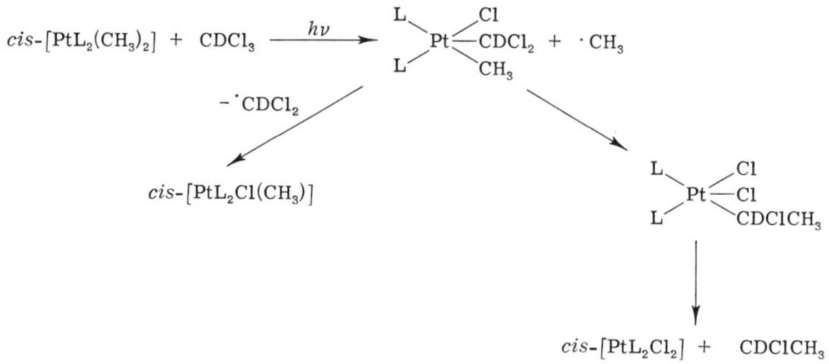

$$\tag{8-37}$$

but the absence of fast reaction in an inert solvent such as C_6D_6 suggested that simple homolysis of the Pt—CH_3 bond did not occur. (However, see the following discussion.) The intermediate (LXXVII) was invoked to account for the eventual formation of propane derivatives.

The cis-$[Pt(C_2H_5)_2(PPh_3)_2]$ and cis-$[Pt(C_2H_5)_2(Ph_2PCH_2CH_2PPh_2)_2]$ complexes were subsequently examined by other workers [45] in somewhat more detail. ESR and deuterium-labeling experiments clearly demonstrated that in benzene solution irradiation does lead to homolysis of the Pt—ethyl bonds. For example, photolysis of a solution containing cis-$[Pt(C_2H_5)_2 \cdot$

$(PPh_3)_2$] and the spin-trap nitrosodurene in an ESR cavity gave the characteristic signal of the ethylnitrosodurene adduct. Ethane was the principal organic product in the absence of a spin-trap with $<5\%$ butane observed. The initial reaction thus appears to be that shown in Eq. (8-38). The ethyl radical subsequently scavenges hydrogen either from solvent or a PPh_3 ligand to yield

$$cis\text{-}[Pt(C_2H_5)_2(PPh_3)_2] \xrightarrow{h\nu} [Pt(C_2H_5)(PPh_3)_2] + C_2H_5 \qquad (8\text{-}38)$$

ethane. The fate of the Pt complex in the absence of a suitable trapping agent is unknown, but a complex analyzing approximately for $[Pt(PPh_3)_2]_x$ was obtained. In the presence of carbon monoxide, the $cis\text{-}[Pt(C_2H_5)_2(PPh_3)_2]$ complex yielded the known $[Pt_3(CO)_3(PPh_3)_4]$ cluster upon photolysis, whereas $cis\text{-}[Pt(C_2H_5)_2(Ph_2PCH_2CH_2PPh_2)]$ gave $[Pt_4(CO)_4(Ph_2PCH_2\cdot CH_2PPh_2)_3]$. The $[Pt_3(CO)_3(PPh_3)_4]$ cluster can also be obtained by treating $[Pt(PPh_3)_3]$ with CO. The CO scavenging experiments indicate that both ethyl groups are photolabile and that $[Pt(PPh_3)_2]$ may be a reasonable intermediate.

C. $[Pt(i\text{-}C_3H_7)_2(COD)]$

Muller and Goser [46] prepared $[Pt(i\text{-}C_3H_7)_2(COD)]$ (COD = 1,5-cyclooctadiene) by treating $[PtCl_2(COD)]$ with $i\text{-}C_3H_7MgBr$. Irradiation of a solution of the complex under N_2 in the presence of excess COD led to production of $[Pt(COD)_2]$ in 22% yield (Eq. 8-39). Although not stated,

$$[Pt(i\text{-}C_3H_7)_2(COD)] \xrightarrow[\text{COD}]{h\nu} [Pt(COD)_2] \qquad (8\text{-}39)$$

this reaction could proceed by photoinduced β-hydride elimination according to the mechanistic scheme in Eq. (8-40). A similar photoreaction to produce $[Ni(COD)_2]$ from $[Ni(i\text{-}C_3H_7)_2(COD)]$ was observed, although irradiation of $[Pt(i\text{-}C_3H_7)_2(NBD)]$ (NBD = norbornadiene) did not lead to $[Pt(NBD)_2]$.

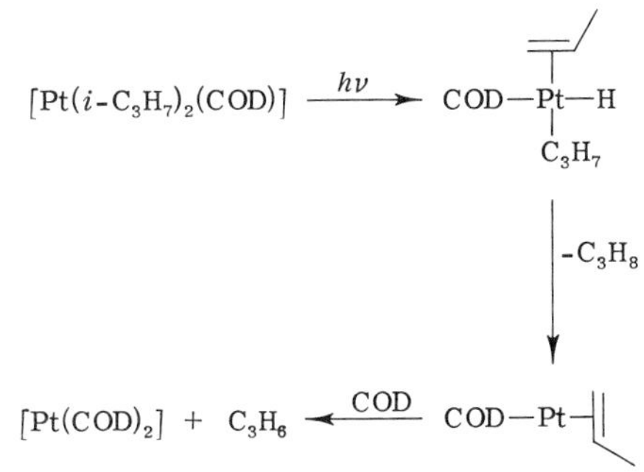

$$ (8\text{-}40) $$

D. $[Pt\, X_2(C_3H_6)(L—L)]$

UV irradiation of the trimethyleneplatinum(IV) compounds $[PtX_2(C_3H_6)\cdot(L—L)]$ ($X = Cl$, Br; $L—L = 2,2'$-bipyridine, 1,10-phenanthroline) has been reported [47] to lead to elimination of C_3H_6 predominantly as cyclopropane and formation of the corresponding platinum(II) compounds (Eq. 8-41). The

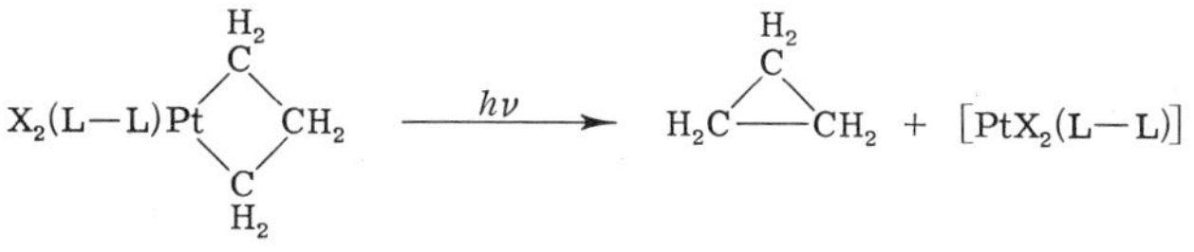

$$X_2(L—L)\,Pt \underset{\underset{H_2}{C}}{\overset{\overset{H_2}{C}}{\diamond}} CH_2 \quad \xrightarrow{h\nu} \quad H_2C \overset{\overset{H_2}{C}}{\diamond} CH_2 \; + \; [PtX_2(L—L)] \qquad (8\text{-}41)$$

kinetics of the reaction suggested simple concerted elimination of C_3H_6 from the excited complex.

VIII. SILVER AND GOLD COMPLEXES

A. $[Au(CH_3)L]$ ($L = PPh_3$, PPh_2Me)

Mitchell and Stone [48] have reported that UV irradiation of $[Au(CH_3)L]$ ($L = PPh_3$, PPh_2Me) in the presence of the fluoroolefins C_2F_4, C_3F_6, and C_2F_3Cl led to insertion products as exemplified by Eq. (8-42). Hexafluoro-2-butyne reacted similarly to give the insertion product (LXXVIII). The

$$[Au(CH_3)L] + C_2F_4 \longrightarrow [Au(CF_2CF_2CH_3)L] \qquad (8\text{-}42)$$

authors favored a free-radical mechanism for the reaction rather than a

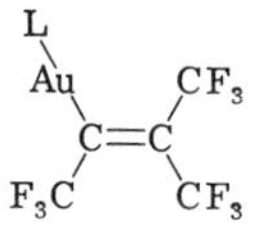

$$\underset{F_3C}{\overset{L}{\underset{\diagdown}{Au}}} \underset{\diagup}{\overset{\diagdown}{\underset{C=C}{}}} \overset{CF_3}{\underset{CF_3}{}}$$

(LXXVIII)

possible oxidative-addition, reductive-elimination sequence.

Van Leeuwen and co-workers [49] subsequently found that excitation of the complex in $CDCl_3$ solution led to formation of $[AuCl(PPh_3)]$. They suggested that this product formed through reaction of the excited complex with $CDCl_3$ to give $[AuCl(PPh_3)]$ directly and production of the triplet radical pair $CH_3 + CDCl_2$. However, they also demonstrated that photo-induced homolysis of the $Au—CH_3$ bond occurs, and they observed the methyl exchange process shown in Eqs. (8-43) and (8-44) [49]. It should be

$$[Au(CH_3)(PPh_3)] \xrightarrow{h\nu} [AuPPh_3] + \dot{C}H_3 \qquad (8\text{-}43)$$

$$[Au(CD_3)(PPh_3)] + \dot{C}H_3 \longrightarrow [Au(CH_3)(PPh_3)] + \dot{C}D_3 \qquad (8\text{-}44)$$

noted that formation of $[AuCl(PPh_3)]$ could also arise through $Au\text{—}CH_3$ homolysis since the $[AuPPh_3]$ product should be reactive enough to abstract Cl from $CDCl_3$.

B. $[Ag(n\text{-}C_4H_9)P(n\text{-}C_4H_9)_3]$

The photolysis of $[Ag(n\text{-}C_4H_9)P(n\text{-}C_4H_9)_3]$ was studied by Whitesides and co-workers [50] who observed that Pyrex-filtered irradiation gave the organic products shown in Eq. (8-45). The product distribution strongly suggested that n-butyl radicals are produced, and photolysis apparently

$$[Ag(n\text{-}C_4H_9)P(n\text{-}C_4H_9)_3] \xrightarrow{h\nu}$$
$$1\text{-}C_4H_8 \ (7\%) + C_4H_{10} \ (39\%) + C_8H_{18} \ (35\%) + C_4H_9 \cdot CH(CH_3)C_3H_5 \ (1\%) \quad (8\text{-}45)$$

leads to homolytic cleavage of the silver–alkyl bond.

IX. THORIUM COMPLEXES

$[Th(\eta^5\text{-}C_5H_5)_3(i\text{-}C_3H_7)]$

Irradiation of this thermally stable thorium complex has been shown by Marks and co-workers [51] to give $[Th(\eta^5\text{-}C_5H_5)_3]$ in yields $>92\%$ (Eq. 8-46). The production of almost equimolar amounts of propylene and

$$[Th(\eta^5\text{-}C_5H_5)_3(i\text{-}C_3H_7)] \xrightarrow{h\nu} [Th(\eta^5\text{-}C_5H_5)_3] + \underset{47\%}{C_3H_6} + \underset{53\%}{C_3H_8} \qquad (8\text{-}46)$$

propane led the authors to suggest that the reaction proceeds by β-elimination of hydride to give an intermediate such as (LXXIX) which then eliminates

$$(\eta^5\text{-}C_5H_5)_3 Th\text{—}H$$
$$| \atop C_3H_7$$

(LXXIX)

C_3H_8. It was further proposed that the reaction occurs through an excited state which induces partial weakening of one of the Th–cyclopentadienyl bonds which in turn lessens the steric crowding and permits the occurrence of the favorable β-hydride elimination reaction [51].

X. SUMMARY

Too few photochemical studies of metal–alkyl complexes have been conducted to allow any broad generalizations to be made except that homolysis of the metal–alkyl bond can be photoinduced. The detailed mechanism by which many alkyl complexes decompose, however, has not been unambiguously determined. It does appear from the $[M(\eta^5\text{-}C_5H_5)_2(CH_3)_2]$ (M = Ti, Zr, Hf) studies that photoinduced reductive elimination of alkane through coupling of the two alkyl groups is not a favored pathway for bis(alkyl) complexes, even though an analogous pathway is the dominant photoreaction mode for polyhydride complexes. The importance of the photoinduced β-hydride elimination reaction has yet to be determined.

REFERENCES

1. H. G. Alt, *J. Organomet. Chem.* **124**, 167 (1977).
2. K. W. Barnett and P. M. Treichel, *Inorg. Chem.* **6**, 294 (1967).
3. H. G. Alt, *J. Organomet. Chem.* **127**, (1977).
4. P. M. Treichel, R. L. Shubkin, K. W. Barnett, and D. Reichard, *Inorg. Chem.* **5**, 1177 (1966).
5. A. N. Nesmeyanov, Y. A. Chapovsky, and Y. A. Ustynyuk, *J. Organomet. Chem.* **9**, 345 (1967).
6. M. L. H. Green and A. N. Stear, *J. Organomet. Chem.* **1**, 230 (1964).
7. M. L. H. Green and P. L. I. Nagy, *J. Chem. Soc.* p. 189 (1963).
8. B. K. Bower and H. G. Tennent, *J. Am. Chem. Soc.* **94**, 2512 (1972).
9. W. Mowat, A. Shortland, G. Yagupsky, N. J. Hill, M. Yagupsky, and G. Wilkinson, *J. Chem. Soc., Dalton Trans.* p. 533 (1972).
10. R. W. Harrigan, G. S. Hammond, and H. B. Gray, *J. Organomet. Chem.* **81**, 79 (1974).
11. H. Alt and M. D. Rausch, *J. Am. Chem. Soc.* **96**, 5936 (1974).
11a. M. D. Rausch, W. H. Boon, and H. G. Alt, *J. Organomet. Chem.* **141**, 299 (1977).
12. E. Samuel, H. G. Alt, D. C. Hrncir, and M. D. Rausch, *J. Organomet. Chem.* **113**, 331 (1976).
12a. C. H. Bamford, R. J. Puddephatt, and D. M. Slater, *J. Organomet. Chem.* **159**, C31 (1978).
13. E. Samuel, P. Millard, and C. Giannotti, *J. Organomet. Chem.* **142**, 289 (1977).
14. W. H. Boon and M. D. Rausch, *Chem. Commun.* p. 397 (1977).
14a. H. G. Alt and M. D. Rausch, *Z. Naturforsch., Teil B* **30**, 813 (1975).
14b. E. Samuel and C. Giannotti, *J. Organomet. Chem.* **113**, C17 (1976).
15. M. Peng and C. H. Brubaker, Jr., *Inorg. Chim. Acta* **26**, 231 (1978).
15a. M. D. Rausch, W. H. Boon, and E. A. Mintz, *J. Organomet. Chem.* **160**, 81 (1978).

16. G. Erker, *J. Organomet. Chem.* **134**, 189 (1977).

17. U. Zucchini, E. Albizzati, and U. Giannini, *J. Organomet. Chem.* **26**, 357 (1971).

18. D. G. H. Ballard and P. W. van Lienden, *Makromol. Chem.* **154**, 177 (1972).

19. H. De Vries, *Recl. Trar. Chim. Pays-Bas* **80**, 866 (1961).

20. H. G. Alt, *Angew. Chem., Int. Ed. Engl.* **15**, 759 (1976); *J. Organomet. Chem.* **124**, 167 (1977); H. G. Alt, J. A. Schwarzle, and C. G. Kreiter, *ibid* **153**, C7 (1978); H. G. Alt and J. A. Schwarzle, *ibid* **162**, 45 (1978).

20a. M. D. Rausch, T. E. Gismondi, H. G. Alt, and J. A. Schwarzle, *Z. Naturforsch., Teil B* **32**, 998 (1977).

21. R. G. Severson and A. Wojcicki, *J. Organomet. Chem.* **157**, 173(1978).

22. K. Ofele and M. Herberhold, *Z. Naturforsch., Teil B* **28**, 306 (1973).

22a. K. Ofele, E. Roos, and M. Herberhold, *Z. Naturforsch., Teil B* **31**, 1070 (1976).

23. R. D. Rieke, H. Kojiana, and K. Ofele, *J. Am. Chem. Soc.* **98**, 6735 (1976).

24. E. O. Fischer and H. Fischer, *Chem. Ber.* **107**, 657 (1974).

25. M. Ephritikhine and M. L. H. Green, *Chem. Commun.* p. 926 (1976).

26. A. Hudson, M. F. Lappert, P. W. Lednor, and B. K. Nicholson, *Chem. Commun.* p. 966 (1974).

27. E. O. Fischer and J. Muller, *Z. Naturforsch., Teil B* **17**, 776 (1962); **18**, 413 and 1137 (1963).

28. E. O. Fischer and J. Muller, *Chem. Ber.* **96**, 3217 (1963).

29. E. O. Fischer and J. Muller, *J. Organomet. Chem.* **1**, 89 (1963); **1**, 464 (1964); **5**, 275 (1966).

30. A. Albini and H. Kisch, *J. Am. Chem. Soc.* **98**, 3869 (1976).

31. J. Evans, S. J. Okrasinski, A. J. Pribula, and J. R. Norton, *J. Am. Chem. Soc.* **99**, 5835 (1977).

32. A. Vogler and R. Hirschmann, *Z. Naturforsch., Teil B* **31**, 1082 (1976).

33. J. R. Sheats and H. M. McConnell, *J. Am. Chem. Soc.* **99**, 7091 (1977).

34. E. A. Koerner von Gustorf, L. H. G. Leenders, I. Fischler, and R. N. Perutz, *Adv. Inorg. Chem. Radiochem.* **19**, 65 (1976).

35. D. Crowfoot-Hodgkin, J. Pickworth, J. H. Robertson, K. N. Trueblood, R. J. Prosen, and J. G. White, *Nature (London)* **175**, 325 (1955).

36. G. N. Schrauzer, J. W. Sibert, and R. J. Windgassen, *J. Am. Chem. Soc.* **90**, 6681 (1968).

37. J. F. Endicott and G. J. Ferraudi, *J. Am. Chem. Soc.* **99**, 243 (1977).

38. K. N. Joblin, A. W. Johnson, M. F. Lappert, and B. K. Nickolson, *Chem. Commun.* p. 441 (1975).

39. C. Giannotti, C. Fontaine, and B. Septe, *J. Organomet. Chem.* **71**, 107 (1974).

40. C. Giannotti, C. Fontaine, A. Chiaroni, and C. Riche, *J. Organomet. Chem.* **113**, 57 (1976).

41. C. Y. Mok and J. F. Endicott, *J. Am. Chem. Soc.* **99**, 1276 (1977).

42. T. Inoue and J. F. Endicott, private communication.

43. J. M. Brown, J. A. Conneely, and K. Mertis, *J. Chem. Soc., Perkin Trans.* 2 p. 905 (1974).

44. P. W. N. M. Van Leeuwen, C. F. Roobeek, and R. Huis, *J. Organomet. Chem.* **142**, 233 (1977).

45. R. Pierantozzi, C. McLaren, and G. L. Geoffroy, to be submitted.

46. J. Muller and P. Goser, *Angew. Chem., Int. Ed. Engl.* **6**, 364 (1967).

47. G. Phillips, R. J. Puddephatt, and C. F. H. Tipper, *J. Organomet. Chem.* **131**, 467 (1977).

48. C. M. Mitchell and F. G. A. Stone, *J. Chem. Soc., Dalton Trans.* p. 102 (1972).

49. P. W. N. M. Van Leeuwen, R. Kaptein, R. Huis, and C. F. Roobeek, *J. Organomet. Chem.* **104**, C44 (1976).

50. G. M. Whitesides, D. E. Bergbreiter, and P. E. Kendall, *J. Am. Chem. Soc.* **96**, 2806 (1974).

51. D. G. Kalina, T. J. Marks, and W. A. Wachter, *J. Am. Chem. Soc.* **99**, 3877 (1977).

Subject Index

A

Alkyl–metal bond homolysis, photoinduced in
[M(1-norbornyl)$_4$], 301
[M(η^5-C$_5$H$_5$)$_2$R$_2$] (M = Ti, Zr, Hf), 301–304
[M(η^5-C$_5$H$_5$)$_2$(Ar)$_2$] (M = Ti, Zr), 304–305
[M(CH$_2$Ph)$_4$] (M = Ti, Zr), 305–307
[W(η^5-C$_5$H$_5$)$_2$(—CHR$_1$CHR$_2$CH$_2$—)], 310
[MnR(CO)$_5$], 311
[Os(CH$_3$)$_2$(CO)$_4$], 312
[Co(CN)$_5$(CH$_2$C$_6$H$_5$)]$^{3-}$, 313–314
Methyl cobalamin and related compounds, 314–319
[Co(CH$_3$)([14]aneN$_4$)(OH$_2$)]$^{2+}$, 317–319
[Rh(C$_2$H$_5$) (NH$_3$)$_5$]$^{2+}$, 319
[Rh(C$_2$H$_5$)(OH$_2$)(NH$_3$)$_4$]$^{2+}$, 319
[MR$_2$L$_2$] (M = Pd, Pt), 320–321
[Au(CH$_3$)L], 322–323
[Ag(n-C$_4$H$_9$)P(n-Bu)$_3$], 323
Arene ligand exchange, photoinduced in
[Cr(η^6-C$_6$H$_6$)(CO)$_3$] derivatives, 225
[Mo(η^6-C$_6$H$_5$CH$_3$)(CO)$_3$], 226
[W(η^6-C$_6$H$_5$CH$_3$)(CO)$_3$], 226
[RuCl$_2$(η^6-arene)PR$_3$], 228

B

Bimolecular excited-state processes, 122–128
Bonding, ligand–metal
alkyl, 7, 8
arene, 5, 6
aryl, 7, 8
carbon monoxide, 2, 3
cyclopentadienyl, 6, 7
hydride, 7, 8
isocyanide, 3, 4, 258–261
olefin, 4, 5
metal–metal, 20–22

C

Chromium compounds
[Cr(CO)$_6$], 13, 26, 40, 45–50, 68–71, 77, 173–178, 180, 182, 184, 262
[Cr(CO)$_5$], 48–49, 69
[Cr(CO)$_5$L], 40, 50
[Cr(CO)$_4$(ethylenediamine)], 56
[Cr(CO)$_4$(carbene)$_2$], 309–310
[Cr(CO)$_4$(diene)], 173, 175–178, 180
[Cr(CO)$_3$(NCCH$_3$)$_3$], 176
[Cr(CO)$_3$(NBD)(PPh$_3$)], 177–178
[Cr$_2$(CO)$_{10}$]$^{2-}$, 61–62
[CrMn(CO)$_{10}$]$^-$, 61–62
[Cr(η^6-arene)(CO)$_3$], 59, 60, 78–81, 176, 225–226
[Cr(η^6-arene)(CO)$_2$L], 59, 60, 78–81
[Cr(η^6-arene)$_2$], 219–225
[Cr(η^6-arene)$_2$]$^+$, 222–225
[Cr$_2$(η^5-C$_5$H$_5$)$_2$(CO)$_4$], 61, 62, 82
[Cr$_2$(η^5-C$_5$H$_5$)$_2$(CO)$_6$], 61–62, 82, 308
[Cr(η^5-C$_5$H$_5$)(CO)$_3$(CH$_3$)], 82, 300, 307–308
[Cr(η^5-C$_5$H$_5$)(COCH$_3$)(CO)$_2$(PPh$_3$)], 308
[Cr(η^5-C$_5$H$_5$)Cl$_2$]$_2$, 308
[Cr(CNR)$_6$], 259, 261–267
[Cr(CNR)$_6$]$^+$, 266
[Cr(CNR)$_5$(Py)], 262, 265
[Cr(1-norbornyl)$_4$], 301
[Cr(bipy)$_3$]$^+$, 224–225
[Cr(bipy)$_3$]$^{2+}$, 224–225
[Cr(H$_2$O)$_6$]$^{3+}$, 223

I